BIOSENSOR PRINCIPLES AND APPLICATIONS

Bioprocess Technology

Series Editor

W. Courtney McGregor

Xoma Corporation
Berkeley, California

Volume 1 Membrane Separations in Biotechnology, *edited by W. Courtney McGregor*

Volume 2 Commercial Production of Monoclonal Antibodies: A Guide for Scale-Up, *edited by Sally S. Seaver*

Volume 3 Handbook on Anaerobic Fermentations, *edited by Larry E. Erickson and Daniel Yee-Chak Fung*

Volume 4 Fermentation Process Development of Industrial Organisms, *edited by Justin O. Neway*

Volume 5 Yeast: Biotechnology and Biocatalysis, *edited by Hubert Verachtert and René De Mot*

Volume 6 Sensors in Bioprocess Control, *edited by John V. Twork and Alexander M. Yacynych*

Volume 7 Fundamentals of Protein Biotechnology, *edited by Stanley Stein*

Volume 8 Yeast Strain Selection, *edited by Chandra J. Panchal*

Volume 9 Separation Processes in Biotechnology, *edited by Juan A. Asenjo*

Volume 10 Large-Scale Mammalian Cell Culture Technology, *edited by Anthony S. Lubiniecki*

Volume 11 Extractive Bioconversions, *edited by Bo Mattiasson and Olle Holst*

Volume 12 Purification and Analysis of Recombinant Proteins, *edited by Ramnath Seetharam and Satish K. Sharma*

Volume 13 Drug Biotechnology Regulation: Scientific Basis and Practices, *edited by Yuan-yuan H. Chiu and John L. Gueriguian*

Volume 14 Protein Immobilization: Fundamentals and Applications, *edited by Richard F. Taylor*

Volume 15 Biosensor Principles and Applications, *edited by Loïc J. Blum and Pierre R. Coulet*

Additional Volumes in Preparation

BIOSENSOR PRINCIPLES AND APPLICATIONS

edited by
Loïc J. Blum
Pierre R. Coulet

Université Claude Bernard Lyon 1
Villeurbanne, France

Marcel Dekker, Inc. New York • Basel • Hong Kong

Library of Congress Cataloging--in--Publication Data

Biosensor principles and applications/edited by Loïc J. Blum and
Pierre R. Coulet
p. cm. -- -- (Bioprocess technology; v. 15)
Includes bibliographical references and index.
ISBN 0-8247-8546-0
1. Biosensors. I. Blum, Loïc J. II. Coulet, Pierre
R. III. Series
R857.B54B54 1991
660'.6-- --dc20 91-23803
CIP

This book is printed on acid-free paper.

Marcel Dekker, Inc.
270 Madison Avenue, New York, New York 10016

Current printing (last digit):
10 9 8 7 6 5 4 3 2 1

PRINTED IN THE UNITED STATES OF AMERICA

Series Introduction

Bioprocess technology encompasses all of the basic and applied sciences as well as the engineering required to fully exploit living systems and bring their products to the marketplace. The technology that develops is eventually expressed in various methodologies and types of equipment and instruments built up along a bioprocess stream. Typically in commercial production, the stream begins at the bioreactor, which can be a classical fermentor, a cell culture perfusion system, or an enzyme bioreactor. Then comes separation of the product from the living systems and/or their components followed by an appropriate number of purification steps. The stream ends with bioproduct finishing, formulation, and packaging. A given bioprocess stream may have some tributaries or outlets and may be overlaid with a variety of monitoring devices and control systems. As with any stream, it will both shape and be shaped with time. Documenting the evolutionary shaping of bioprocess technology is the purpose of this series.

Now that several products from recombinant DNA and cell fusion techniques are on the market, the new era of bioprocess technology is well established and validated. Books of this series represent developments in various segments of bioprocessing that have paralleled progress in the life sciences. For obvious proprietary reasons, some developments in industry, although validated, may be published only later, if at all. Therefore, our continuing series will follow the growth of this field as it is available from both academia and industry.

W. Courtney McGregor

Preface

A strong demand exists for improving the control and automation of industrial processes and for monitoring specific analytes in hospital critical care services, environmental control, or even defense. Following key parameters in real time remains difficult, and on-site analysis, which is an alternative to time-consuming conventional analysis performed in central laboratories, is raising a great amount of interest.

New generations of sensors able to directly provide direct and immediate information on the composition of their surroundings are promising tools in this area. Among them, biosensors, which can be thought of as highly sophisticated chemical sensors incorporating some kind of biological material in a sensing layer intimately associated with a transducer, are very attractive. Based on the highly specific and sensitive biomolecular recognition of target analytes, they provide a transduced signal, generally electrical, that can be correlated, after calibration, to the concentration of the analyte present in a complex medium.

Enzyme electrodes are the archetype of the first generation of biosensors, and some of them are now commercially available.

New generations of biosensors are emerging. They are based on novel and promising transducers such as field-effect transistors or optoelectronic devices. Efforts have been made by different groups to improve the selectivity and sensitivity of the sensing layer, to explore new concepts in transduction modes, and to miniaturize both the probes and related smart-signal processing systems.

Even though the literature published on this subject during the last two decades is abundant, it is still quite difficult to develop a clear idea of the future of such devices by reading randomly gathered papers.

Our aim in preparing this book was to ask authors—all active in the field and selected for their expertise—to contribute chapters based on experimental facts, presenting the state of the art in their domain that can serve as a reliable basis for other researchers.

As a matter of fact, the main bottleneck in such an interdisciplinary area is the common language that has to be developed to bridge the gap between readers from different disciplines. For this purpose, we asked authors to be didactic and comprehensive using examples from their own work and presenting major and relevant contributions they have identified in their field. Although it is particularly difficult in such a book to avoid overlapping as well as omissions, emphasis in each chapter is on the principle of measurement, the biological material used, the description of transducers, and characteristics and performances of the biosensors thus designed.

In our opinion, groups gathering specialists of biomolecular engineering, microelectronics, optronics, computer sciences, and automation capable of developing a comprehensive interdisciplinary approach may soon claim a decisive leadership in the field.

Finally, our hope is that reading *Biosensor Principles and Applications* will boost the constitution of such groups and will encourage young scientists to join this challenging and promising area in the near future.

Loïc J. Blum
Pierre R. Coulet

Contents

Series Introduction *iii*
Preface *v*
Contributors *ix*

1. What Is a Biosensor? 1

Pierre R. Coulet

2. Amperometric Enzyme Electrodes for Substrate and Enzyme Activity Determinations 7

Gilbert Bardeletti, Florence Séchaud and Pierre R. Coulet

3. Development of Amperometric Biosensors for Enzyme Immunoassay 47

J. V. Bannister, I. John Higgins and Anthony P. F. Turner

4. Potentiometric Enzyme Electrodes 63

Jean-Michel Kauffman and George G. Guilbault

5. Enzyme Thermistor Devices 83

Bengt Danielsson

6. Analytical Applications of Piezoelectric Crystal Biosensors 107

John H. T. Luong and George G. Guilbault

7. FET-Based Biosensors 139

Toshihide Kuriyama and Jun Kimura

8. Chemically Mediated Fiberoptic Biosensors 163

Bernhard P. H. Schaffar and Otto S. Wolfbeis

9. Fluorophore- and Chromophore-Based Fiberoptic Biosensors 195

Mark A. Arnold

10. Bioluminescence- and Chemiluminescence-Based Fiberoptic Sensors 213

Loïc J. Blum and Sabine M. Gautier

11. Immunosensors 249

Masuo Aizawa

12. Microbial Biosensors 267

I. Karube and M. E. SangMok Chang

13. In Vivo Biosensors 303

P. Vadgama and M. A. Desai

14. Trends and Prospects 339

Loïc J. Blum and Pierre R. Coulet

Index 345

Contributors

Masuo Aizawa Department of Bioengineering, Tokyo Institute of Technology, Tokyo, Japan

Mark A. Arnold Department of Chemistry, University of Iowa, Iowa City, Iowa

J. V. Bannister Biotechnology Centre, Cranfield Institute of Technology, Cranfield, Bedfordshire, England

Gilbert Bardeletti Laboratoire de Génie Enzymatique, CNRS-Université Claude Bernard Lyon 1, Villeurbanne, France

Loïc J. Blum Laboratoire de Génie Enzymatique, CNRS-Université Claude Bernard Lyon 1, Villeurbanne, France

M. E. SangMok Chang Research Center for Advanced Science and Technology, University of Tokyo, Tokyo, Japan

Pierre R. Coulet Laboratoire de Génie Enzymatique, CNRS-Université Claude Bernard Lyon 1, Villeurbanne, France

Bengt Danielsson Department of Pure and Applied Biochemistry, Chemical Center, University of Lund, Lund, Sweden

M. A. Desai Department of Medicine (Clinical Biochemistry), University of Manchester, Hope Hospital, Salford, England

Sabine M. Gautier Laboratoire de Génie Enzymatique, CNRS-Université Claude Bernard Lyon 1, Villeurbanne, France

George G. Guilbault Department of Chemistry, University of New Orleans, New Orleans, Louisiana

I. John Higgins Biotechnology Centre, Cranfield Institute of Technology, Cranfield, Bedfordshire, England

I. Karube Research Center for Advanced Science and Technology, University of Tokyo, Tokyo, Japan

Jean-Michel Kauffman Institut de Pharmacie, Université Libre de Bruxelles, Brussels, Belgium

Jun Kimura Resources and Environmental Protection Research Laboratories, NEC Corporation, Kanagawa, Japan

Toshihide Kuriyama Microelectronics Research Laboratory, NEC Corporation, Kanagawa, Japan

John H. T. Luong Biotechnology Research Institute, National Research Council of Canada, Montreal, Quebec, Canada

Bernhard P. H. Schaffar Biomedical Research and Development, AVL LIST GmbH, Graz, Austria

Florence Séchaud Laboratoire de Génie Enzymatique, CNRS-Université Claude Bernard Lyon 1, Villeurbanne, France

Anthony Peter Francis Turner Biotechnology Centre, Cranfield Institute of Technology, Cranfield, Bedfordshire, England

P. Vadgama Department of Medicine (Clinical Biochemistry), University of Manchester, Hope Hospital, Salford, England

Otto S. Wolfbeis Analytical Division, Institute of Organic Chemistry, Karl Franzens University, Graz, Austria

BIOSENSOR PRINCIPLES AND APPLICATIONS

1

What Is a Biosensor?

Pierre R. Coulet

CNRS-Université Claude Bernard Lyon 1
Villeurbanne, France

DEFINITION AND BACKGROUND

The abundant literature that can be related to the keyword *Biosensor* proves without doubt that the field is attractive. It is an interdisciplinary area for which sharp limits cannot be defined easily. The concept of biosensor has evolved; for some authors it is a self-contained analytical device that responds selectively and reversibly to the concentration or activity of chemical species in biological samples. No mention is made here of a biologically active material involved in the device; thus any sensor physically or chemically operated in biological samples can be considered a biosensor. This definition is obviously too broad but may involve, for instance, microelectrodes implanted in animal tissues, like brain. For most of the authors in this book the association of a biological sensing material with a transducer is compulsory, and even if different definitions are given, a biosensor can be simply defined, in a first approach as a device that intimately associates a biological sensing element and a transducer.

The first biosensor described, even if the term was not used at the time, was the combination of the Clark amperometric oxygen electrode serving as transducer and the enzyme glucose oxidase as sensing element for glucose monitoring. In 1962 Clark and Lyons (1) took advantage of the fact that an analyte like glucose could be enzymatically oxidized with, in parallel, consumption of the coreactant O_2 or the appearance of a product, H_2O_2, which could be electrochemically monitored. The enzyme, retained by a perm-selective membrane, thus added to the amperometric detector a high selectivity that could not be obtained without the bioelement.

During the following decade a lot of effort was devoted to obtaining bioconjugates for enzyme immunoassays. Various methods for enzyme immobilization were also described, including adsorption, entrapment in a gel lattice, covalent binding trough activated groups on the support, or the use of a cross-linking reagent (2).

In 1967 Updike and Hicks (3) gave the name *enzyme electrode* to a device comprising a polyacrylamide gel with entrapped glucose oxidase coating an oxygen electrode for the determination of glucose. Besides amperometry, potentiometric electrodes were also proposed by Guilbault and Montalvo in 1969 (4).

Since the early 1970s various combinations of biological material associated with different types of transducers gave birth to the larger concept of "biosensor." As a matter of fact, as exemplified in Figure 1, a biosensor associates a bioactive sensing layer with any suitable transducer giving a usable output signal. Biomolecular sensing can be defined as the possibility of detecting analytes of biological interest, like metabolites, but also including drugs and toxins, using an affinity receptor (enzymes being the simplest and historically the first employed), which can be a natural system or an artificial one mimicking a natural one, able to recognize a target molecule in a complex medium among thousands of others.

To obtain a usable output signal that can be correlated with the amount or concentration of analyte present in the medium, multiple events must take place sequentially. Briefly, a first chemical or physical signal consecutive to molecular recognition by the bioactive layer is converted by the transducer into a second signal, generally electrical, with a transduction mode that can be electrochemical, thermal, optical, or based on mass variation. The selective molecular recognition of the target molecule can theoretically be achieved with various kinds of affinity systems, for example (but not exclusively),

Enzyme for substrate
Antibody for antigen
Lectin for sugar
Nucleic acid for complementary sequence

The first problem we must face is the degree of bioamplification obtained when molecular recognition occurs. If the bioactive molecule present in the sensing layer is a biocatalyst a reaction takes place in the presence of the specific target analyte, and a variable amount of coreactant or product may be either consumed or produced, respectively, in a short time depending on the turnover. Biocatalysis thus corresponds to an amplification step generating a chemical signal.

In contrast, the use of antibodies for the detection of antigens is not normally a

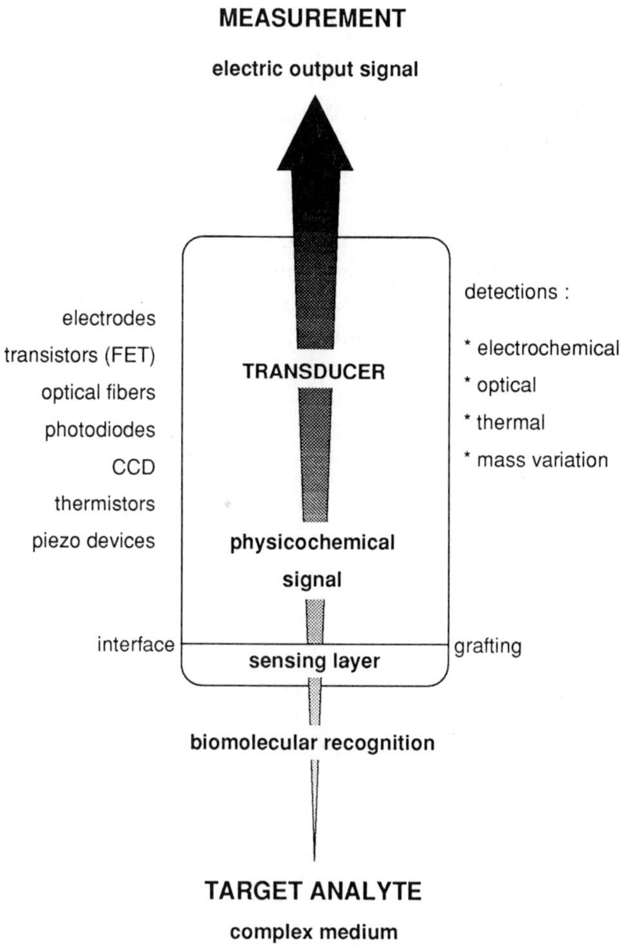

Figure 1 Configuration of a biosensor.

biocatalysis phenomenon, and different approaches can be considered. A bioconjugate involving a bound enzyme can be prepared, and the presence of the target antigen is determined through the related enzyme reaction. Conventional detection has already been described with enzyme electrodes. New approaches are now being intensively explored, and antibody-antigen binding can be sensed directly through transducers sensitive to mass variation for example.

Another key point to which attention must be paid is the intrinsic specificity of the biological material involved in the recognition process. Some enzymes, for instance, may be strictly specific, like urease, or highly specific, like glucose oxidase. Others, like alcohol oxidase or amino acid oxidases, recognize a large spectrum of alcohols and amino acids, respectively. This intrinsic specificity is difficult to modify, and problems of interferences may arise. For antibodies the specificity can be strongly enhanced by using the monoclonal antibodies now widely produced in many laboratories.

Numerous attempts to find a universal transducer that matches any kind of reaction have been reported. Heat variation, for instance, appears as a signal following practically any chemical or biochemical reaction. Mass variation, consecutive to molecular recognition, also appears very attractive as a universal signal for transduction, especially for antigen-antibody reaction in which no biocatalysis occurs. Piezoelectric devices sensitive to mass, density, or viscosity variations can be used as transducers. The change in oscillation frequency can be correlated to the change in interfacial mass. Quartz-based piezoelectric oscillators and surface acoustic wave-detectors are the two types currently used.

IMPORTANCE OF THE BIOLOGICAL SENSING SYSTEM

Operating Conditions

It must be kept in mind that most of these biological systems have extraordinary potentialities but are also fragile and must be used under strictly defined conditions. For instance, most enzymes have an optimal pH range in which their activity is maximal; this pH zone must be compatible with the characteristics of the transducer. Except for rare enzymes capable of undergoing, for a short period, temperatures higher than 100°C, most biocatalysts must be used in a quite narrow range of temperature (15–40°C).

There is also a demand for measurements in gaseous or solvent phases, however, and until now biosensors work mostly in aqueous media.

Immobilization of the Biological System: Role of the Sensing Layer

The simplest way to retain bioactive molecules on the tip of a transducer is to trap them behind a perm-selective membrane. This has been mainly used in addition to embedding procedures, in polyacrylamide gels for the design of enzyme electrodes. The trend now is to use disposable membranes with bound bioactive material. The availability of preactivated membranes suitable for the immediate preparation of any bioactive membrane thus appeared as a real improvement.

The removal of interference is a prerequisite for the wide use of biosensors in industrial processes. This can be achieved by using multilayer membranes, such as those developed by Yellow Springs Instrument Co. (for glucose or lactate enzyme electrodes), with the enzyme sandwiched between a special cellulose acetate membrane and a polycarbonate nucleopore membrane. The main role of this membrane is to prevent proteins and other macromolecules from passing into the bioactive layer. Cellulose acetate membrane allows only molecules of the size of hydrogen peroxide to cross and contact the platinum anode, thus preventing interference by ascorbic acid or uric acid, for example, at the fixed potential.

For miniaturized biosensors different methods for selectively depositing bioactive layers have been described, and details can be found in corresponding chapters in this book.

WHAT ARE THE CRITERIA FOR BELONGING TO THE BIOSENSOR FAMILY?

A biosensor should respond selectively, continuously, rapidly, specifically, and ideally without added reagent, and then different criteria must also be considered.

What about *vicinity*? The biosensing element must be either intimately connected to or integrated within a physicochemical transducer. In most cases the systems described fulfill this requirement, but more and more systems are composed of microcolumn on which molecular recognition occurs and of transducer separated from the column for different reasons, mainly the incompatibility of the biosensing reaction (pH for urease in the urea detector) and of the detector itself (pNH_3 sensor). Microcolumns can sometimes be replaced by nylon coils, especially in luminescent reactions, with the luminescence enzyme system bound onto the internal wall of the tube, allowing the reaction medium containing the sample to flow through.

Generally a biosensor is described as a small probe or pen-sized device. This is true for a few, but generally the probe must be connected to a signal-processing system for the readout of results. A question then arises, where does the sensor stop and the instrument begin? Can we consider, for instance, a spectrophotometer operating an enzyme-based fiberoptic probe as a biosensor, or is this simply remote analysis that can be performed outside the standard cuvette?

What is *rapidly*? For enzyme electrodes a short response time may be between a few seconds and 30 sec, for instance. For immunosensors 15 min is acceptable and considered short compared to alternative techniques that are far more time consuming. For microbial sensors 20–30 min is very short, if we understand that biological oxygen demand (BOD) measurements, for instance, may last 5 days using conventional microbiological methods.

The problem of specificity has already been addressed, but we must keep in mind that interference may occur on the sensing element or on the transducer as well.

It is often claimed that a biosensor must ideally work without added reagent, but what is the exact meaning of a reagentless technique? If we are very dogmatic this means that the biosensor tip can be placed directly in contact with crude samples, undiluted and not pretreated, a drop of whole blood for biomedical analysis or a few milliliters of fermentation broth or raw sewage, for example.

More generally the sample is injected into a reaction medium in contact with the bioactive tip. Do we then consider the buffer used an added reagent, or are only cofactors or coreactants, if required, concerned? Indeed, it is possible to imagine a system with cofactors or coreactants entrapped within the sensing layer with a slow release, allowing the reaction to take place when the target analyte is present. In this case the property of reversibility is still questionable since a system of regeneration *in situ* must be conceived.

What about *reusability?* If a disposable sensing tip for a one-shot determination is considered, what is the difference, in terms of cost and efficiency, when taking into account the preparation of the bioactive part, from a device in which reagents must be added for each measurement?

In fact, the design of a biosensor depends mainly on the area in which it operates. To be reliable and of practical use there may be no need for all the qualities of the ideal biosensor. Finally, only one criterion can be retained: *convenience*.

The following chapters deal with the main types of biosensors based on the concept and principles evoked here. What appears the most difficult but also the most promising is the interdisciplinary aspect of this field and the competency associated with designing these new analytical tools, which are capable of working in real and practical conditions and thus able to challenge well-established methods.

Undoubtedly, the extraordinary capabilities of biomolecules in selective recognition allied to the still advancing evolution of electronics or optics will lead in the near future to unexpected breakthroughs in the analytical field.

REFERENCES

1. Clark, L. C. Jr., and Lyons, C. (1962). Electrode systems for continuous monitoring in cardiovascular surgery, *Ann. NY Acad. Sci.*, *102*:29.
2. Silman, I. H., and Katchalski, E. (1966). Water-insoluble derivatives of enzymes, antigens and antibodies, *Annu. Rev. Biochem.*, 35:873.
3. Updike, S. J., and Hicks, G. P. (1967). The enzyme electrode, *Nature*, *214*:986.
4. Guilbault, G. G., and Montalvo, J. (1969). A urea specific enzyme electrode, *J. Am. Chem. Soc.*, *91*:2164.

2

Amperometric Enzyme Electrodes for Substrate and Enzyme Activity Determinations

Gilbert Bardeletti, Florence Séchaud, and Pierre R. Coulet

CNRS-Université Claude Bernard Lyon 1
Villeurbanne, France

INTRODUCTION

The use of an enzyme in association with an electrode was the first type of biosensor proposed. This approach combines the specificity of an enzymatic reaction with the analytical power of an electrochemical detection (1), applied for the first time to glucose determination (2). The enzyme is maintained in close contact with the electrochemical transducer, and the electrical signal obtained can be correlated with the concentration of the substrate to be detected. Since then a strong interest in such analytical devices arose, and numerous papers have appeared in the literature describing enzyme electrodes (3–6).

These sensors appear very elegant tools in different domains: biotechnology, biomedical engineering, environment, and others, because a real need exists for the determination of various metabolites in complex mixtures. The characteristics and performance of such devices appear attractive: sensitivity, reliability, handiness, simplicity, and rapidity (saving of reactants and cheapness). Some biosensor-based analyzers are now commercially available or are under development by major companies.

However, several drawbacks must still be overcome for the wider utilization of such devices. Significant improvements have recently been proposed by using new types of enzyme-sensor associations, novel supports or immobilization procedures, more appropriate electrochemical detection techniques, or by purifying novel enzymes.

FUNDAMENTALS

In an amperometric enzyme electrode the function of the enzyme is to generate (or to consume) an electroactive species in a stoichiometric relationship with its substrate or target analyte. The amperometric transducer allows the electrochemical reaction (oxidation or reduction) to proceed at the electrode surface, giving rise to a current. This current is directly related to the bulk substrate concentration.

It is important that electron transfer between the biological molecules and electrode material be facilitated. In this context we examine in this section some fundamental features of the biosensor components as they influence the response of the device.

Enzymes and Immobilization Procedures

Enzymes

The most common enzymes used in monoenzymatic systems are oxidases, which catalyze the model oxidation reaction:

$$\text{Substrate } S + O_2 \xrightarrow{\text{oxidase}} \text{product } P + H_2O_2$$

Oxidases are usually flavoproteins that use O_2, the natural electron acceptor, to regenerate the reduced enzyme during the reaction.

New systems have recently been developed in which a chemical mediator replaces oxygen. Among them, ferrocene-ferricinium ion couples or derivatives are the most frequently employed.

If the compound X to be assayed is not a substrate of an oxidase then an enzyme sequence can be used, generally involving a hydrolase as auxiliary enzyme, leading to a compound S that is a substrate of an oxidase:

$$X + H_2O \xrightarrow{\text{hydrolase}} S + Y$$

Indeed, the use of an enzyme sequence generally leads to a longer response time and the reaction may be subject to interference. The enzymatic sequence is responsible for the recognition of the analyte and thus determines the specificity of the overall system.

Immobilization Procedures

Enzymes have mostly been immobilized in membranes or gels, but direct coupling to the transducer can also be achieved as exemplified in modified electrodes. Factors that affect enzyme electrode performances include:

The type of immobilization as well as the thickness of the enzymic layer
The enzyme loading
The conditions for optimum enzyme activity

A fast response of the sensor requires intimate contact between the enzyme and the transducer. This is achieved by confining the enzyme in a thin layer as close as possible to the transducer.

One of the main factors affecting the performance of the biosensor is the enzyme immobilization procedure itself, and different approaches to enzyme immobilization have been used in the design of enzyme electrodes:

Confinement of soluble enzyme with a perm-selective membrane
Physical or chemical adsorption
Entrapment in gels (eventually behind a semipermeable membrane)
Cross-linking by means of bifunctional reagents (sometimes in combination with adsorption and entrapment)
Covalent binding to an activated insoluble support

These procedures have recently been thoroughly reviewed by Barker (7) and Kennedy and White (8).

Adsorption of enzymes on insoluble supports results from ionic, polar, and hydrogen bonding or hydrophobic interactions. The major advantage of the method is its simplicity using mild conditions. However, the tightness of the enzyme linkages is highly dependent on pH, solvent, and temperature, and a slight change in one of these factors causes desorption. Enzyme electrodes using this immobilization procedure give poor performance.

The first example of an immobilized enzyme electrode was reported by Updike and Hicks (9). Glucose oxidase was entrapped in a polyacrylamide gel placed on the surface of an oxygen electrode. Other materials besides polyacrylamide gels, such as starch gel, silicone rubber (3), and polyvinyl alcohol (10) have been used since then. However, the gel entrapment procedure suffers from major drawbacks:

Large diffusional barriers to the transport of substrate and product, which increase the response time
Denaturation of part of the enzyme molecules during polymerization as a result of the use of free radicals
Loss of enzyme activity by leakage

Bifunctional (especially glutaraldehyde) or multifunctional reagents that induce intermolecular cross-linking can be used to insolubilize enzymes. Some disadvantages of the method are:

The difficulty in controlling the reaction
The gelatinous nature of the final product
The need for a large amount of enzyme

Generally, a neutral protein is added to the mixture to improve the mechanical performance of the membrane and to lower the amount of enzyme. For example, Thomas and colleagues (11,12) produced an enzymic membrane in the following: A cross-linking reaction occurred in a solution containing albumin (30 mg/ml) and glutaraldehyde (2 mg/ml). By spreading uricase (0.5 mg/ml) or urease (4 mg/ml) on a glass surface, a transparent proteic film was obtained after about 1 hr. More recently Romette and Cooney (13) described a L-glutamine enzyme electrode using a selective gas film (6 μm thick polypropylene film) coated by a gelatin-glutaminase solution. The bilayer was air dried and then immersed in a 1.25% glutaraldehyde solution for 3 min. This membrane was fixed with an O-ring on the oxygen probe tip.

The enzyme layer can be sandwiched between two permeable membranes. An outer cellophane, polycarbonate, or cellulose acetate membrane served to exclude larger molecules or molecules like ascorbate. Enzyme electrodes can be prepared either by trapping powdered enzyme with a few microliters of buffer or a thin layer of chemically bound or entrapped enzyme between the electrode and a perm-selective membrane (cellophane or porous polycarbonate membrane, for example) secured with a rubber O-ring (14,15). Some examples of enzyme immobilization performed by covalent attachment using a preexisting membrane are given here.

Collagen Membrane. Membranes prepared by electrodeposition of fibrous proteins, such as collagen, were developed by Karube and Suzuki (16). This approach was very useful for enzyme immobilization and for the design of different types of enzyme electrodes. Coulet et al. (17) proposed a mild method of enzyme coupling on collagen membranes that had undergone an acylazide activation process. The technique was very convenient for oxidases, and other types of enzymes were immobilized on the surface of collagen films. Different types of coupling could be performed using one or several enzymes either at random or in an asymmetric way, with different enzymes on each face of the membrane (18–22).

Synthetic Preactivated Membranes. To simplify the immobilization procedure and to prepare easily tailor-made biosensors, Biodyne immunoaffinity polyamide membranes, provided in a preactived form by Pall, France, were further selected in our group for glucose oxidase immobilization (23). Enzyme coupling was obtained simply by wetting the membrane with 10 μl of a concentrated enzyme solution, applied on one or both sides of the membrane. The reaction was achieved within 1 min at room temperature. Immobilization can also be performed by immersing the preactivated membrane in 2–5 ml of a stirred enzyme solution for some hours. Bioactive disks of a size matching the electrode tip are cut out of the enzymic membrane and immediately mounted on the transducer or stored in a suitable buffer. This procedure was extended to numerous substrate and enzyme detections (24–27).

Nylon Nets. Enzymes have been chemically immobilized on nylon nets through lysine spacers and glutaraldehyde after treatment with dimethylsulfate and 2,4,6-trinitrobenzenesulfonic acid. The resulting thin enzyme membrane was held using an O-ring over a platinum electrode (28). This procedure was used for a single or two enzyme immobilizations. The best results were obtained with an asymmetric coupling, as previously shown on collagen membranes (29). Nylon mesh was also used by Moody et al. (30).

Cellulose Acetate Membranes. Sternberg et al. (31) designed microsensors with multilayered structures in which glucose oxidase was covalently bound to a cellulose acetate layer using bovine serum albumin and parabenzoquinone linkages. These microsensors were implanted subcutaneously in anesthetized rats for blood glucose monitoring (32).

Graphite Disk. Glucose oxidase immobilization could also be achieved on graphite disks (4.5 mm in diameter and 1 mm thick) using adipic dihydrazide to cross-link the enzyme. The enzyme-coated disks were stored in 10 mM sodium phosphate buffer (pH 7) at 4°C. Before use a disk was placed on the probe and firmly maintained by a 0.2 μm pore size polycarbonate membrane (33).

Detection Modes and Transducers: Enzyme Electrode Design

Definition and Fundamentals

The first type of transducer used was an amperometric electrode. Since then other types of transducers have been developed and are described in other chapters. Amperometry is a classic method in electrochemistry, like potentiometry, polarography, voltammetry, chronoamperometry, and coulometry. The application of a potential between a reference and an indicating electrode allows one to measure a current when an electroactive analyte is oxidized or reduced, depending on the voltage, at the indicating electrode. The current is related to the rate of the electrochemical reaction that occurs. All compounds exhibiting electroactivity in aqueous solution (standard range 1.0 to −0.6 V *versus* NHE, the normal hydrogen electrode) may normally be detected and measured using an amperometric technique.

Transduction of the signal due to the enzymatic reaction in a biosensor can operate in one of three ways. First, the electrode simply detects either a consumed natural coreactant (O_2) or a product (H_2O_2) appearing during the biocatalysis. Hydrogen peroxide and oxygen sensors are the most usual devices. At an appropriate fixed potential (about ±700 mV, respectively, *versus* a reference electrode, for example, Ag/AgCl) the electrode transduces the electrochemical oxidation or reduction in an anodic or cathodic current respectively. This current is directly proportional to the analyte concentration in a definite range.

More recently systems involving mediators have been described in the literature. These redox species (ferrocene and its derivatives), replacing oxygen in oxidase reactions, require a lower operating potential with regard to hydrogen peroxide detection and prevent the occurrence of electrochemical interference at this potential. Moreover, these new systems have the considerable advantage of overcoming the problem of classic sensors in which the response depends not only on the substrate concentration but also on the oxygen partial pressure.

Enzyme reacting directly and rapidly at the electrode itself still constitutes a challenge for the design of amperometric biosensors. As a matter of fact, under ordinary conditions most proteins exhibit no direct electroactivity. To achieve a feasible direct electron transfer the electrode must be used as a "cofactor." A promising approach to this problem is to employ more appropriate sensors, with conducting polymers deposited on the electrode surface (34,35), modified electrodes (6,36), carbon paste sensors, and rotating ring-disk electrodes (37–39), or finally to employ more sophisticated electrochemical methods. The numerous voltammetric techniques allow one to choose between conventional dc cyclic voltammetry, adsorptive stripping voltammetry, and square-wave and pulse techniques (40). The development of these techniques has now greatly increased but has not yet led to reliable sensors.

Until now these amperometric transducers related to enzyme electrode design described in the literature mostly concern O_2 or H_2O_2 monitoring (Fig. 1), with an increasing interest in mediated systems.

Oxygen and Hydrogen Peroxide Detection

Oxygen Electrode. The Clark electrode was historically the first transducer associated with soluble glucose oxidase by Clark and Lyons in 1962 for glucose monitoring (2). The sensor consisted of a platinum or gold disk cathode polarized at a potential of about −0.7 V *versus* a silver/silver chloride anode having a ring shape. The set was covered with a film of polyethylene, PTFE (Teflon), polypropylene, rubber, or polyvinyl chloride (PVC) with a high permeability toward oxygen.

This arrangement allows oxygen reduction at the cathode to proceed. Different steps are involved in this electrochemical process, leading finally to H_2O as shown in the overall reaction:

$$1/2O_2 + 2H^+ + 2e^- \xrightarrow{Pt} H_2O$$

The measured cathodic current resulting from this electrochemical reduction is directly proportional to the oxygen level in the solution (O_2 partial pressure).

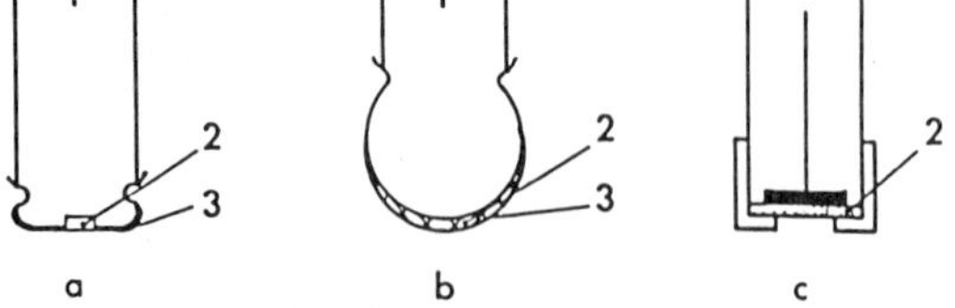

Figure 1 Amperometric enzyme electrodes. (a) Dialysis membrane electrode with soluble enzyme, or (b) entrapped enzyme and (c) enzyme membrane electrode: (1) transducer, (2) enzyme, (3) dialysis membrane.

Thus the depletion of oxygen at the biosensor tip due to the oxidase reaction can be easily correlated with the substrate concentration. A linear relationship between the current variation and the analyte concentration can be established in a definite range. It must be noted that sometimes the term "polarographic oxygen detector" is wrongly used because the term "polarography" should be reserved for electroanalytical techniques based on the dropping mercury electrode.

Hydrogen Peroxide Detection. Detection of hydrogen peroxide can also be performed using a platinum disk anode polarized at about 0.7 V *versus* a silver/silver chloride reference. At this potential H_2O_2 oxidation occurs at the platinum surface, according to the reaction

$$H_2O_2 \xrightarrow{Pt} O_2 + 2H^+ + 2e^-$$

The enzymic membrane is positioned on the sensor tip confining a buffered and chlorinated solution. The sensor responds linearly to the hydrogen peroxide generated by the enzyme reaction, the output current being correlated with the analyte concentration.

Electrochemical Aspects. In amperometry electrochemical processes are generally complex and may be considered a succession of electron transfers and chemical events (41). For example, the overall oxygen reduction (stoichiometrically, four electrons for one oxygen molecule) involves different steps (such as oxygen reduction to hydrogen peroxide, hydrogen peroxide reduction to H_2O, and dismutation of hydrogen peroxide) (42), and numerous parameters influence the rates of these reactions (the potential, the nature of electrode metal, and the operating conditions).

The overall sensor current is thus dependent on many factors, including charge transfer, adsorption, chemical kinetics, diffusion, convection, and substrate mass transport. The electrode response may be studied and characterized particularly by evaluation of the behavior of the rotating ring-disk electrode and investigation of theoretical aspects (5,37,38). Numerous relationships, each giving the generalized equation of current as a function of these parameters, have been elaborated (43–50).

Background Current. When switching on the sensor the potential is applied between the two electrodes and under defined operating conditions the output current decreases and reaches a low stable value after a variable time. This so-called background current can be attributed to the possible electrochemical reaction of impurities or to the buffer itself (42).

For the H_2O_2-detecting sensor (equipped with a membrane), the value of the noise current is generally very low (5–20 nA) and depends not only on the platinum area and sensor geometry but also on the operating conditions.

For the Clark electrode the definition of background current is different. It should be determined in a deaerated solution (complete lack of oxygen). Its value

(0.1–0.2 μA under the best conditions) depends on sensor configuration and geometry, particularly on the platinum or gold cathode area. A second reference can be obtained with the current resulting from an oxygen-saturated solution, which thus represents 100% of available current, for example 30μA/cm Pt area (42). Calibration of the Clark electrode is established between these two references (0–100%). Under operating conditions, measurements are then realized in subtraction mode with regard to this maximum value; that is, a current decrease is recorded, in contrast to the H_2O_2 sensor in which a current increase occurs.

Sensor Response. For each step, change in O_2 or H_2O_2 concentration in the medium, variation in the output current allows one to define the intrinsic characteristics of response time and intensity, transient current, and so on.

This is a result of the establishment, each time, of steady-state conditions: a "pseudoequilibrium" between the appearance of the species close to the sensor and its consumption at the indicating electrode. In the usual case, when the determining step is mass transfer, the overall sensor current is directly related to the flux of reduced (or oxidized) species at the electrode surface. Thus the output current is proportional to the analyte concentration, and this linear relationship represents the main interest of an amperometric transducer.

For the Clark electrode various models have been elaborated to account for this situation (49,51). The relationship giving the resulting steady-state current of a rotating disk electrode is certainly the expression closest to reality for an amperometric membrane sensor (5).

Stability and Influence of Various Parameters. O_2 and H_2O_2 detectors show good long-term stability provided that the operating conditions remain constant. An overall variation in the O_2 probe signal, lower than ±2.5% over periods of many months, was reported by Hitchman (42).

The influence of several parameters, such as temperature, stirring, and the salting-out effect, can give rise to real difficulties in the use of the O_2 probe.

The Clark electrode shows a large dependence on temperature, and the signal increases by 1–6% for a rise of 1°C, in contrast to oxygen solubility, which decreases with temperature (42). This dependence may be due to variations in membrane permeability, and various methods of temperature correction have been proposed leading to the design of automatic temperature compensation systems.

In the same way, the effect of stirring on sensitivity is important. Control of mixing conditions is required to maintain a constant bulk concentration of oxygen at the membrane surface.

In high ionic strength solutions oxygen solubility falls with increasing salt concentration: this is the salting-out effect. Nevertheless, as the Clark electrode responds to the O_2 partial pressure across the membrane and because this O_2 partial pressure remains constant even under highly saline conditions, the current detection level gives the same value. This phenomenon must be carefully taken into account when using the Clark electrode-based bioprobe in complex mixtures.

Other parameters influencing the behavior of the two types of transducers, that is, the Clark electrode and the H_2O_2 sensor, can be mentioned. First, the two metallic elements (Pt or Au for the working electrode; Ag for the reference electrode) are susceptible to contamination and poisoning by various compounds, such as gases (H_2S, SO_2), thioorganic materials, and dissolved or solid impurities. The membrane plays the rolé of an effective barrier, which is sometimes insufficient. For this reason membrane properties (such as integrity and solidity) should be periodically controlled. Otherwise, mechanical (by polishing) or electrochemical (by pulse) treatment of poisoned electrodes should be done.

The reference electrode must also exhibit a stable potential. It is also important that its area be large enough that polarization can be kept constant. In the case of a Ag/AgCl reference electrode, a 1 cm^2 area seems adequate with a saturated electrolyte solution. Another requirement is that the reference electrode has a sufficient mass so that its consumption is negligible.

Likewise, attention must be paid to the intrinsic characteristics of the working electrode, such as electrolyte composition (42,52).

When all these parameters are well controlled, the electrochemical process proceeds successfully.

Mediated Detection and Modified Electrodes

Mediated Detection. Replacing oxygen by an alternative electron donor-acceptor results in a mediated enzyme sensor. The ideal mediator should react rapidly with the reduced enzyme, and to avoid the problem of interference in real samples the oxidation potential of the mediators should be low and independent of pH (53).

First-Generation Mediators. This category includes water-soluble electron acceptors, such as ferricyanide, various quinones, and organic dyes, generally used with soluble enzymes.

In 1970 Williams et al. (54) described an enzyme electrode in which O_2 was replaced by benzoquinone and the hydroquinone, produced by the glucose oxidase (GOD)-catalyzed reaction, was measured amperometrically (400 mV *versus* SCE). A second nonenzymic electrode was used to compensate for the current produced by the nonspecific oxidation of other electroactive species present in the sample. The use of this mediator for glucose determination was recently compared by Asperger et al. (55) with two classic methods based on O_2 or H_2O_2 detection, with a serious advantage for the mediated reaction.

Hexacyanoferrate(III) can be used as an electron acceptor with some dehydrogenase-based enzymatic systems, and the hexacyanoferrate(II) produced is oxidized at a fixed potential on a platinum electrode.

$$\underset{\text{Hexacyanoferrate(III)}}{Fe(CN)_6^{3-}} + e^- \longrightarrow \underset{\text{hexacyanoferrate(II)}}{Fe(CN)_6^{4-}}$$

Comtat and his group (56) described such a device permitting continuous assays of L-lactate. The polarization potential of Pt was 250 mV *versus* Ag/AgCl, and the reaction chamber contained yeast lactate dehydrogenase (LDH) and Fe $(CN)_6^{3-}$. A bienzyme electrode for L-carnitine was recently described based on the same principle (57).

In the same way Karube et al. (58) used a similar system to develop an ethanol sensor based on membrane-bound alcohol dehydrogenase (ADH), with the mediator present in the bulk solution, using a polarization potential of 350 mV *versus* Ag/AgCl.

Hexacyanoferrate(III) can be reduced at a platinum electrode at-100 mV *versus* Ag/AgCl. Cosgrove et al. (59) described sensors for H_2O_2 using peroxidase and catalase. Peroxidase catalyzes the reaction between hexacyanoferrate(II) and H_2O_2 according to the reaction:

$$H_2O_2 + 2Fe(CN)_6^{4-} + 2H^+ \xrightarrow{\text{peroxidase}} 2H_2O + 2Fe(CN)_6^{3-}$$

and the hexacyanoferrate(III) thus generated is reduced at the Pt electrode.

These first mediated sensors were not widely adopted because the mediator should be added too often to the sample. However, an analyzer based on this approach is now on the market (Microzym-L from SGI, France).

Mediators of the Second Generation. Some years ago Cass et al. (60,61) demonstrated that ferrocene (biscyclopentadienyliron) and some of its derivatives could be used as mediators in enzyme electrodes.

Most ferrocenes are hydrophobic and may be retained on carbon electrodes by simple adsorption (62). They are well characterized electrochemically and operate at a redox potential that is generally low (160 mV *versus* SCE), which can be varied by changing their structure (63).

As illustrated by Cardosi and Turner (53), a glucose sensor can be constructed by covalently coupling GOD to a carbon foil previously modified with 1,1′-dimethylferrocene. The reaction sequence is

$$\text{Glucose} + 2\text{ferricinium}^+ + H_2O \xrightarrow{\text{glucose oxidase}} \text{gluconic acid} + 2\text{ferrocene}$$

$$2\text{Ferrocene} \xrightarrow{\text{160mV/SCE}} 2\text{ferricinium}^+ + 2e^-$$

The oxidation potential of 1,1′--dimethylferrocene is low enough to avoid interference from other reduced substances at physiological pH. Other enzyme sensors for glycolate, L-amino acid, and galactose utilizing 1,1′-dimethylferrocene showed a response to the substrate at the optimum pH of the corresponding enzyme (64).

Green and Hill (65) also developed the use of ferrocene derivatives of lidocaine and theophylline (acting as electron acceptors from reduced GOD) in rapid homogeneous enzyme immunoassays. Other mediator compounds such as tetrathiafulvalenes (TTF) or tetracyanoquinodimethane (TCNQ) have been used at a low working potential (220mV *versus* Ag/AgCl) (53).

Modified Electrodes. New approaches are now being investigated by other groups. To achieve a feasible direct electron transfer between the enzyme and the electrode material, various techniques are now being developed. The electrode material itself can be modified, generally by chemical or electrochemical means, to obtain electron transfer at an adequate potential. For this purpose the enzyme or its cofactor can be adsorbed or entrapped (with polymerization or not) inside the sensor.

Electrodes have been described that exhibit good performance with the enzyme adsorbed on the electrode material. The carbon paste technique particularly (64) is of strong interest for such a purpose.

Jönsson and Gorton (66) described a glucose sensor made of a carbonaceous electrode covered with a catalytic layer of a mixture of palladium and gold: Pd and Au were sputtered on the graphite. Such a preparation allowed them to drastically decrease the voltage to 300 mV *versus* Ag/AgCl to detect H_2O_2. Hintsche and Scheller (67) described a glucose oxidase/peroxidase sensor by coadsorbing the two enzymes on a nonporous carbon. In a similar way the enzyme can be incorporated into the micropores of platinum particles (68). The platinized platinum, which is very porous, is utilized as a matrix and the metal used to detect H_2O_2. This design is especially useful for the construction of microelectrodes.

Benetto et al. (69) immobilized GOD by direct cross-linking on a platinized carbon electrode (not requiring a mediator) with good results.

For determining the substrates of NAD-dependent dehydrogenases, the direct electrooxidation of NADH on carbonaceous materials or platinum appeared an attractive method. For example, Kobos (70) developed a L-lactate electrode by retaining the cofactor, NAD, on the surface of a Pt electrode. Palleschi et al. (71), immobilizing 3-hydroxybutyrate dehydrogenase on a graphite electrode and detecting NADH, designed a 3-hydroxybutyrate probe, allowing measurements in the range 5–100 μmol/L with good precision.

The necessary overpotential (from 1.1 to 1.3 V *versus* SCE) was too high for practical use; however its decrease may be obtained by the use of modified electrodes, particularly by quinoidal compounds (72).

Methods for the oxidation of the cofactor, NADH, are discussed by Gorton (73). To choose successful transducers, the kinetics of electron transfer and a decrease in overvoltage on these chemically modified electrodes are analyzed in detail by this author.

A theoretical approach and a comparative study were proposed by Bergel and Comtat (74) for modeling and optimizing electrochemical cofactor regeneration in electroenzymatic reactors.

Covalent linkage of the enzyme, for example to graphite or vitreous carbon, would ideally allow a direct-exchange flux of electrons between the prosthetic group of the immobilized enzyme and the electrode surface. Generally, a redox molecule (the cofactor or cosubstrate) acts as a shuttle between the active site of the enzyme and the electrode surface. Bourdillon et al. (75) thus demonstrated the efficiency of the coupling between enzymatic catalysis and coenzyme electrochemical regeneration (Fig. 2).

As quoted by Oyama et al. (76), enzymes can be fixed onto electrode surfaces, generally by covalent attachment via intermediate linkages using cyanuric chloride (77), glutaraldehyde (37,78), carbodiimide (39,60,79), and boronate (80). They can also be entrapped in a polymer matrix electrochemically prepared on electrodes (34,81,82).

Cenas et al. (83) reported the use of redox polymers, dopamine-modified dextran, and hydroquinone-piperazine copolymer, which were electrocatalytically efficient for the oxidation of reduced GOD.

Gunasingham and Tan (84) reported a platinum-dispersed Nafion film-modified glassy carbon as an electrocatalytic surface for an amperometric glucose enzyme electrode.

Ikeda et al. (85) proposed a kinetic model of the biocatalysis observed at the film-coated GOD-benzoquinone carbon paste electrode.

Bartlett and Whitaker (34,86) studied an electrochemical polymerization of *N*-methylpyrrole with GOD, producing an adherent film containing the active

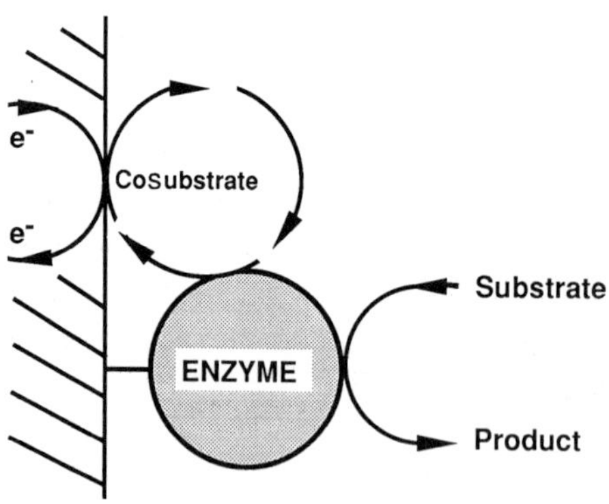

Figure 2 A pathway for enzymatic electrocatalysis using a redox molecule. Cosubstrate acts as a shuttle between enzyme-active site and electrode surface. (*From Ref. 75.*)

enzyme at the electrode surface. This approach has numerous advantages: the spatial distribution of enzyme and thickness of the enzymic film can be easily controlled, and by changing the redox state of polymer the enzymic activity can be modulated. Aizawa et al. (87) also designed an electroconductive enzyme membrane by electrochemical polymerization of pyrrole.

Oyama et al. (76) described a graphite carbon electrode modified by an inner electropolymerized cobalt tetrakis(*o*-aminophenyl)porphyrin. This electrode was coated with a bovine serum albumin (BSA) matrix and GOD cross-linked with glutaraldehyde. By monitoring the current due to O_2 reduction, an amperometric determination of glucose was achieved with results comparable to those reported with conventional GOD sensors.

Enzymes can be chemically immobilized directly onto platinum electrodes after anodization or platinization (88). Ikariyama et al. (89) showed that the nature of pretreatment of the Pt electrode plays a very important role in improving the selectivity of enzyme microsensors. The surface state of such a transducer should be carefully controlled.

A sensor responding to glucose in whole blood was described by Genshaw (90,91). In this device GOD is directly coupled to the platinum by silane and cross-linked with glutaraldehyde. A layer of rubber is cast over the enzyme layer. The polarization potential is maintained at 0.7 V *versus* Ag/AgCl to detect H_2O_2.

Moody et al. (92) described a miniature enzyme electrode consisting of GOD covalently attached to a silanized and anodized Pt surface via glutaraldehyde. The determination of glucose was realized by monitoring the anodic oxidation of H_2O_2 at 600 mV *versus* Ag/AgCl.

Watanabe et al. (93) recently developed a sensor for hydrogen peroxide employing a horseradish peroxidase (HRP) monolayer covalently attached by silane treatment to a tin oxide electrode and a dissolved electron mediator. As an extension, a GOD/HRP bilayer-modified electrode exhibited good performance in glucose determinations.

Investigation of the properties of electrode surfaces derivatized with hydrophobic materials using the Langmuir-Blodgett technique are underway in different laboratories. Garcia et al. (94) presented preliminary results on the behavior of glassy carbon electrodes modified by cast layers of phosphatidylcholine.

These chemically modified electrodes with immobilized enzyme are very attractive. Their relative simplicity of assembly, response characteristics, and the possibility of miniaturization compared to conventional bioprobes favor the increased development of these devices. However, they still suffer from a short lifetime and poor stability.

Other Detection Methods

Among the different amperometric detection methods modern voltammetric techniques like direct current cyclic voltammetry, square-wave voltammetry, and pulse techniques (40,95) have been recently reported for the determination of various analytes. These electrochemical methods have been already used in pharmaceutical analysis, especially for the detection of metals in various samples (40). In association with an enzymatic system this approach can now be extended to the development of convenient biosensors with significant advantages.

Davis et al. (96) developed an ATP sensor for monitoring creatine kinase activity with an enzyme-mediated electrode using cyclic voltammetry. The probe could be used to assay creatine kinase activities over the range 0.01–10 IU/ml.

Suaud-Chagny and Gonon (97) described an immobilized lactate dehydrogenase (LDH)-carbon fiber microelectrode sensitive to pyruvate (detection limit lower than 1 μM). The enzyme is embedded in an inert protein electrochemically deposited on the tip of the sensor, and the principle of measurement is based on the voltammetric detection of NADH. The applications of such a tool could be very extensive, especially in the field of clinical assay, by measuring enzymatic reactions involving the NAD/NADH system.

Wang et al. (98,99) recently developed a biologically modified carbon paste electrode for dopamine detection. A plant tissue preparation containing polyphenol oxidase was incorporated into the carbon paste matrix, and square-wave and differential pulse voltammetric techniques were used successfully. The stability of the biocatalytic activity and the voltammetric sensing allowed the determination of dopamine from 1.3×10^{-6} to 9×10^{-5} M. The detection limit was close to 4×10^{-7} M and the analytical throughput was 60 assays per hr.

Ikeda et al. (100) proposed a D-gluconate sensor using D-gluconate dehydrogenase immobilized on a benzoquinone-carbon paste electrode and the cyclic voltammetric detection of this mediator. This probe provides a linear current response to D-gluconate from 0.1 to 4 mM.

It must be pointed out that recent advances in microprocessor-controlled voltammetric analyzers now facilitate the development of these new types of biosensors.

Biosensor Characteristics

We focus in this section on the performance of enzymatic probes based on O_2 or H_2O_2 detection.

Sensitivity and Detection Limit

Two modes for the characterization of the response of the amperometric transducer are possible: steady-state or dynamic.

The steady-state response is defined as the step variation in current (Δi) between the baseline and the plateau reached after addition of the analyte-containing sample. This step is proportional to the variation in the analyte concentration (ΔC). The slope of the calibration curve, $\Delta i/\Delta C$, represents the sensor sensitivity. It can be expressed, for example, with H_2O_2 detection in mA/M or MA/M $\cdot cm^2$ when taking into account the active area of the electrode tip.

The dynamic response is obtained by monitoring the first derivative of the signal, as with the glucose electrode developed in the laboratory. The height of the peak was found to be proportional to the analyte concentration in the same range of concentration (101,102).

The inherent sensitivity of a sensor is partly influenced by mass-transfer limitations and particularly by the membrane permeability and thickness. For example, the $\Delta i/\Delta C$ value is inversely proportional to the thickness of membranes positioned on the electrode tip. Thus the presence of a significant diffusional barrier becomes a drawback to sensor sensitivity.

The detection limit is defined as the lowest analyte concentration that can be detected with an acceptable signal-to-noise ratio. With a Clark electrode the lowest limit of detection for oxygen is about 0.5 ppm (mg O_2/L) for a usually air-saturated solution corresponding to about 8 ppm at room temperature (25°C) and normal pressure (1 atm) (42). With an hydrogen peroxide sensor as previously described the detection limit for H_2O_2 is in the nanomolar range.

For a bioprobe a threshold toward detected metabolite, when clearly reported, not merely stated as zero, allows interesting comparisons of different system performances (Tables 1 and 2).

Response Time

The transient current in the sensor response depends not only on kinetic parameters of the electrochemical reaction but also essentially on diffusion conditions. Models to elucidate the effects of sensor design parameters and predict the characteristic responses have been developed (Fig. 3) (46–50).

The time necessary to reach the plateau defines the sensor steady-state response time. When the dynamic response is considered, the response time is advantageously shortened. Commonly, for a Clark electrode, 90% of the full response is obtained after 20 sec and 99% within 40–50 sec. In this case it has been demonstrated (42) that the response time t is proportional to the square of the membrane thickness b and inversely proportional to the oxygen diffusion coefficient D_m through this membrane: $t = b^2/D_m$.

For a bioprobe, in most cases dealing with monoenzyme systems, the response time is generally reported to be about 1–3 min. With a rate mode as developed in the laboratory, the dynamic response time was about 45 sec using a two-electrode–based analyzer (Glucoprocesseur).

The multienzyme systems have a longer response time because of the enzymic sequence. Numerous other factors can influence the response time, including the nature and thickness of the membrane, the amount of immobilized enzyme, the substrate concentration, the geometry of the electrode tip or of the measurement cell, and stirring and temperature.

Linearity and Calibration

The main interest of an amperometric transducer is the linear relationship existing between the output current and the analyte concentration, in contrast to

Table 1 Characteristics and Performances of Amperometric Enzyme Electrodes Used in Biologic Fluid Analyses (Glucose and L-Lactate Determinations Appear in Table 3)

Analytes in biologic fluids	Biosensors		
	Enzyme	Sensor	Immobilization procedure
Oxalate	Oxalate oxidase	H_2O_2	Sandwiched; Gore-tex dialysis membrane
Oxalate	Oxalate oxidase	O_2	Glutaraldehyde, BSA; pig intestine type H
Oxalate	Oxalate oxidase	H_2O_2	Covalently bound; preactivated polyamide
Salicylate	Salicylate hydroxylase	O_2	Glutaraldehyde, BSA; pig intestine
Urate	Uricase + peroxidase	$FeCN_6^{4-}$	Glutaraldehyde, BSA; dialysis membrane
Urea	Urease	O_2	Entrapped; dialysis membrane
Urea	Urease	O_2	Poly-γ-methyl-L-glutamate; CCl_3 or NH_2
Cholesterol	Cholesterol oxidase	H_2O_2	Covalently bound; collagen membrane
Cholesterol	Cholesterol oxidase + cholesterol esterase + peroxidase	$H_2O_2/FeCN_6^{4-}$	Glutaraldehyde-methacrylate gel/silk-cellulose acetate
Choline	Choline oxidase	H_2O_2	Glutaraldehyde; cellulose acetate, nylon net
Choline (phospholipids)	Choline oxidase	O_2	Entrapment; cellulose triacetate polymer, Teflon
Choline + acetylcholine	Choline oxidase + choline esterase	H_2O_2	Covalently bound; preactivated polyamide
Choline + acetylcholine	Choline oxidase + choline esterase	H_2O_2	Entrapped photo-crosslinkage
Acetylcholine	Choline oxidase + choline esterase	H_2O_2	Glutaraldehyde; cellulose acetate/nylon net
Phosphatidylcholine	Phospholipase D + choline oxidase	H_2O_2	Covalently bound; agarose gel
Bilirubin	Glucose oxidase + horse radish peroxidase	H_2O_2	Entrapped; gelatin dialysis membrane
Creatinine	Creatinine amidohydrolase + creatine amidinohydrolase + sarcosine oxidase	H_2O_2	Glutaraldehyde; cellulose acetate
Glycerophosphate	L-α-glycerophosphate oxidase + peroxidase	$FeCN_6^{3-}$ + silanized glassy carbon	BSA + glutaraldehyde; covalent fixation of enzymes directly onto silanized glassy carbon

[a] Not indicated
[b] Steady-state response
[c] Kinetic, dynamic response

Performances					
Detection limit	Linearity	Response time	Stability	Notes	Ref.
$2x10^{-6}$ M	Up to $5x10^{-4}$M	1-4 min	> 2 weeks	Urine, dilution 1:40	108
NI[a]	$1.3x10^{-5}$-$2.5x10^{-4}$ M	7-25 min	2 months/ 200 assays	Urine	109
10^{-8} M	10^{-7}M-10^{-4}M	<1 min	4 months/ 150 assays	Tacussel analyzer	25
NI	10^{-5}M-$1.9x10^{-3}$M	<6 min	NI	NAD(P)H in buffer	110
NI	Up to $3.5x10^{-4}$M	3 min (std)[b]/ 15 sec (kin)[c]	NI		111
$8x10^{-4}$ M	10^{-3}-$8x10^{-2}$M	NI	20 days		112
2.5 mg/dl	2.5/100 mg/dl	1.5 min	18 days	With immobilized nitrifying bacteria	113
10^{-8} M	10^{-7}-$8x10^{-5}$M	<5 min (std) <1 min (kin)	several months 150 assays	Free cholesterol	19
$4x10^{-4}$ M	$4x10^{-4}$-$1.2x10^{-3}$M	2min (std)/ 30 sec (kin)	10 days	Free and total cholesterol (within 5 min)	114
NI	Up to 10^{-4} M	1-2 min (std)	20 days		115
NI	$2.5x10^{-5}$-$1.2x10^{-4}$ M	2.5 min	20 days	Phospholipase D in solution	116, 117
$5x10^{-9}$ M	Up to $2x10^{-5}$ M	2-3 min	3 months (-20°C)		27
NI	$25x10^{-7}$-$15x10^{-5}$ M choline $2x10^{-5}$-$7.5x10^{-4}$M acetylcholine	2-4 min	60 days (dry state), 4°C		118
NI	10^{-6}-10^{-5} M	NI	NI		115
NI	Up to 3 g/dm	4 min	NI		119
NI	$5x10^{-6}$- 5 10^{-5} M	2 min	NI		120
1 mg/l	Up to 100 mg/l	20 sec (kin)	500 assays 9 months, 4°C	Storage (air, 4°C) 25 μl serum	121
NI	$5x10^{-6}$-10^{-4} M	20 sec 5 min time cycle	60 days (95%), 4°C 70 days (80%), 4°C	Chemical amplification for biologic fluids	122, 123

Table 2 Characteristics and Performances of Amperometric Enzyme Electrodes Used in Foodstuff Analyses (Glucose and L-Lactate Determinations Appear in Table 3)

Analytes in food industry	Biosensors		
	Enzyme	Sensor	Immobilization procedure
Galactose	Galactose oxidase	H_2O_2	Covalently bound on collagen membrane
Fructose	Hexokinase + $NADP^+$ + glucose-6-phosphate dehydrogenase	O_2	Gelatin entrapment sandwiched; polypropylene/dialysis membrane
Maltose	GOD + glucoamylase	H_2O_2	Covalently bound on collagen membrane
Sucrose	GOD + invertase + mutarotase	O_2	Glutaraldehyde and BSA; Teflon / Spectrapor 2
Sucrose	GOD + invertase + mutarotase	O_2	Glutaraldehyde; collagen membrane
Sucrose	GOD + invertase + mutarotase	H_2O_2	Covalently bound on collagen membrane
Lactose	β-Galactosidase + GOD	O_2	Glutaraldehyde and BSA; magnetic film
Lactose	β-Galactosidase + GOD	H_2O_2	Covalently bound on collagen membrane
Pyruvate	Pyruvate oxidase	O_2	Acetylcellulose
Pyruvate	Pyruvate oxidase	H_2O_2	Covalently bound on preactivated polyamide
Pyruvate	Pyruvate oxidase	H_2O_2	Carbodiimide; polyamide and nylon 6.6
Isocitrate	Horseradish peroxidase + isocitrate dehydrogenase	O_2	Gelatin + glutaraldehyde ; polyethylene dialysis membrane
Ascorbate	Ascorbate oxidase	O_2	Glutaraldehyde; collagen paste/Teflon
Ethanol	Alcohol dehydrogenase	Pt (+350 mV)/ $FeCN_6^{4-}$	Glutaraldehyde; cellulose triacetate
Aspartame	α-Chymotrypsin ('A') + alcohol oxidase	O_2	Gel of AOD sandwiched between teflon and retaining membrane; α-chymotrypsine in solution
Monoamines	Monoamine oxidase	O_2	Glutaraldehyde; collagen
Hypoxanthine	Xanthine oxidase	O_2	Glutaraldehyde; cellulose triacetate
IMP[e]	5'-Nucleotidase + xanthine oxidase + nucleoside phosphorylase	O_2	Glutaraldehyde; triacetylcellulose; 1,8 diamino-4-aminomethyloctane
ATP/ADP	Pyruvate kinase + hexokinase + LDH + lactate monooxygenase	O_2	Gelatin; polyethylene dialysis membrane

[a]Steady-state response.
[b]Kinetic, dynamic response.
[c]Not indicated.
[d]Glutamate pyruvate transaninase.
[e]Inosine 5′ monophosphate.

Performances					
Detection limit	Linearity	Response time	Stability	Notes	Ref.
$5x10^{-7}$ M	$5x10^{-7}$-$6x10^{-4}$M	5-6 min (std)[a] 1 min (kin)[b]	NI		20
$3x10^{-4}$ M	Up to $3x10^{-3}$ M	3-5 min	NI		124
$5x10^{-8}$ M	Up to 10^{-3} M	2-3 min (kin) 2-20 min (std)	2 months (air dried)	Air-dried enzyme, room θ storage	18, 29
NI[c]	$3.3x10^{-6}$-$7x10^{-3}$ M	15 sec (kin) 6 min (std)	>300 assays	Soft drink	125
NI	$2.5x10^{-3}$-10^{-2} M	3 min	10 days, 85 assays		126
10^{-4} M	10^{-4}- $2.5x10^{-3}$ M	8-10 min	NI	No interference with glucose up to 2 mM	20
NI	10^{-3}-$4x10^{-3}$ M	NI	NI		127
$3.5x10^{-5}$ M	$6.7x10^{-5}$-10^{-3} M	4-10 min	NI		20
$6x10^{-5}$ M	Up to $8x10^{-4}$ M	2 min	50 assays	Determinations of GPT[d]	128
10^{-5} M	$2x10^{-5}$-10^{-1} M	2-3 min	1 month	Foodstuffs, analyzer	105
NI	10^{-6}-10^{-4} M	NI	-13% after 30 days		129
$2x10^{-4}$ M	10^{-4}-$2x10^{-3}$ M	50 sec; cycle time 6 min	140 assays		130
NI	$5x10^{-5}$-$5x10^{-4}$ M	1-5 min	3 weeks, 100 assays	Interferes with catecholamine, D-isoascorbate, hydroquinone, food	131
NI	10^{-4}- $5x10^{-3}$ M	<5 min	80% 1 month, 30°C	Interferes with propane-1-ol, butane-1-ol, alcoholic beverages	58
NI	80-800 ppm	3-5 min	2 weeks (θ room) > 7 weeks (4°C)	Diet soft drinks, methanol determinations (degradation result) without ('A')	132
NI	$5x10^{-5}$-$2x10^{-4}$ M	NI	NI	Meat freshness	133
8.5 mg/dm	$6.2x10^{-5}$-$1.5x10^{-3}$ M	100 sec	>100 assays, 30 days	Fish meat	134
NI	10^{-3}- $5x10^{-3}$ M	NI	15 days	Fish meat	135
$2.5x10^{-7}$ M	Up to $6x10^{-6}$ M	4-10 min	NI	Amplified reaction	136

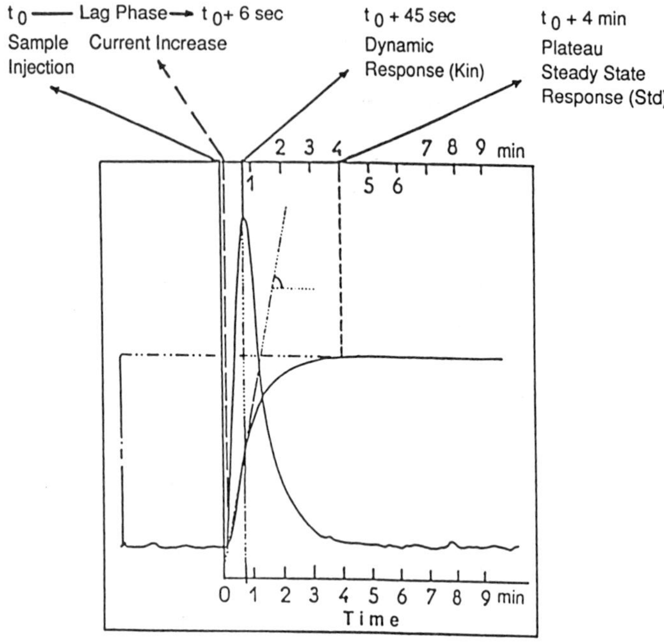

Figure 3 Dynamic and steady-state response time of an amperometric enzyme electrode based on H_2O_2 detection.

potentiometric detection, which leads to a logarithmic calibration. For the hydrogen peroxide transducer the extent of the linear range depends on the characteristics of the sensor geometry. Calibration is very easily performed by adding aliquots of H_2O_2 standard concentrations in measurement solutions. A range from 100 nM to 10 mM is obtained with probes used in the laboratory.

For the Clark electrode the response is linear over nearly four orders of magnitude (42). Generally a two-point calibration is used: zero oxygen concentration obtained by using a pure inert gas (N_2) or a sodium sulfite solution and air-saturated water or simply air as a standard maximum value. It is possible to calibrate an O_2 detector with more points by using gas mixtures of controlled composition or dissolved oxygen standard solutions measured by well-accepted Winkler procedures (103). This calibration should be made with great care to avoid systematic error in the measurements. When these precautions are observed good agreement is reported between classic methods and this biosensor (42). This difficulty may be considered a drawback with regard to the simplicity of employing a hydrogen peroxide sensor.

With a bioprobe the linear calibration range obtained depends on the type of enzyme involved in the sensor design and may vary greatly. Linearity is generally observed up to substrate concentrations lower than the Michaelis constant value, except in some diffusion-limited systems (46,104).

Selectivity and Interferences

The amperometric detection is a selective analytic method in of itself. At a fixed and determined potential only some molecules can electrochemically react on the sensor. The presence of the membrane allows one to increase this selectivity. Particularly in the O_2 probe, the hydrophobic film, which is highly permeable to gases, avoids interference due to ionic or organic species. Moreover, a membrane with different permeabilities to different gases in fact acts as a filter and gives this sensor good selectivity for O_2. In the H_2O_2 sensor the membrane prevents deposition of impurities on the electrode tip, thus poisoning the electrochemical element. The ultimate selectivity is obviously provided by the specificity of the enzymic system.

When an interfering species is present the choice, when possible, of an adapted polarization potential minimizes to a relatively large degree the contribution of this interference (105), as exemplified by the recent development of mediated or modified electrodes.

In some cases, devices using multilayer membranes (106,107) allow one to avoid the influence of electroactive interferent species, but the response time may be increased.

Another method for continuously eliminating the effect of interfering species is the differential system constituted by two sensors as developed in the laboratory with the microprocessor analyzer Glucoprocesseur. One of the sensors is covered by the enzymic membrane; the other, called compensating electrode, is covered by the same membrane without enzyme. The result is obtained by continuous subtraction of the two currents (101).

SUBSTRATE DETERMINATIONS

Most of the papers published thus far deal with analyte determinations in the biomedical field or the food industry. It mut be pointed out that experiments have been mainly conducted in the laboratory in reconstituted media as close as possible to real operating conditions on site. The main results are presented in Table 1 for biological fluids and in Table 2 for foodstuff analysis. Glucose and L-lactate determinations are presented in Table 3 which deals with commercially available analyzers.

ENZYME ACTIVITY DETERMINATIONS

The determination of various enzyme activities in physiological media is of clinical importance for the diagnosis of myocardial or liver damage (hepatitis and cirrhosis), cardiac infection, cancer, and other diseases. Routinely, enzyme activities in biologic media (including serum, whole blood, and urine) are determined by spectrophotometric methods in which expensive reagents, such as coenzymes [NAD(P)H], are necessary. These methods usually require sample pretreatment, and the procedure is time consuming. Only recently have a few reports dealt with the use of biosensors for the measurement of enzyme activities.

The determination of α-amylase (EC 3.2.1.1) activity using a single enzyme sensor has been reported by Osawa et al. (137) using an analyzer (Gluco-20, Fuji, Japan) that requires only 20 μl sample. Complete measurement was achieved in 110 sec. The coefficients of variation (CV) for the α-amylase activity determination were 5.3 and 4.4% at 727 and 1030 IU/L, respectively.

$$\text{Maltopentaose} + H_2O \xrightarrow{\alpha\text{-amylase}} \text{maltose} + \text{maltotriose}$$

$$\text{Maltose} + \text{maltotriose} + H_2O \xrightarrow{\alpha\text{-glucosidase}} \text{D-glucose}$$

with α-glucosidase added to the buffer.

The determination of α-amylase activity using starch as substrate and a bienzyme sensor (glucoamylase and glucose oxidase) was reported by Renneberg et al. (138) and Yoda and Tsuchida (139).

Nikoleis and Mottola (140), using sample injection in closed flow-through systems with glucose oxidase containing maltase as an impurity and amperometric measurement of oxygen decrease, measured α-amylase activity with starch as substrate. Determinations were achieved from 500 to 5000 IU/L with a relative standard deviation (RSD) of about 2% and a maximum determination rate of 120 injections per hr.

Blum et al. (141) described a sensitive method for the rapid determination of the activity of several oxidases (glucose oxidase, EC 1.1.3.4; galactose oxidase, EC 1.1.3.9; and cholesterol oxidase, EC 1.1.3.6) and hydrolases (invertase, EC 3.2.1.26; β-galactosidase, EC 3.2.1.23; and glucoamylase, EC 3.2.1.3), for either soluble or immobilized enzymes, based on the electrochemical detection of hydrogen peroxide. When associated with an immobilized glucose oxidase membrane this sensor allows the determination of the activity of any enzyme which generates glucose and no preincubation step is required. Good reproducibility was found, and the results compared well with those obtained with other techniques.

A pyruvate sensor was used to determine lactate dehydrogenase (EC 1.1.1.27) activity in human sera (142). Lactate dehydrogenase (LDH) catalyzes the pyruvate-producing reaction

$$\text{L-lactate} + NAD^+ \xrightarrow{\text{LDH}} \text{pyruvate} + NADH + H^+$$

Mizutani et al. (143) thus described the sequential determination of L-lactate and LDH activity using a lactate oxidase sensor according to the reaction

$$\text{L-lactate} + O_2 \xrightarrow{\text{lactate oxidase}} \text{pyruvate} + H_2O_2$$

A simple glucose oxidase enzyme electrode was used for determination of the activity of alkaline phosphatase (AP; EC 3.1.3.1) in very turbid media based on the hydrolysis of glucose-6-phosphate and measurement of increasing glucose level:

$$\text{Glucose-6-phosphate} + H_2O \xrightarrow{\text{AP}} \text{glucose} + \text{phosphate}$$

Application of such an electrode to a commercially available routine analyzer (GMK 01, GDR) was proposed (144).

Weigelt et al. (145) described a pyruvate-sensing bienzyme electrode based on lactate dehydrogenase, lactate monooxygenase (LMO), and an oxygen probe for pyruvate kinase (EC 2.7.1.40) determination in erythrocytes. Pyruvate kinase (PK) can be measured in the range 0.12–1.85 μmol per sample.

$$\text{Phosphoenolpyruvate} + \text{ADP} \xrightarrow{\text{PK}} \text{pyruvate} + \text{ATP}$$

$$\text{Pyruvate} + \text{NADH} + H^+ \xrightarrow{\text{LDH}} \text{L-lactate} + \text{NAD}^+$$

$$\text{L-lactate} + O_2 \xrightarrow{\text{LMO}} \text{acetate} + CO_2 + H_2O$$

Coimmobilization of PK with the two other enzymes, LDH and LMO, leads to a trienzyme sensor for creatine kinase (EC 2.7.3.2) determination in the range 0.1–1.76 μmol per sample:

$$\text{Creatine phosphate} + \text{ADP} \xrightarrow{\text{CK}} \text{creatine} + \text{ATP}$$

The measurement is based on the ability of this electrode to sense ADP. In this system CK competes with PK for ADP, and the overall process is limited by ADP diffusion in the trienzyme membrane. The sensor appeared suitable for human serum determination, but the sum of L-lactate and pyruvate concentrations should not exceed 6.5 mM.

The MB isoenzyme of serum creatine kinase was determined by Fonong (146) using an immunoinhibition amperometric technique. Glycerokinase (GK) and glycerol-3-phosphate oxidase (G-3-POD) were coimmobilized on a collagen membrane at the surface of a platinum electrode polarized at 700 mV:

$$\text{Phosphocreatine} + \text{ADP} \xrightarrow{\text{CK (MB)}} \text{creatine} + \text{ATP}$$

$$\text{ATP} + \text{glycerol} \xrightarrow{\text{GK}} \text{glycerol-3-phosphate} + \text{ADP}$$

$$\text{Glycerol-3-phosphate} + O_2 \xrightarrow{\text{G-3-POD}} \text{dihydroxyacetone phosphate} + H_2O_2$$

One of the reaction products, H_2O_2, was measured amperometrically and related to the activity of creatine kinase. Calibration curves using human standards were linear from 10 to 500 IU/L with a detection limit of 8 IU/L. The precision of the method (RSD) was 3.1% at a level of 10 IU/L of CK. The method requires 30 min incubation and then about 10 min for measurement and needs an injection of only 100 μl serum.

The determination of transaminases, glutamate pyruvate transaminase (EC 2.6.1.2) and glutamate oxaloacetate transaminase (EC 2.6.1.1), is also very important. Normal activities in serum are 5–24 IU/L for glutamate pyruvate transaminase (GPT) and 5–20 IU/L for glutamate oxaloacetate transaminase (GOT) but they can increase 100 to 1000-fold in some diseases. Fonong and Barber (147) described a method for the separate determination of these transaminases based on the use of lactate dehydrogenase and malate dehydrogenase (MDH) and on the oxidation of NADH at a platinum electrode polarized at 700 mV:

$$\text{L-alanine} + \alpha\text{-ketoglutarate} \xrightarrow{\text{GPT}} \text{pyruvate} + \text{L-glutamate}$$

$$\text{Pyruvate} + \text{NADH} + H^+ \xrightarrow{\text{LDH}} \text{L-lactate} + \text{NAD}^+$$

$$\text{L-aspartate} + \alpha\text{-ketoglutarate} \xrightarrow{\text{GOT}} \text{oxaloacetate} + \text{L-glutamate}$$

$$\text{Oxaloacetate} + \text{NADH} + H^+ \xrightarrow{\text{MDH}} \text{L-malate} + \text{NAD}^+$$

The NADH consumption rate can be related to the activity of the dehydrogenases. Using LDH and MDH coimmobilized on a collagen membrane placed on the surface of a platinum anode, GPT and GOT could be separately determined using a Yellow Springs Instrument Co. (YSI) Model 2510 oxidase probe and a YSI Model 25 current meter. Calibration curves were almost linear from 0 to 40 IU/L for GPT with a detection limit of 2 IU/L. The calibration graph for GOT was linear and extended from 0 to 50 IU/L with a detection limit of 2 IU/L. No significant decrease in the activity of the immobilized enzymes was observed after about 140 determinations in 14 days. After myocardial or liver damage values in serum were much higher and dilution was needed.

Wollenberger et al. (148) used a glutamate electrode with a glutamate oxidase membrane prepared by gelatin entrapment sandwiched between two dialysis membranes and attached to a hydrogen peroxide electrode for the cumulative detection of both transaminases in a 20–2000 IU/L concentration range:

$$\text{L-glutamate} + O_2 + H_2O \xrightarrow{\text{glutamate oxidase}} \alpha\text{-ketoglutarate} + NH_3 + H_2O_2$$

Kihara et al. (149) applied a bienzyme electrode comprising oxaloacetate decarboxylase and pyruvate oxidase to the successive measurement of both transminases based on alanine addition after determination of GOT activity. The measuring range was up to 1500 IU/L for each enzyme. GOT and GPT activities were determined sequentially. The addition of alanine dilutes the GOT in the measurement cell. The method does not provide an optimal concentration of this substrate for GPT and consequently does not lead to an accurate measurement of GPT activity.

We recently reported transminase determinations performed separately, in sequence, or simultaneously (105). A microprocessor-based analzyer (Glucoprocesseur), including two electrodes and involving the oxidation of hydrogen peroxide produced by oxidases, was used for the fast and reliable determination of GOT and GPT over a wide range of concentrations. GPT and GOT were determined either separately or successively with a specific enzyme electrode using pyruvate oxidase alone or mixed with oxaloacetate decarboxylase. With a two-electrode–based analyzer using pyruvate oxidase immobilized on one electrode and oxaloacetate decarboxylase and pyruvate oxidase coimmobilized and set on the second electrode (Fig. 4), GOT and GPT activities were simultaneously determined. Results were obtained in less than 4 min under the same conditions at optimal substrate concentrations. A linear range of 6–30,000 IU/L in the sample, a determination limit of 3 IU/L, with RSD values of 4.1% and correlation coefficients with the commercial kit of 0.997 and 0.996 for GPT and GOT, respectively, were obtained. This procedure avoids the drawback of the addition of alanine after GOT determination.

A biosensor system based on an oxygen electrode for the continuous-flow determination of glucose oxidase and lactate dehydrogenase activities (in this case associated with a pyruvate oxidase membrane) was recently developed by Okuma et al. (150). Calibration curves for glucose oxidase and lactate dehy-

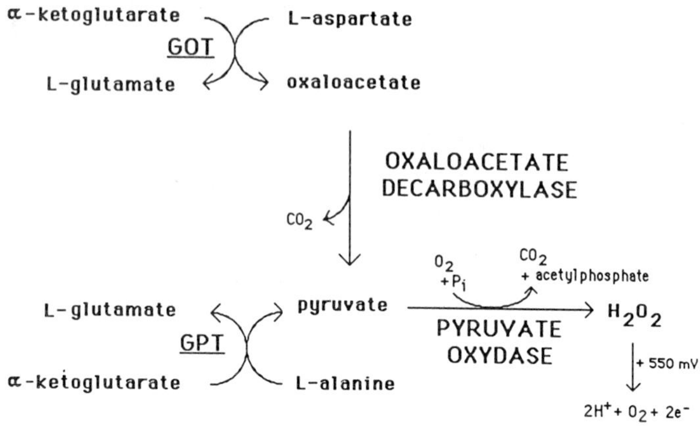

Figure 4 General scheme for determination of transaminase activity. (*From Ref. 105*.)

drogenase activities were linear up to 81 and 300 units (50 μl sample injection), respectively. One assay could be completed within 15 min, and sensors were stable for more than 25 days at 5°C.

The determination of serum cholinesterase activity plays an important role in anesthesiology.

$$\text{Acetylcholine or choline esters} + H_2O \xrightarrow{\text{choline esterase}} \text{choline} + \text{acetate or organic acid}$$

$$\text{Choline} + 2O_2 \xrightarrow{\text{choline oxidase}} \text{betaïne} + 2H_2O_2$$

Biosensors based on O_2 and H_2O_2 determinations have been proposed by Mizutani and Tsuda (151) and Yao (152). Although these methods offer good accuracy and precision, they lack sensitivity and suffer from some interference.

More recently, Palleschi et al. (153) determined serum cholinesterase activity using a hydrogen peroxide electrode and choline oxidase immobilized on a nylon net, 30 determinations per hr can be performed using only 5 μl serum samples. The lifetime of the probe is about 1–3 months (400 analyses).

COMMERCIAL DEVICES

Only a few biosensor-based instruments are commercially available. The most commonly used biosensors are the glucose and L-lactate electrodes using O_2 or H_2O_2 detection associated with glucose oxidase and lactate oxidase, respectively. The main characteristics and performance of these devices are presented in Table 3.

Table 3 Commercial Devices Based on Amperometric Enzyme Electrodes

Manufacturer	Instrument	Analyte	Technology	Sample (μl)	Linear Range	Frequency	Precision	Stability	Remarks
AIC, Anal. Inst. Corp., Japan	Glucoroder E	Glucose	Immobilized GOD[a], oxygen electrode	5-40	0-55 mM serum 0-330 mM urine	120-150 samples per hr	ND[b]	6000 assays or 2 months use	
Baxter Travenol Lab., United States	Exactech	Glucose	GOD + ferrocene, (test strips)	Drop of whole blood	2.2-25 mM	30 sec	3.3%, 15 mM 8%, 3 mM	One-shot test	No replaceable battery; world's smallest blood glucose meter no larger than a pen; technology easy to use, fast and accurate.
DKK, Denki Kagaku Kaku, Japan	BAS 1	Glucose L-ascorbic acid L-lactic acid	Enzyme membrane, oxygen electrode	ND	0-200 mg/L 0-150 mg/L 0-60 mg/L	Response time 1 min	ND	1-3 months in use, 2-6 months on storage, 10°C	
Fuji Electric Co., Japan	Gluco 20 and amylase glucose analyzer	Glucose	Immobilized GOD, H_2O_2 electrode	20	0-27 mM	80 samples per hr	1-7%	>500 samples	
	UA-300	Uric acid	Immobilized uricase, H_2O_2 electrode	20	10-900 mg/L	60 samples per hr	0.75%, urine; 2.6%, plasma	ND	Whole blood, plasma, urine serum; not interfered by up to 200 mg/L of ascorbic acid, 100 mg/L of bilirubin
Kyoto Daiichi Kagaku, Japan	Glucose Auto and Stat GA-1120	Glucose	Immobilized GOD, H_2O_2 electrode	100-200	10-1000 mg/dl	35 samples per hr	ND	1000 assays	Standardized periodically
Markwell Medical, United States	Direct 30/30	Glucose	GOD membrane, H_2O_2 electrode	7-50	Up to 500 mg/dl	Response time 30 sec	2-5%	enzyme membrane cartridge, 30 days	Whole blood interferes with plasma (<5 %), self-monitor pocket portable

Table 3 (*Continued*)

Manufacturer	Instrument	Analyte	Technology	Sample (μl)	Linear Range	Frequency	Precision	Stability	Remarks
Omron, Tateisi Electronics Co., Japan	HER-100	Lactate	Immobilized LOD[c], H_2O_2 electrode	100	0-80 mg/dl	Response time < 80 sec	< 5% (50 mg/dl)	ND	
Provesta Corporation, United States	Bioanalyzer	Glucose L-lactate Alcohols Lactose	Enzyme gel maintained with a semi-permeable membrane, oxygen electrode	10-100	10 mg/L glucose 5 mg/L L-lactate 1 mg/L alcohols 20 mg/L lactose	4-5 min for a sample	5%	ND	
Seres, France	Enzymat	Glucose Lactate Lysine	Co-cross-linking of oxidase + gelatin, oxygen electrode		0.1-4 g/L glucose 0.1 g/L L-lactate 0.1 g/L lysine	2 min for a sample	ND	ND	
Setric SGI, France	Microzym-L	L-Lactate	Yeast lactate dehydrogenase + hexacyanoferrate (III /II)	60-100 10-fold diluted	1-10 mM threshold 0.05 mM	30-60 samples per hr	0.2 mM	200 samples or 2 weeks with one refill	Energetic autonomy (battery): 6 hr, medicine, food and fermentation, cell culture
Solea-Tacussel, France	Glucoprocesseur	Glucose	GOD membrane, H_2O_2 electrode	5-500	0.05-500 mM	40 samples per hr	1-2 %	2000 assays or 12 months	30 or 37°C; usable for L-lactate, oxalate or transaminase determinations
Toa Electronics Ldt., Japan	FGA-1	Glucose	GOD membrane, H_2O_2 electrode	2 ml	0-20 g/dl	6 min.	ND	ND	Combination of enzyme sensor and flow injection analysis; on-line; automatic (every 1-9 hr) or manual calibration

Universal Sensors, United States	ABD	Glucose	Immobilized GOD, H_2O_2 or oxygen electrode		0.005 - 100 mg/dl	30-60 samples per hr	<3 %	500 assays 6 months	Enzyme membranes system for sucrose, lactate, oxalate, alcohol and son on
YSI (Yellow Springs Instrument Co.), United States	YSI 27	Glucose Sucrose Alcohol L-lactate Lactose	Kits with imobilized oxidase enzymes, H_2O_2 electrode	25 25 5 10-25 25	0-20 g/L glucose 0-20 g/L sucrose 0- g/L alcohol 0-5 g/L lactose	20 samples per hr	2% (glucose) Sensitivity, 1 mg/dl (glucose)	10 days	For sucrose in free glucose medium
	YSI 2000	Glucose + L-lactate	Immobilized oxidase enzymes, H_2O_2 electrode	550	0-20 g/L glucose 0-2 g/l L-lactate	60 samples per hr	0.04 g/L (2.5 %)	3 weeks (glucose) 2 weeks (L-lactate)	Fermentation broths, cell culture, control on-line, quality and production control, manually or automatic sampler
	YSI 2300 G YSI 2300 L YSI 2300 GL	Glucose L-lactate Glucose + L-lactate	Glucose oxidase Lactate oxidase	25	0-5 g/L glucose 0-1.35 g/L L-lactate	Response time 90 sec	ND	ND	Autocalibration every 15 min
ZWG, Germany	GM K 01	Glucose	Immobilized GOD	20-50	0.5-50 mM glucose	60 - 90 samples per hr	1.5%	>1000 assays	
		Uric acid	Immobilized uricase	100	0.2-1 mM uric acid	40 samples per hr	1.2 %		

[a] Glucose oxidase
[b] Not done
[c] Lactate oxidase

CONCLUSION

Among the different types of biosensors that have been described, amperometric enzymes electrodes are the first to be commercialized. A few, however, led to the design of reliable commercial devices. Depending on the enzyme used, different analytes, such as glucose, oxalate, lactose, L-lactate, and pyruvate (20,24–27), have been monitored in the laboratory with an analyzer (Glucoprocesseur) using a very simple and fast procedure allowing the appropriate enzyme membranes to be prepared cheaply and reliably (23). The differential two-electrode system allows the detection and subtraction of interference in complex media (foodstuffs, fermented milk, and blood). The use of the same analyzer equipped with two enzyme electrodes could be extended to the accurate and simultaneous determination of both GOT and GPT. This approach, also chosen by Scheller and his group (145,154,155), who used different bioactive membranes with another analyzer (Glukometer), may make such biosensors more widely acceptable in analytical and industrial processes.

Special mention must be made of another approach involving miniaturization and a new design as exemplified by the Exactech pen-sized probe for whole-blood glucose determinations developed at Cranfield (156).

Provided an increase in stability could be obtained, the development of highly reliable enzyme electrodes can be expected.

REFERENCES

1. Clark, L. C. Jr. (1959). Electrochemical device for chemical analysis, U.S. Patent 2,913,386.
2. Clark, L. C. Jr., and Lyons, C. (1962). Electrode systems for continuous monitoring in cardiovascular surgery, *Ann. NY Acad. Sci., 102*:29.
3. Guilbault, G. G. (1984). *Analytical Uses of Immobilized Enzymes*, Marcel Dekker, New York.
4. Turner, A. P. F., Karube, I., and Wilson, G. S. (1987). *Biosensors, Fundamentals and Applications*, Oxford University Press, New York.
5. Wilson, G. S. (1987). Fundamentals of amperometric sensors. In: *Biosensors, Fundamentals and Applications* (A. P. F. Turner, I. Karube, and G. S. Wilson, Eds), Oxford University Press, New York, p. 165.
6. Albery, W. J., and Craston, D. H. (1987). Amperometric enzyme electrodes: theory and experiment. In: *Biosensors, Fundamentals and Applications* (A. P. F. Turner, I. Karube, and G. S. Wilson, Eds), Oxford University Press, New York, p. 180.
7. Barker, S. A. (1987). Immobilization of the biological component of biosensors. In: *Biosensors, Fundamentals and Applications* (A. P. F. Turner, I. Karube, and G. S. Wilson, Eds), Oxford University Press, New York, p. 85.
8. Kennedy, J. F., and White, C. A. (1985). Data on techniques of enzyme immobilization and bioaffinity procedures. In: *Handbook of Enzyme Biotechnology* (A. Wiseman, II), John Wiley & Sons, Chichester, p. 380.

9. Updike, J. W., and Hicks, J. P. (1967). The enzyme electrode, *Nature (Lond.), 214*: 986.
10. Kirstein, D., Kirstein, L., and Scheller, F. (1985). Enzyme electrode for urea with amperometric indication. Part I. Basic principle, *Biosensors, 1*: 117.
11. Thomas, D., Barbotin, J. N., David, A., Hervagault, J. F., and Romette, J. L. (1977). Experimental evidence for a kinetic and electrochemical memory in enzyme membranes, *Proc. Natl. Acad. Sci. USA, 74*: 5314.
12. Romette, J. L., Kernevez, J. P., and Thomas, D. (1974). Memory in enzyme membranes, *Nature, 249*: 490.
13. Romette, J. L., and Cooney, C. L. (1987). L-glutamine enzyme electrode for on-line mammalian cell culture process control, *Anal. Lett., 20*: 1069.
14. Guilbault, G. G., and Lubrano, G. J. (1973). A enzyme electrode for the amperometric determination of glucose, *Anal. Chim. Acta, 64*: 439.
15. Tsuchida, T., and Yoda, K. (1981). Immobilization of D-glucose oxidase onto a hydrogen peroxide permselective membrane and application for an enzyme electrode, *Enzyme Microb. Technol., 3*:326.
16. Karube, I., and Suzuki, S. (1972). Electrochemical aggregation of tropocollagen, *Biochem. Biophys. Res. Commun., 48*: 320.
17. Coulet, P. R., Julliard, J. H., and Gautheron, D. C. (1974). A mild method of general use for covalent coupling of enzymes to chemically activated collagen films, *Biotechnol. Bioeng., 16*: 1055.
18. Coulet, P. R., and Bertrand, C. (1979). Asymmetrical coupling of enzymatic systems on collagen membranes. Application to multienzyme electrodes, *Anal. Lett., 12*: 581.
19. Bertrand, C., Coulet, P. R., and Gautheron, D. C. (1979). Enzyme electrode with collagen-immobilized cholesterol oxidase for the microdetermination of free cholesterol, *Anal. Lett., 12*: 1477.
20. Bertrand, C., Coulet, P. R., and Gautheron, D. C. (1981). Multipurpose electrode with different enzyme systems bound to collagen films, *Anal. Chim. Acta, 126*: 23.
21. Coulet, P. R., and Blum, L. J. (1983). Fast electroenzymatic measurement of total cholesterol in serum, *Anal. Lett., 16*: 541.
22. Guilbault, G. G., and Coulet, P. R. (1983). Creatinine-selective enzyme electrodes, *Anal. Chim. Acta, 152*: 223.
23. Assolant-Vinet, C. H., and Coulet, P. R. (1986). New immobilized enzyme membranes for tailor-made biosensors, *Anal. Lett., 19*: 875.
24. Bardeletti, G., Sechaud, F., and Coulet, P. R. (1986). A reliable L-lactate electrode with a new membrane for enzyme immobilization for amperometric assay of lactate, *Anal. Chim. Acta, 187*: 47.
25. Assolant-Vinet, C. H., Bardeletti, G., and Coulet, P. R. (1987). A novel enzyme membrane electrode for oxalate determination in foodstuffs, *Anal. Lett., 20*: 513.
26. Sechaud, F., Peguin, S., Coulet, P. R., and Bardeletti, G. (1989). Fast and reliable organic acid determination in fermented milk using an enzyme-electrode based analyser, *Proc. Biochem., 24*: 33.
27. Morelis, R. M., and Coulet, P. R. (1990). Sensitive biosensor for choline and acetylcholine involving fast immobilization of a bienzyme system on a disposable membrane, *Anal. Chim. Acta, 231*: 27.
28. Mascini, M., Iannello, M., and Palleschi, G. (1983). Enzyme electrodes with

improved mechanical and analytical characteristics obtained by binding enzymes to nylon nets, *Anal. Chim. Acta, 146*: 135.
29. Bardeletti, G., and Coulet, P. R. (1984). Air-dried enzyme electrodes, *Anal. Chem., 56*: 591.
30. Moody, G. J., Sanghera, G. S., and Thomas, J. D. R. (1988). Factors concerning the design and calibration of an amperometric enzyme electrode system for the flow injection analysis of cholesterol, *Analyst, 113*: 1419.
31. Sternberg, R., Bindra, D. S., Wilson, G. S., and Thevenot, D. R. (1988). Covalent enzyme coupling on cellulose acetate membranes for glucose sensor development, *Anal. Chem., 60*: 2781.
32. Sternberg, R., Barrau, M. B., Gangiotti, L., Thevenot, D. R., Bindra, D. S., Wilson, G. S., Vehlo, G., Froguel, P., and Reach, G. (1988). Study and development of multilayer needle-type enzyme-based glucose microsensors, *Biosensors*, 4: 27.
33. Bradley, J., Kidd, A. J., Anderson, P. A., Dear, A. M., Ashby, R. E., and Turner, A. P. F. (1989). Rapid determination of glucose content of molasses using a biosensor, *Analyst, 114*: 375.
34. Bartlett, P. N., and Whitaker, R. G. (1987). Electrochemical immobilisation of enzymes. Part I. Theory, *J. Electroanal. Chem., 224*: 27.
35. Bartlett, P. N., and Whitaker, R. G. (1988). Strategies for the development of amperometric enzyme electrodes, *Biosensors, 3*: 359.
36. Johnson, D. C., Ryan, M. D., and Wilson, G. S. (1986). Dynamic electrochemistry: Methodology and application, *Anal. Chem., 58*: 33R.
37. Shu, F. R., and Wilson, G. S. (1976). Rotating ring-disk enzyme electrode for surface catalysis studies, *Anal. Chem., 48*: 1679.
38. Kamin, R. A., and Wilson, G. S. (1980). Rotating ring-disk enzyme electrode for biocatalysis kinetic studies and characterization of the immobilized enzyme layer, *Anal. Chem., 52*: 1193.
39. Bourdillon, C., Bourgeois, J. P., and Thomas, D. (1980). Covalent linkage of glucose oxidase on modified glassy carbon electrodes. Kinetic phenomena, *J. Am. Chem. Soc., 102*: 4231.
40. Junter, G. A. (1988). *Electrochemical Detection Techniques in the Applied Biosciences*, Vols. 1 and 2, Ellis Horwood Ltd., Chichester.
41. Bard, A. J., and Faulkner, L. R. (1983). *Electrochimie—Principes, Méthodes et Applications*, Masson, Paris.
42. Hitchman, M. L. (1978). *Measurement of Dissolved Oxygen*, John Wiley and Sons, New York.
43. Mell, L. D., and Maloy, J. T. (1976). Amperometric response enhancement of the immobilized glucose oxidase enzyme electrode, *Anal. Chem., 48*: 1597.
44. Bradley, C. R., and Rechnitz, G. A. (1984). Kinetic analysis of enzyme electrode response, *Anal. Chem., 56*: 664.
45. Bergel, A., and Comtat, M. (1984). Theoretical evaluation of transient responses of an amperometric enzyme electrode, *Anal. Chem., 56*: 2905.
46. Tse, P. H. S., and Gough, D. A. (1987). Transient response of an enzyme electrode sensor for glucose, *Anal. Chem., 59*: 2339.
47. Schulmeister, T. (1987). Mathematical treatment of concentration profiles and

anodic current of amperometric enzyme electrodes with chemically-amplified response, *Anal. Chim. Acta, 201*: 305.
48. Ylilammi, M., and Lehtinen, L. (1988). Numerical analysis of a theoretical one-dimensional amperometric enzyme sensor, *Med. Biol. Eng. Comput., 26*: 81.
49. Lemke, K. (1988). Mathematical simulation of an amperometric enzyme-substrate electrode with a pO_2 basic sensor. Part 1 and Part 2, *Med. Biol. Eng. Comput., 26*: 523.
50. Kulys, J. J., Sorochinskii, V. V., and Vidziunaite, R. A. (1986). Transient response of bienzyme electrodes, *Biosensors, 2*: 135.
51. Mancy, K. H., Okun, D. A., and Reilley, C. N. (1962). A galvanic cell oxygen analyzer, *J. Electroanal. Chem., 4*: 65.
52. Hoare, J. P. (1968). *The Electrochemistry of Oxygen*, Interscience, New York.
53. Cardosi, M. F., and Turner, A. P. F. (1987). Glucose sensors for the management of diabetes mellitus. In: *The Diabetes Annual, Vol. 3* (K. G. M. M. Alberti and L. P. Krall, Eds), Elsevier, Amsterdam, p. 560.
54. Williams, D. L., Doig, A. R., and Korosi, A. (1970). Electrochemical enzymatic analysis of blood glucose and lactate, *Anal. Chem., 42*: 118.
55. Asperger, L., Geppert, G., and Krabish, C. H. (1987). Comparison of amperometric measuring principles for determinations of glucose with electrodes based on glucose oxidase, *Anal. Chim. Acta, 201*: 281.
56. Durliat, H., Comtat, M., Mahenc, J., and Baudras, A. (1975). Dosage ampérometrique du lactate à l'aide d'une électrode à enzyme, *J. Electroanal. Chem., 66*: 73.
57. Comtat, M., Galy, M., Goulas, P., and Souppe, J. (1988). Amperometric bi-enzyme electrode for L-carnitine, *Anal. Chim. Acta, 208*: 295.
58. Kitagawa, Y., Kitabatake, K., Kubo, I., Tamiya, E., and Karube, I. (1989). Alcohol sensor based on membrane-bound alcohol dehydrogenase, *Anal. Chim. Acta, 218*: 61.
59. Cosgrove, M., Moody, G. J., and Thomas, J. D. R. (1988). Chemically immobilized enzyme electrodes for hydrogen peroxide determination, *Analyst, 113*: 1811.
60. Cass, A. E. G., Davis, G., Francis, G. D., Hill, H. A. O., Aston, W. J., Higgins, I. J., Plotkin, E. V., Scott, L. D. L., and Turner, A. P. F. (1984). Ferrocene-mediated enzyme for amperometric determination of glucose, *Anal. Chem., 56*: 667.
61. Cass, A. E. G., Davis, G., Green, M. J., and Hill, H. A. O. (1985). Ferricinium ion as an electron acceptor for reductases, *J. Electroanal. Chem., 190*: 117.
62. Brooks, S. L., Ashby, R. E., Turner, A. P. F., Calder, M. R., and Clarke, D. J. (1987). Development of an on-line glucose sensor for fermentation monitoring, *Biosensors, 3*: 45.
63. Szentrimay, R., Yeh, P., and Kuwana, T. (1977). Evaluation of mediator-titrants for the indirect coulometric titration of biocomponents, *Am. Chem. Soc. Symp., 38*: 143.
64. Dicks, J. M., Aston, W. J., Davis, G., and Turner, A. P. F. (1986). Mediated amperometric biosensors for D-galactose, glycolate and L-amino acids based on a ferrocene-modified carbon paste electrode, *Anal. Chim. Acta, 182*: 103.

65. Green, M. J., and Hill, H. A. O. (1986). Amperometric enzyme electrodes, *J. Chem. Soc., 82*: 1237.
66. Jönsson, G., and Gorton, L. (1987). An amperometric glucose electrode based on adsorbed glucose oxidase on palladium/gold modified graphite, *Anal. Lett., 20*: 839.
67. Hintsche, R., and Scheller, F. W. (1987). Enzyme adsorption at electrode surfaces: an enzyme sensor with adjustable sensitivity, *Stud. Biophys., 119*: 179.
68. Ikariyama, Y., Yamauchi, S., Yukiashi, T., and Ushioda, H. (1987). Micro-enzyme electrode prepared on platinized platinum, *Anal. Lett., 20*: 1407.
69. Bennetto, H. P., DeKeyzer, D. R., Delaney, G. M., Koshy, A., Mason, J. R., Razack, L. A., Stirling, J. L., and Thurston, C. F. (1987). An amperometric biosensor, *Int. Anal. 8*: 22.
70. Kobos, R. K. (1987). Enzyme-based electrochemical biosensors, *Trends Anal. Chem. 6*: 6.
71. Palleschi, G., Rathore, H. S., and Mascini, M. (1988). Amperometric probe for 3-hydroxybutyrate with immobilized 3-hydroxybutyrate dehydrogenase, *Anal. Chim. Acta., 209*: 223.
72. Kulys, J. J., Cenas, N. K., and Kanapieniene, J. J. (1987). Electrochemical regeneration of NAD^+ on chemically modified electrodes and its application to biosensors, *Stud. Biophys., 119*: 175.
73. Gorton, L. (1986). Chemically modified electrodes for the elctrocatalytic oxidation of nicotinamide coenzymes, *J. Chem. Soc., 82*: 1245.
74. Bergel, A., and Comtat, M. (1986). Electroenzymatic reactors with coenzyme regeneration: A theoretical approach, *Biotechnol. Bioeng. 28*: 728.
75. Bourdillon, C., Laval, J. M., and Thomas, D. (1986). Enzymatic electrocatalysis: Controlled potential electrolysis and cosubstrate regeneration with immobilized enzyme modified electrode, *J. Electrochem. Soc., 133*: 706.
76. Oyama, N., Ohsaka, T., Mizunuma, M., and Kobayashi, M. (1988). Electro-polymerized cobalt tetrakis(*o*-aminiphenyl)porphyrin film mediated enzyme electrode for amperometric determination of glucose, *Anal. Chem., 60*: 2534.
77. Ianniello, R. M., and Yacynych, A. M., (1981). Immobilized enzyme chemically modified electrode as an amperometric sensor, *Anal. Chem., 53*: 2090.
78. Yao, T. (1983). A chemically-modified enzyme membrane electrode as an amperometric glucose sensor, *Anal. Chim. Acta, 148*: 27.
79. Ianniello, R. M., Lindsay, T. J., and Yacynych, A. M. (1982). Immobilized xanthine oxidase chemically modified electrode as a dual analytical sensor, *Anal. Chem., 54*: 1980.
80. Narasimhan, K., and Wingard, L. B. Jr. (1986). Immobilization of flavins on electrode surfaces. Part I and Part II, *J. Mol. Catal., 34*: 253.
81. Foulds, N. C., and Lowe, C. R. (1986). Enzyme entrapment in electrically conducting polymers immobilization of glucose oxidase in polypyrrole and its application in amperometric glucose sensors, *J. Chem. Soc., 82*: 1259.
82. Shinohara, H., Chiba, T., and Alizawa, M. (1988). Enzyme microsensor for glucose with an electrochemically synthetized enzyme polyaniline film, *Sens. Actuat., 13*: 79.
83. Cenas, N. K., Pocius, A. K., and Kulys, J. J. (1983). Electron exchange between

flavin- and heme-containing enzymes and electrodes modified by redox polymers, *Bioelectrochem. Bioenerg.*, *11*: 61.

84. Gunasingham, H., and Tan, C. B. (1989). Platinum-dispersed Nafion film modified glassy carbon as an electrocatalytic surface for an amperometric glucose enzyme electrode, *Analyst*, *114*: 695.
85. Ikeda, T., Hamada, H., and Senda, M. (1986). Electrocatalytic oxidation of glucose at a glucose oxidase-immobilized benzoquinone-mixed carbon paste electrode, *Agr. Biol. Chem.*, *50*: 883.
86. Bartlett, P. N., and Whitaker, R. G. (1987). Electrochemical immobilisation of enzymes. Part II. Glucose oxidase immobilised in poly-*N*-methylpyrrole, *J. Electroanal. Chem.*, *224*: 37.
87. Yabuki, S., Shinohara, H., and Aizawa, M. (1989). Electro-conductive enzyme membrane, *J. Chem. Soc.*, *14*: 945.
88. Beh, S. K., Moody, G. J., and Thomas, J. D. R. (1989). Effect of pre-treatment of platinum for modified platinum wire glucose oxidase amperometric electrodes, *Analyst*, *114*: 29.
89. Ikariyama, Y., Yamauchi, S., Yukiashi, T., and Ushioda, H. (1988). Surface control of platinized platinum as a transducer matrix for micro-enzyme electrodes, *J. Electroanal. Chem.*, *251*: 267.
90. Genshaw, M. A. (1988). Enzyme electrode for determining glucose in whole blood, *Clin. Chem.*, *34*: 1717.
91. Genshaw, M. A., and Jones, J. E. (1989). Whole blood glucose enzyme electrode, *J. Electrochem. Soc.*, *136*: 414.
92. Moody, G. J., Sanghera, G. S., and Thomas, J. D. R. (1986). Modified platinum wire glucose oxidase amperometric electrode, *Analyst*, *111*: 1235.
93. Tatsuma, T., Okawa, Y., and Watanabe, T. (1989). Enzyme monolayer- and bilayer-modified tin oxide electrodes for the determination of hydrogen peroxide and glucose, *Anal. Chem.*, *61*: 2352.
94. Garcia, O. J., Quintela, P. A., and Kaifer, A. E. (1989). Electrodes modified with a film of phosphatidylcholine: electrochemistry inside a lipid layer, *Anal. Chem.*, *61*: 979.
95. Davis, G. (1985). Electrochemical techniques for the development of amperometric biosensors, *Biosensors*, *1*: 161.
96. Davis, G., Green, M. J., and Hill, H. A. O. (1986). Detection of ATP and creatine kinase using an enzyme electrode, *Enzyme Microb. Technol.*, *8*: 349.
97. Suaud-Ghagny, M. F., and Gonon, F. G. (1986). Immobilization of lactate dehydrogenase on a pyrolytic carbon fiber microelectrode, *Anal. Chem.*, *58*: 412.
98. Wang, J., and Lin, M. S. (1988). Mixed plant tissue-carbon paste bioelectrode, *Anal. Chem.*, *60*: 1545.
99. Wang, J., and Brennsteiner, A. (1988). Miniature tissue based voltammetric bioelectrodes, *Anal. Lett.*, *21*: 1773.
100. Ikeda, T., Miki, K., Fushimi, F., and Senda, M. (1988). Amperometric D-gluconate sensor using D-gluconate dehydrogenase from bacterial membranes, *Agr. Biol. Chem.*, *52*: 1557.
101. Thevenot, D. R., Sternberg, R., Coulet, P. R., Laurent, J., and Gautheron, D. C. (1979). Enzyme collagen membrane for electrochemical determination of glucose, *Anal. Chem.*, *51*: 96.

102. Coulet, P. R., Sternberg, R., and Thevenot, D. R. (1980). Electrochemical study of reactions at interfaces of glucose oxidase collagen membranes, *Biochim. Biophys. Acta, 612*: 317.
103. Winkler, L. W. (1888). The determination of dissolved oxygen in water, *Ber. Dtsch. Chem. Ges., 21*: 2843.
104. Gough, D. A., and Leypoldt, J. K. (1981). Theoretical aspects of enzyme electrode design, *Appl. Biochem. Bioeng., 3*: 175.
105. Peguin, S., Coulet, P. R., and Bardeletti, G., (1989). Pyruvate oxidase and oxaloacetate decarboxylase enzyme electrodes. Simultaneous determination of transaminases with a two-electrode-based analyzer, *Anal. Chim. Acta, 222*: 83.
106. Clark, L. C. Jr. (1970). Membrane polarographic electrode system and method with electrochemical compensation, U.S. Patent 3,539,455.
107. Scheller, F., and Renneberg, R. (1983). Glucose-eliminating enzyme electrode for direct sucrose determination in glucose-containing samples, *Anal. Chim. Acta, 152*: 265.
108. Bradley, C. R., and Rechnitz, G. A. (1986). Comparison of oxalate oxidase enzyme electrodes for urinary oxalate determinations, *Anal. Lett., 19*: 151.
109. Nabi Rahni, M. A., Guilbault, G. G., and De Oliveira, N. G. (1986). Immobilized enzyme electrode for the determination of oxalate in urine, *Anal. Chem., 58*: 523.
110. Nabi Rahni, M. A., Guilbault, G. G., and De Oliveira, G. N. (1986). Immobilized enzyme electrode for the determination of salicylate in blood serum, *Anal. Chim. Acta, 181*: 219.
111. Kulis, Y. Y., Pesliakene, M. V., Laurinavichyus, V. S. A., Tatikyan, S. S., and Simonyan, A. L., (1985). Determination of uric acid in biological solutions using bienzyme electrodes, *J. Anal. Chem., 40*: 1652.
112. Kirstein, D., Scheller, F., Olsson, B., and Johansson, G. (1985). Enzyme electrode for urea with amperometric indication, *Anal. Chim. Acta, 171*: 345.
113. Kubo, I., and Karube, I. (1986). Modified polymer for enzyme immobilization and characterization of immobilized urease, *Anal. Lett., 19*: 667.
114. Wollenberger, U., Kühn, M., Scheller, F., Deppmeyer, V., and Jänchen, M. (1983). Amperometric enzyme sequence electrodes for cholesterol, *Bioelectrochem. Bioenerg., 11*: 307.
115. Mascini, M., and Moscone, D. (1986). Amperometric acetylcholine and choline sensors with immobilized enzymes, *Anal. Chim. Acta, 179*: 439.
116. Campanella, L., Tomassetti, M., De Angelis, G., Sammartino, M. P., and Cordatore, M. (1987). A new assay for choline-containing phospholipids in amniotic fluid by an enzyme sensor, *Clin. Chim. Acta, 169*: 175.
117. Campanella, L., Tomassetti, M., De Angelis, G., and Sammartino, M. P. (1988). Enzyme sensor for the determination of choline-containing phospholipids in some biological fluids, *Analyst, 113*: 77.
118. Marty, J. L., Sode, K., and Karube, I. (1990). Amperometric determination of choline and acetylcholine with enzymes immobilized in a photo-crosslinkable polymer, *Anal. Chim. Acta, 228*: 49.
119. Matsumoto, K., Seijo, H., Karube, I., and Suzuki, S. (1980). Amperometric determination of choline with use of immobilized choline oxidase, *Biotechnol. Bioeng., 22*: 1071.

120. Renneberg, R., Pfeiffer, D., Scheller, F., and Jänchen, M. (1982). Enzyme sequence and competition electrodes based on immobilized glucose oxidase, peroxidase and catalase, *Anal. Chim. Acta, 134*: 359.
121. Tsuchida, T., and Yoda, K. (1983). Multi-enzyme membrane electrodes for determination of creatinine and creatine in serum, *Clin. Chem., 29*: 51.
122. Takagi, M., Kojima, J., and Morita, N. (1988). An enzyme electrode sensor for glycerophosphate, *Anal. Sci., 4*: 361.
123. Kojima, J., Morita, N., and Takagi, M. (1988). Enzyme electrode for determination of glycerophosphate by combined use of glycerophosphate oxidase and peroxidase, *Anal. Sci., 4*: 497.
124. Schubert, F., Kirstein, D., Scheller, F., Abraham, M., and Boross, L. (1986). Enzyme electrode for fructose, glucose-6-phosphate and ATP determination, *Anal. Lett., 19*: 2155.
125. Xu, Y., Guilbault, G. G., Kuan, S. S. (1989). Sucrose enzyme electrode, *Anal. Chem., 61*: 782.
126. Satoh, I., Karube, I., and Suzuki, S. (1976). Enzyme electrode for sucrose, *Biotechol. Bioeng., 18*: 269.
127. Cordonnier, M., Lawny, F., Chapot, D., and Thomas, D. (1975). Magnetic enzyme membranes as active elements of electrochemical sensors. Lactose, saccharose, maltose bienzyme electrodes, *FEBS Lett., 59*: 263.
128. Mizutani, F., Tsuda, K., Karube, I., Suzuki, S., and Matsumoto, K. (1980). Determination of glutamate pyruvate transaminase and pyruvate with an amperometric pyruvate oxidase sensor, *Anal. Chim. Acta, 118*: 65.
129. Mascini, M., and Mazzei, F. (1987). Amperometric sensor for pyruvate with immobilized pyruvate oxidase, *Anal. Chim. Acta, 192*: 9.
130. Shubert, F., Kirstein, D., Abraham, M., Scheller, F., and Boross, L. (1985). Horseradish peroxidase based bienzyme electrode for isocitrate, *Acta Biotechnol., 5*: 375.
131. Matsumoto, K., Yamada, K., and Osajima, Y. (1981). Ascorbate electrode for determination of L-ascorbic acid in food, *Anal. Chem., 53*: 1974.
132. Smith, V. J., Green, R. A., and Hopkins, T. R. (1989). Determination of aspartame in beverages using an alcohol oxidase enzyme electrode, *J. Assoc. Offic. Anal. Chem., 72*: 30.
133. Karube, I., Satoh, I., Araki, Y., Suzuki, S., and Yamada, H. (1980). Monoamine oxidase electrode in freshness testing of meat, *Enzyme Microb. Technol., 2*: 117.
134. Watanabe, E., Ando, K., Karube, I., Matsuoka, H., and Suzuki, S. (1983). Determination of hypoxanthine in fish meat with an enzyme sensor, *J. Food Sci., 48*: 496.
135. Watanabe, E., Endo, H., and Toyama, K. (1988). Determination of inosine-5′-monophosphate in the presence of inosine and hypoxanthine with an enzyme sensor, *Appl. Microbiol. Biotechnol., 29*: 341.
136. Wollenberger, U., Schubert, F., Scheller, F., Danielsson, B., and Mosbach, K. (1987). A biosensor for ADP with internal substrate amplification, *Anal. Lett., 20*: 657.
137. Osawa, H., Akiyama, S., and Hamada, T. (1981). Determination of uric acid, glucose and amylase in whole blood using enzyme electrode. In: Proceedings 1st International Sensor Symposium, Tsukuba, Japan, p. 163.

138. Renneberg, R., Scheller, F., Riedel, K., Litschko, E., and Richter, M. (1983). Development of anti-interference enzyme layer for α-amylase measurements in glucose containing samples, *Anal. Lett.*, *16*: 877.
139. Yoda, K., and Tsuchida, T. (1983). Bi-enzyme electrode for determination of α-amylase activity in serum. In: Proceedings 2nd International Sensor Symposium, Fukuoka, Japan, p. 648.
140. Nikoleis, D. P., and Mottola, H. A. (1978). Repetitive determinations of amylase, maltose, sucrose and lactose by sample injection in closed flow-through systems, *Anal. Chem.*, *50*: 1665.
141. Blum, L. J., Bertrand, C., and Coulet, P. R. (1983). Amperometric sensor and immobilized enzyme electrode for the determination of enzymatic activities, *Anal. Lett.*, *16*: 525.
142. Minoura, N., Yamada, S., Karube, I., Kubo, I., and Suzuki, S. (1982). Determination of lactate dehydrogenase with immobilized enzyme electrode, *Anal. Chim. Acta*, *132*: 355.
143. Mizutani, F., Sasaki, K., and Shimura, Y. (1983). Sequential determination of L-lactate and lactate dehydrogenase with immobilized enzyme electrode, *Anal. Chem.*, *55*: 35.
144. Jänchen, M., and Glawion, S. (1987). Quantitative determination of alkaline phosphatase with an amperometric glucose oxidase enzyme electrode, *Stud. Biophys.*, *119*: 187.
145. Weigelt, D., Schubert, F., and Scheller, F. (1988). Determination of pyruvate kinase and creatine kinase using multienzyme electrodes, *Anal. Lett.*, *21*: 225.
146. Fonong, T. (1989). Immobilized enzyme assay of creatine kinase with amperometric detection, *Anal. Biochem.*, *176*: 234.
147. Fonong, T., and Barber, T. (1988). Determination of alanine aminotransferase and aspartate aminotransferase with immobilised enzymes and electrochemical detection, *Analyst*, *113*: 1807.
148. Wollenberger, U., Scheller, F. W., Böhmer, A., Passarge, M., and Müller, H.-G. (1989). A specific electrode for L-glutamate—development and application, *Biosensors*, *4*: 381.
149. Kihara, K., Yasukawa, E., and Hirose, S. (1984). Sequential determination of glutamate-oxalacetate transaminase and glutamate-pyruvate transaminase activities in serum using an immobilized bienzyme-poly(vinyl chloride) membrane electrode, *Anal. Chem.*, *56*: 1876.
150. Okuma, H., Sekimukai, S., Hoshi, M., Toyama, K., and Watanabe, E. (1989). Biosensor system for continuous flow determination of enzyme activities. I. Determination of glucose oxidase and lactic dehydrogenase activities, *Enzyme Microb. Technol.*, *11*: 824.
151. Mizutani, F., and Tsuda, K. (1982). Amperometric determination of cholinesterase with use of an immobilized enzyme electrode, *Anal. Chim. Acta*, *139*: 359.
152. Yao, T. (1983). Flow injection analysis for cholinesterase in blood serum by use of a choline-sensitive electrode as an amperometric detector, *Anal. Chim. Acta*, *153*: 169.
153. Palleschi, G., Lavagnini, M. G., Moscone, D., Pilloton, R., D'Ottavio, D., and Evangelisti, M. E. (1990). Determination of serum cholinesterase activity and

dibucaine numbers by an amperometric choline sensor, *Biosensors Bioelectronics, 5*: 27.

154. Weigelt, D., Schubert, F., and Scheller, F. (1987). Enzyme sensor for the determination of lactate and lactate dehydrogenase activity, *Analyst, 112*: 1155.
155. Weigelt, D., Schubert, F., and Scheller, F. (1987). Enzyme electrodes for lactate, pyruvate and lactate dehydrogenase activity determination, *Fresenius' Z. Anal. Chem., 328*: 259.
156. Matthews, D. R., Holman, R. R., Bown, E., Steemson, A., Watson, A., Hughes, S., and Scott, D. (1987). Pen-sized digital 30-second blood glucose meter, *Lancet, 1*: 778.

3

Development of Amperometric Biosensors for Enzyme Immunoassay

J. V. Bannister, I. John Higgins and Anthony P. F. Turner

Cranfield Institute of Technology
Cranfield, Bedfordshire, England

Immunoassays are based on the interaction between the antigen (Ag) and its specific antibody (Ab) to form an antigen-antibody complex:

$$\mathrm{Ag} + \mathrm{Ab} = \mathrm{AbAg}$$

Equilibrium is reached, and the equilibrium constant for the reaction is defined as

$$K = \frac{[\mathrm{AbAg}]}{[\mathrm{Ab}][\mathrm{Ag}]}$$

This indicates that for a fixed concentration of antibody the ratio of bound to free antigen at equilibrium is quantitatively related to the total amount of antigen present. The reaction forms the basis of all immunoassays. When a fixed amount of labeled antigen is introduced into the assay the concentration of the unknown antigen can then be determined. Similarly, an unknown concentration of antibodies can be determined by using labeled antibodies. The labeling agents currently used are radioisotopes, enzymes, red cells, fluorescent probes, chemiluminescent probes, and metal tags. The predominant label in recent years has been the radioisotope, and the assay is identified as radioimmunoassay (RIA). A significant advantage of RIA is the absence of interfering radioactivity in clinical samples. However, the inconvenience of handling radioactivity has stimulated the development of nonisotopic immunoassays based on other labels. The growing trend away from radioactive labels has led to the development of enzyme-linked immunoassays (EIA).

Enzyme-linked immonassays are commonly categorized as heterogeneous, in which antibody-bound antigen is separated from free antigen, or homogeneous, in which there is no separation step. Heterogeneous EIA as an analytic technique originated about 15 years ago, and four types of assays have been developed. These are competitive EIA for antigen (Fig. 1a), which involves competition between labeled and unlabeled antigen for a limited amount of antibody; the alternative form of this assay is one in which the enzyme labeled antibody is reacted with antigen and then added to excess of solid-phase antigen (Fig. 1b). The enzyme activity bound to the solid phase after separation is inversely proportional to the concentration of free antigen.

The remaining types of heterogeneous EIA are known as the sandwich assays for antigen and antibody. In the sandwich assay for antigen (Fig. 2a) solid-phase antibody is incubated with the antigen to be measured. The solid phase is then

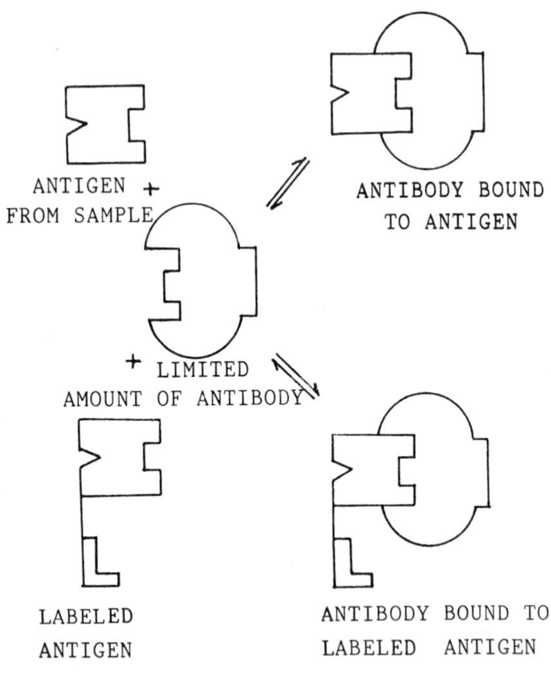

(a)

Figure 1 (a) Competitive enzyme immunoassay for antigen. Separation of bound enzyme is necessary before measurement of enzyme activity. (b) Competitive enzyme-linked immunosorbent assay for antigen. Enzyme-labeled antibody reacts specifically with antigen in the sample and is then added to excess solid-phase antigen. After washing the enzyme label still attached to the solid phase is measured. [Reproduced with permission from Black, C., and Gould, B. J. (1984). *Analyst, 109*: 533.]

washed and enzyme-labeled antibody is added. The enzyme activity attached to the solid support is proportional to the concentration of the antigen. The sandwich assay for antibody detection involves binding of the antigen to the solid phase and a second enzyme-labeled antibody is added (Fig. 2b). The use of solid supports in both competitive and sandwich (noncompetitive) techniques have been named the enzyme-linked immunosorbent assays (ELISA).

A number of enzymes are currently used as labels for heterogeneous and homogeneous EIA. These include alkaline phosphatase, horseradish peroxidase, β-galactosidase, glucose oxidase, glucose-6-phosphate dehydrogenase, and urease. The advantages and disadvantages of enzyme labels compared with radioactivity labels have been described by Blake and Gould (1). These are listed in Table 1. Ideally, the enzyme used as the label should be available in pure form at low cost and it should have a low k_m and a high V_{max} at the pH required for good antibody-antigen interaction. Besides enzymes, molecules acting with or upon enzymes have also been used as immunoassay labels. These include cofactors, prosthetic groups, substrates, and inhibitors. The assay depends on the steric hindrance caused by antibody attaching to the hapten-labeled conjugate

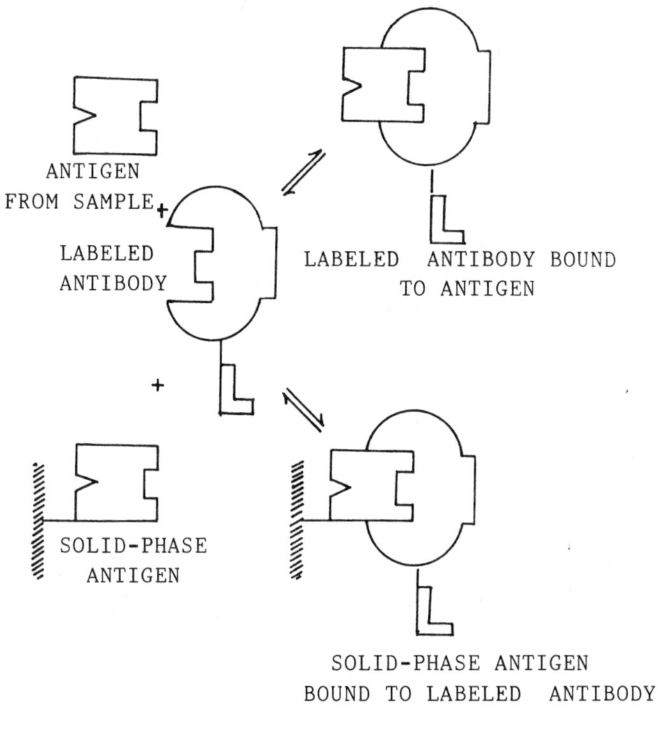

(b)

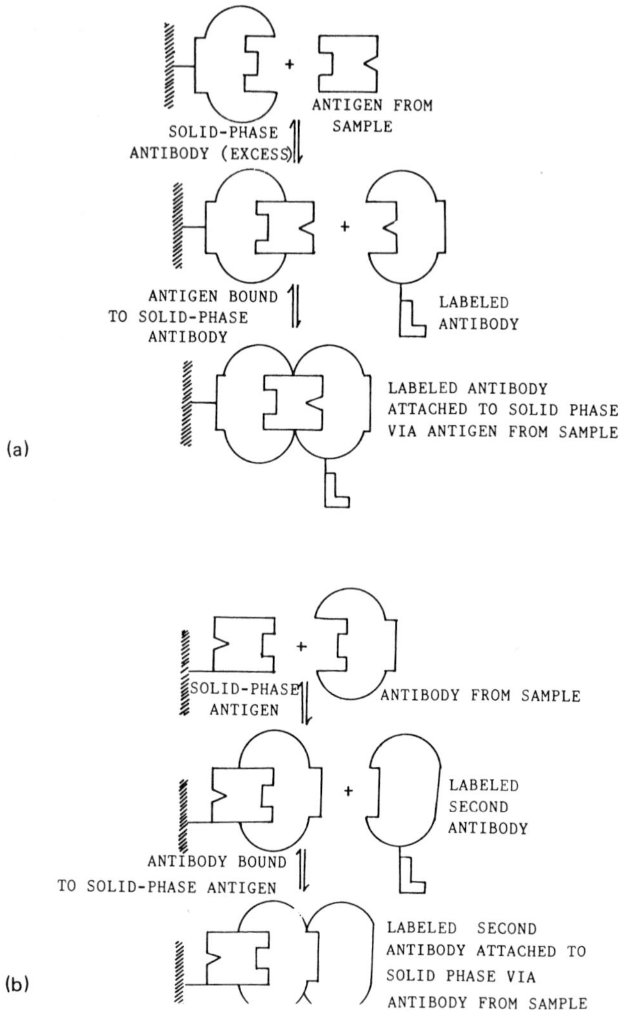

Figure 2 (a) Sandwich assay for antigen. Antigen in the sample is mixed with excess solid-phase antibody. After washing of the solid phase, enzyme-labeled antibody, which is specific for another site on the antigen, is added. The enzyme label, which remains bound after washing, is measured. (b) Sandwich assay for antibody. Antibody in the sample is mixed with excess solid-phase antigen. After washing of the solid phase, enzyme-labeled second antibody is added. Bound enzyme activity is measured after washing. [Reproduced with permission from Blake, C., and Gould, B. J. (1984). *Analyst, 109*: 533.]

and modulating the enzyme activity by altering the natural function of the label with the enzyme (Fig. 3). The model system uses flavine-adenine dinucleotide (FAD) as the prosthetic group and glucose oxidase as the enzyme (2). Enzyme substrates have also been employed as fluorescent immunoassay labels (3,4). In this method neither the labeled ligand nor the antibody-bound ligand is fluorescent. However, when the labeled ligand is cleared by enzymatic hydrolysis it releases that part of the molecule that is able to fluoresce (Fig. 4). β-Galactosidase is normally used as the enzyme label, and the synthetic fluorogenic substrate is a β-galactosylumbelliferone conjugate. Cleavage of the labeled ligand is sterically hindered when it is bound to the specific ligand antibody. Enzyme inhibitors have also been used in the development of homogeneous enzyme-linked immunoassays (5). The inhibitor-hapten complex exerts an inhibitory action when freely present in the reaction solution, but it is sterically hindered from interacting with the enzyme when the antigenic portion of the conjugate is antibody bound (Fig. 5). The system has been developed for phosphonate-hapten complex and acetylcholinesterase (6). Phosphonate irreversibly inhibits the enzyme when it is free but is not inhibitory when bound to antibodies.

Table 1 Comparison of Enzyme Labels with Radiolabels in Immunoassay

Advantages of enzyme labels
No radiation hazards occur during labeling or disposal of waste.
Enzyme-labeled products can have a long shelf life, *e.g.*, 1 year or more.
Equipment for enzyme assay can be inexpensive and is generally available.
Homogeneous assays can be completed in a few minutes and are readily automated.
Heterogeneous assays are ideal for visual qualitative tests.
Multiple simultaneous assays are possible.
Disadvantages of enzyme labels
Plasma constituents may affect enzyme activity.
Assay of enzyme activity can be more complex than measurement of some types of radioisotope.
There is less control of enzyme labeling reactions.
At present homogeneous EIA have limited sensitivity.

Source: Reproduced with permission from Blake, C., and Gould, B.J., *Analyst*, 109: 533. (1984).

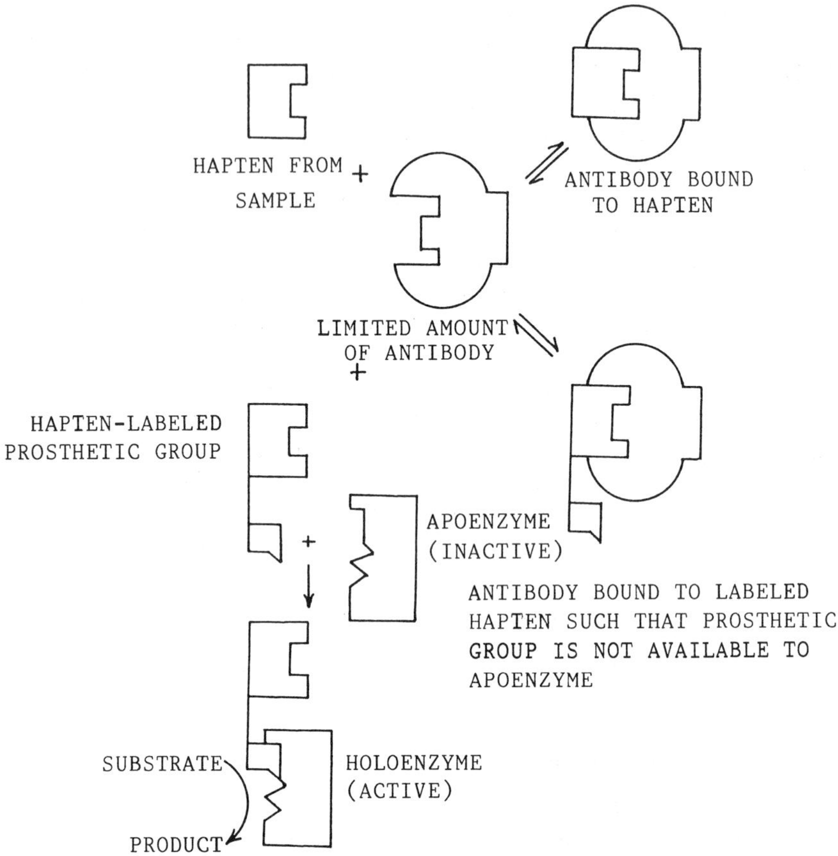

Figure 3 Principle of prosthetic group-labeled immunoassay. The apoenzyme becomes active only when it has been converted into the holoenzyme by binding to the prosthetic group, which is part of the active site of the enzyme. [Reproduced with permission from Blake, C., and Gould, B. J., (1984). *Analyst, 109*: 533.]

The catalytic activity of the enzyme used as label is conventionally detected either spectrophotometrically or fluorimetrically. However, in recent years there has been considerable interest in the development of electrochemical biosensors for immunoassays. Electrochemical biosensors can be either potentiometric or amperometric devices. In the potentiometric devices a potential difference is generated by the sensing element, and this is measured relative to an accurate reference element under conditions of zero current flow. This has the advantage that there is no net consumption of material, and therefore the role of diffusion or mass transport is not important. However, such devices depend upon the rapid development of thermodynamic equilibrium at the electrode surface. Poten-

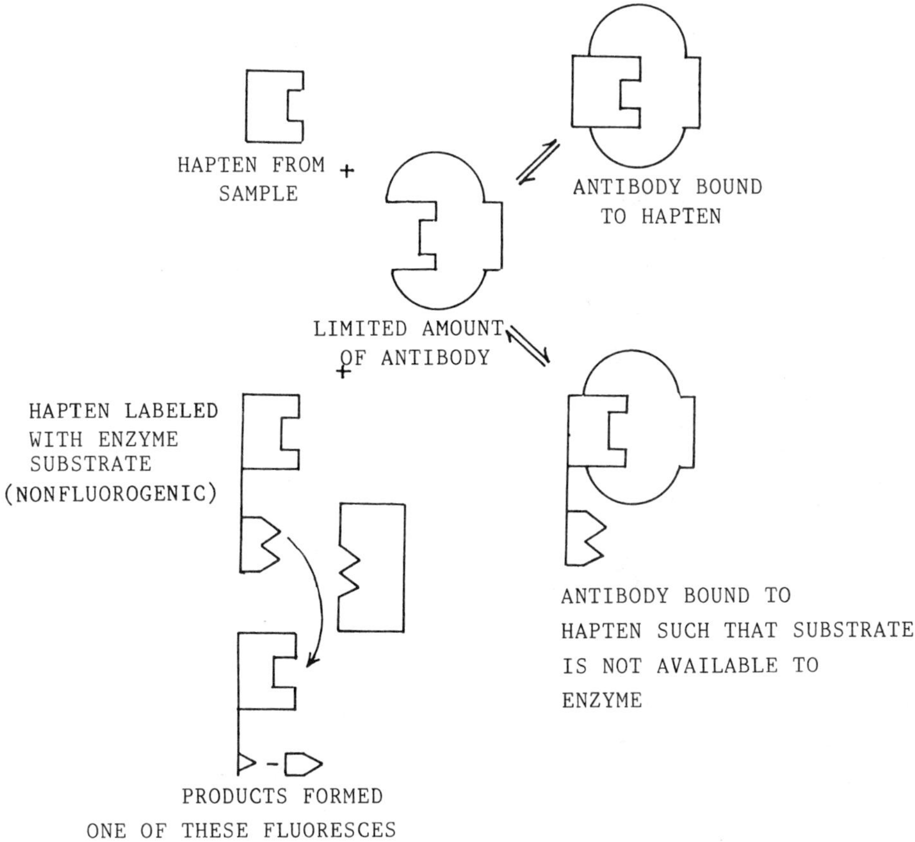

Figure 4 Principle of substrate-labeled fluorescent immunoassay. Only a limited amount of the substrate-labeled hapten can be used, and therefore only a small amount of substrate is available for reaction, which necessitates the formation of a fluorescent product to obtain adequate sensitivity. Either the rate of formation of fluorescent product or the total amount formed can be measured. [Reproduced with permission from Blake, C., and Gould, B. J. (1984). *Analyst, 109*: 533.]

tiometric devices for the detection of such ions as K^+, Na^+, NH_4^+, H^+, and Cl^- are currently in use (7). These devices can be used in conjunction with the following enzymes, which have been used as enzyme labels: acetylcholinesterase, adenosine deaminase, and urease. As knowledge of electrode materials advances, however, it is becoming increasingly clear that amperometric devices are more amenable to routine use. In the amperometric sensor a voltage is applied between the working and reference electrodes. This imposed potential encourages electron transfer to occur, causing a net current flow that is proportional to the concentration of the analyte. In a typical device the current gener-

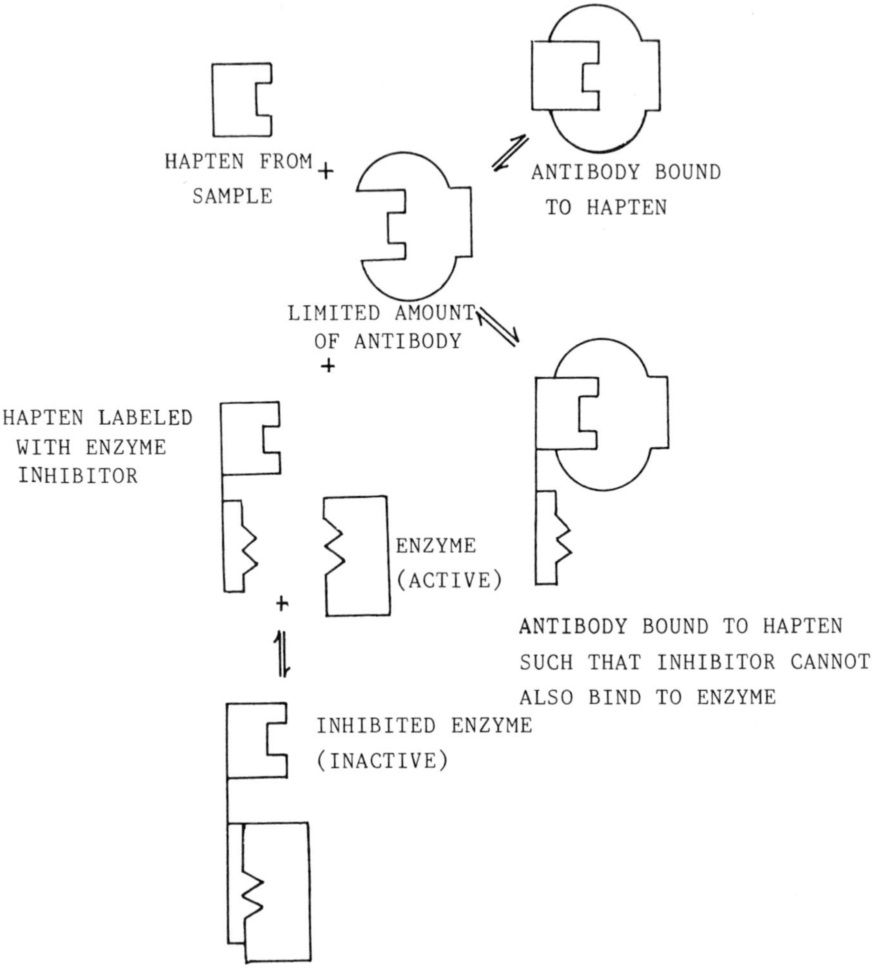

Figure 5 Principle of an inhibitor-based homogeneous enzyme immunoassay. The inhibitor-labeled hapten combines with enzymes that lose activity. The reaction between enzyme and inhibitor can be reversible (as shown) or irreversible. [Reproduced with permission from Blake, C., and Gould, B. J., (1984). *Analyst, 109*: 533.]

ated is directly proportional to the specific binding reaction, whereas in a potentiometric measurement the relationship is logarithmic. The combination of substrate specificity and the sensitivity of electrochemical methods has also been applied to the design of a number of elegant electrochemical immunoassays. The strategy employed is based on labeling the antigen with an enzyme that catalyzes the production of an electrochemically detectable product (8). The rate at which the product is formed is related to the concentration of the analyte in the sample.

A number of heterogeneous enzyme immunoassays with electrochemical detection have been developed. These are based on the ELISA technique. Alkaline phosphatase is used as the enzyme label. It catalyzes the hydrolysis of phenyl phosphate to phenol and phosphate. Phenyl phosphate is electroinactive and hence noninterfering (9). The liberated phenol is detected electrochemically by oxidation at a carbon paste working electrode at 870 mV versus Ag/AgCl following separation by either liquid chromatography or flow injection analysis. High-performance immunoaffinity chromatography with electrochemical detection of the column effluent has also been applied to immunoassays (10).

The detection limits for early enzyme immunoassays with electrochemical measurements were quite high. A detection limit of 1.0 ng/ml was achieved for orosomucoid (11), 5.0 ng/ml for digoxin, and 100 pg/ml for IgG (12). The detection limit for IgG can be lowered from 100 to 7.5 pg/ml following reduction of nonspecific adsorption (13). The use of a separation step eliminates interference from electroactive constituents that may be present in the sample and possible fouling of the electrode by the adsorption of protein films. However, it is possible to eliminate the separation step if the enzyme-generated product is electroactive at a potential below 200 mV. It is above this potential that interference from blood constituents, such as cysteine and ascorbic acid, takes place. One such system involves an electroactive ferrocene-linked substrate for alkaline phosphatase, [*N*-ferrocenoyl]-4-aminophenyl phosphate (14). The cyclic voltammogram of the substrate is consistent with the quasi-reversible ferrocene/ferricinium ion one-electron redox couple (Fig. 6a). A dramatic change in the ferrocene oxidation potential after reaction with alkaline phosphatase is observed. Two new irreversible oxidation peaks at 180 and 590 mV are observed (Fig. 6b). The peak at 590 mV is due to the oxidation of the phenol group; that at 180 mV is due to the oxidation of ferrocene to the ferricinium ion in this compound. Since the substrate is electroinactive at 180 mV, the product of the reaction is detectable without interference from excess substrate in solution, and the change in peak current at 180 mV was found to be related to the enzyme concentration.

The useful oxidation potential of ferrocene and its ability to act as a mediator (15,16) for redox reactions has led to the development of a homogeneous amperometric immunoassay for the drug lidocaine (17). The binding of a ferrocene-lidocaine complex by antibody against the drug was found to inhibit the ability of ferrocene to act as a mediator for the glucose/glucose oxidase reaction. However, this inhibition could be reversed by the addition of free lidocaine, which competes with the ferrocene-lidocaine complex for the available antibody binding sites. The catalytic current produced was found to be directly proportional to the concentration of analyte.

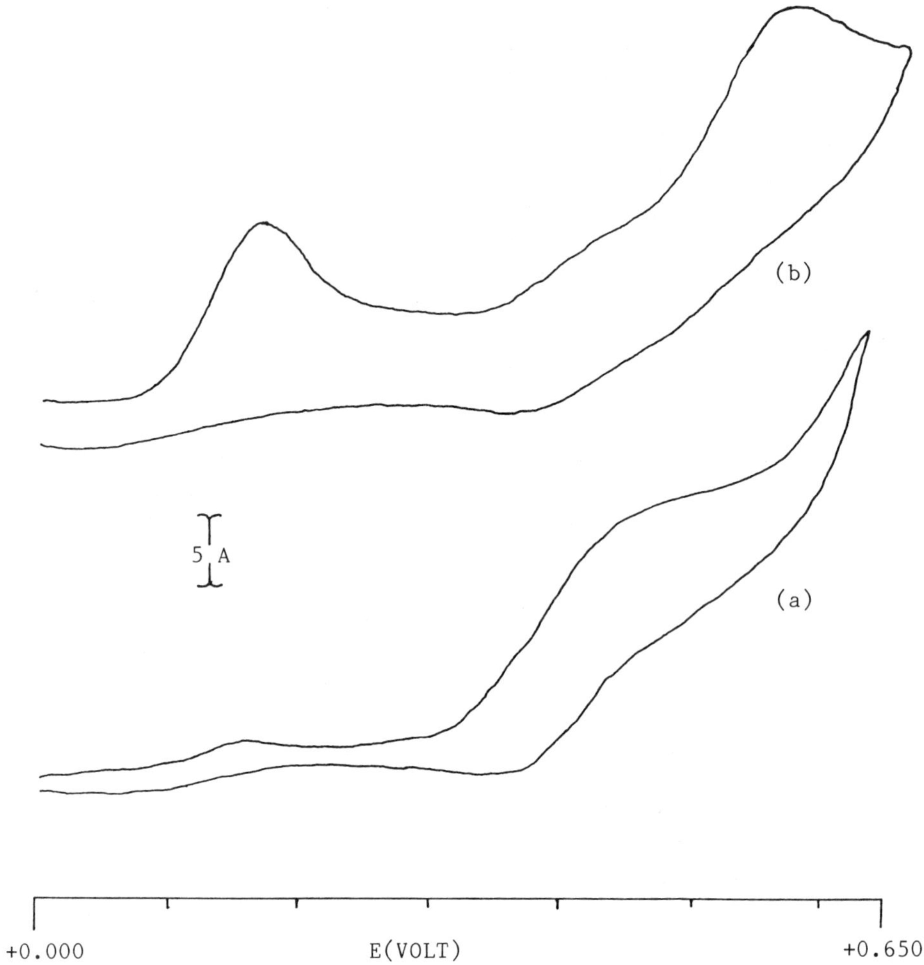

Figure 6 A dc voltammogram of [*N*-ferrocynoyl]-4-aminophenyl phosphate in the absence (a) and presence (b) of alkaline phosphatase. [Reproduced with permission from McNeil, C. J., et al. (1988). *Biosensors, 3*: 199.]

Another type of homogeneous enzyme immunoassay that can be carried out amperometrically involves the conversion of the apo to the holo form of glucose oxidase (18). The assay is carried out spectrophotometrically for the detection of 2,4-dinitrophenyl-aminocaproic acid (Fig. 7). In the first step active holoenzyme is covalently labeled with the analyte. The second step involves acid denaturation of the labeled holoenzyme, resulting in the FAD prosthetic group dissociating from it; gel chromatography of the acidified solution results in separation of the

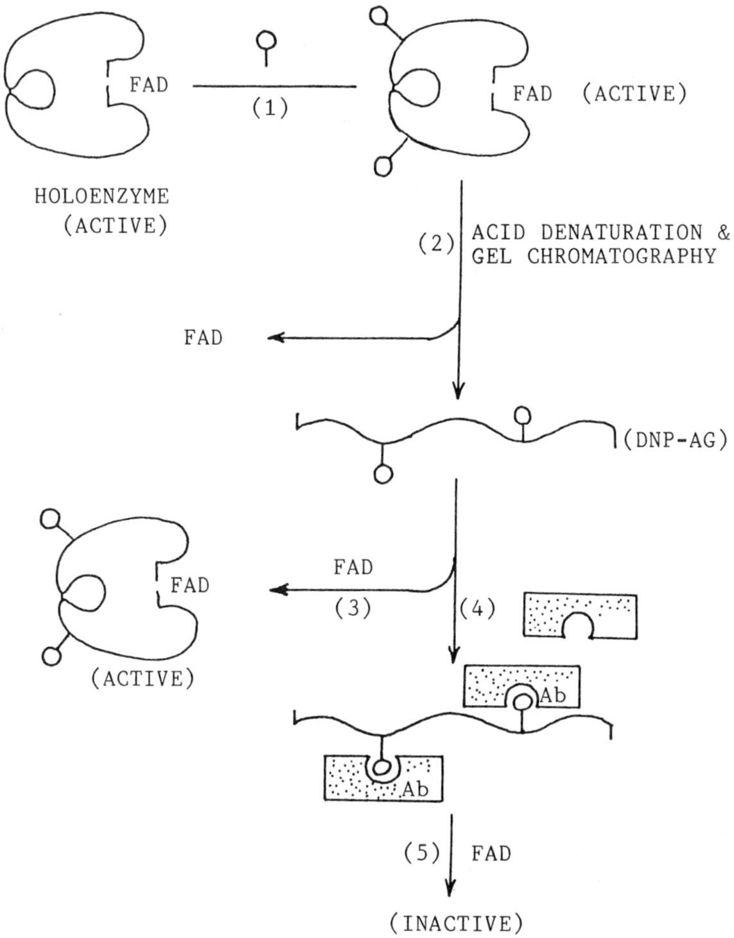

Figure 7 Homogeneous enzyme immunoassay involving the conversion of apo- to hologlucose oxidase. Holoenzyme; glucose oxidase; 2,4-dinitrophenyl (DNP); DNP-AG; 2,4-dintrophenyl-labeled apoglucose oxidase; and Ab; antibody to DNP group. [Reproduced with permission from Ngo, T. T., and Lenhoff, H. M. (1983). *Biochem. Biophys. Res. Commun.*, *114*: 1097.]

released FAD from the labeled apoglucose oxidase. Addition of FAD in the third step to the labeled apoglucose oxidase at neutral pH gives a reconstituted active holoenzyme. However, if instead of FAD antibodies to the analyte are added before the addition of FAD very little active holoenzyme is obtained. The binding of antibody molecules to the labeled denatured apoenzyme prevents reconstitution and binding of FAD. Therefore, reduced electrochemical activity for glucose oxidase is obtained.

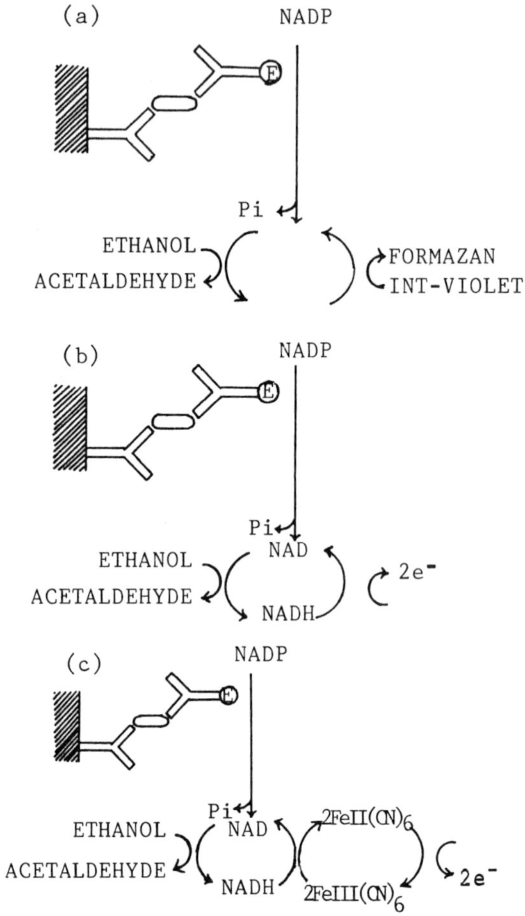

Figure 8 (a) The amplification system for immunoassay. Alkaline phosphatase catalyzes the diphosphorylation of $NADP^+$ to NAD^+, which enters a redox cycle in which ethanol is oxidized and *p*-iodonitrotetrazolium violet (INT violet) is reduced to a colored formazan dye. (b) Electrochemical measurement in which the redox cycle is coupled to $NMP^+/TCNQ^+$. (c) Electrochemical measurement in which ferricyanide acts as an electroacceptor from the diaphorase active site and subsequently reduces at a platinum electrode. [Reproduced with permission from Stanley, C. J., et al. (1988). *J. Immunol. Methods, 112*: 153.]

The reduction in the rate of enzyme catalysis occurring when the antibody binds to an enzyme-labeled antigen has led to the development of immunoassays involving the amperometric measurement of NADH (19). Phenytoin, an anti-Nepileptic drug, was labeled with glucose-6-phosphate dehydrogenase. The enzyme catalyzes the reduction of NAD^+ to NADH. The concentration of the reduced cofactor is monitored at 750 mV versus Ag/AgCl. In the presence of antibodies against the drug the enzyme activity is decreased.

Cofactor has been utilized as the basis of an amplification system for immunoassays (20–23). The enzyme-amplified immunoassay differs from the conventional type in that the product from the enzyme label is not measured but can instead act catalytically on the secondary system, which remains silent until activated. The product is NAD^+, which is generated from $NADP^+$ by alkaline phosphatase (Fig. 8a). In the cycle NAD^+ is reduced by ethanol in the presence of alcohol dehydrogenase and the NADH formed reduces tetrazolium salt in the presence of diaphorase to regenerate NAD^+ and to produce an intensely colored formazan dye. The rate of reduction of the tetrazolium salt is directly proportional to the concentration of the NAD^+ originally formed from $NADP^+$. The amplification reaction has recently been adapted for electrochemical measurement. Three amperometric methods have been described (24,25). The first two methods described by Stanley *et al.* (24) employ either an organic salt electrode composed of *N*-methylphenazinium salt (NMP^+) and 7,7,8,8-tetracyanoquinodimethane (TCNQ) or a platinum wire with ferricyanide as electron-transfer mediator (Fig. 8b and 8c); the third method, described by McNeil *et al.* (25), measures the electrochemical oxidation of hydrogen peroxide generated during the oxidation of NADH to NAD^+ by the enzyme NADH oxidase isolated from the thermophile *Thermus aquaticus* (Fig. 9) (26). The work reported by Stanley *et al.* (24) was later refined to use carbon electrodes in an electrochemical immunoassay for prostatic acid phosphatase (27).

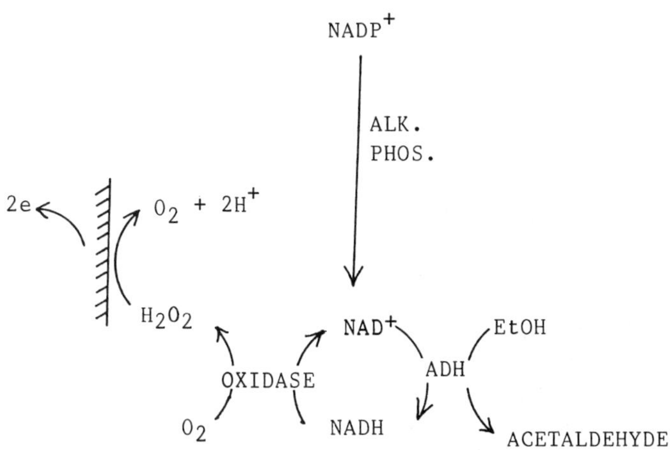

Figure 9 Enzyme cycling system. The enzyme label catalyzes the diphosphorylation of $NADP^+$ to NAD^+, which enters a cycle in which NADH oxidase-catalyzed oxidation of NADH produces H_2O_2, which is detected at the electrode. [Reproduced with permission from McNeil, C. J., et al. (1989). *Anal. Chem., 61*: 25.]

It is clear that electrochemical detectors are proving to be excellent sensors for quantitating enzyme immunoassays. Electrochemical methods are not limited by the dynamic range of optical methods and are essentially free of the various problems associated with absorption and fluorescence methods. The very low detection limits of electrochemical techniques and the use of highly selective monoclonal antibodies will make possible in the future the detection of extremely low levels of antigens in easy-to-use formats. Electrochemical immunoassay may also find application in automated laboratory analyzers, where its sensitivity can rival that of alternative methodologies.

REFERENCES

1. Blake, C., and Gould, B. J. (1984). Use of enzymes in immunoassay techniques. A review, *Analyst, 109*: 533.
2. Morris, D. L., Ellis, P. B., Carrico, R. J., Yeager, F. M., Schroeder, H. R., Albarella, J. P., Boguslaski, R. C., Hornby, W. E., and Rawson, D. (1981). Flavin adenine dinucleotide as a label in homogeneous colorimetric immunoassays, *Anal. Chem., 53*: 658.
3. Burd, J. F., Carrico, R. J., Fetter, M. C., Buckler, R. T., Johnson, R. D., Boguslaski, R. C., and Christner, J. E. (1977). Specific protein-binding reactions monitored by enzymatic hydrolysis of ligand-fluorescent dye conjugates, *Anal. Biochem., 77*: 56.
4. Burd, J. F., Wong, R. C., Feeney, J. E., Carrico, R. J., and Boguslaski, R. C. (1977b). Homogeneous reactant-labelled fluorescent immunoassay for therapeutic drugs exemplified by gentamicin determination in human serum, *Clin. Chem., 23*: 1402.
5. Ngo, T. T., and Lenhoff, H. M. (1980). Enzyme modulators as tools for the development of homogeneous enzyme immunoassays, *FEBS Lett., 116*: 285.
6. Finley, P. R., Williams, R. J., and Lichti, D. A. (1980). Evaluation of a new homogeneous enzyme inhibitor immunoassay of serum thyroxine with use of a bichromatic analyzer, *Clin. Chem., 26*: 1723.
7. Kuan, S. S., and Guilbault, G. G. (1987). In: *Biosensors—Fundamentals and Applications* (A. P. F. Turner, I. Karube, and G. S. Wilson, Eds), pp. 135–152.
8. Heineman, W. R., Halsall, W. B., Wehneyer, K. R., Doyle, M. J., and Wright, D. S. (1987). Immunoassay with electrochemical detection, *Methods Biochem. Anal., 32*: 345.
9. Heineman, W. R., and Halsall, H. B. (1985). Strategies for electrochemical immunoassay, *Anal. Chem., 57*: 1321A.
10. De Alvis, W. V., and Wilson, G. S. (1985). Rapid sub-picomole electrochemical enzyme immunoassay for immunoglobulin G, *Anal. Chem., 57*: 2754.
11. Doyle, M. J., Halsall, H. B., and Heineman, W. R. (1984). Enzyme-linked immunoadsorbent assay with electrochemical detection for $_1$-acid glycoprotein, *Anal. Chem., 56*: 2355.
12. Wehmeyer, K. R., Halsall, H. E., and Heinemann, W. B. (1985). Heterogeneous enzyme immunoassay with electrochemical detection: Competitive and ''sandwich''-type immunoassays, *Clin. Chem., 31*: 1546.

13. Jenkins, S. H., Heineman, W. R., and Halsall, H. B. (1988). Extending the detection limit of solid-phase electrochemical enzyme immunoassay to the attomole level, *Anal. Biochem., 168*: 292.
14. McNeil, C. J., Higgins, I. J., and Bannister, J. V. (1988). Amperometric determination of alkaline phosphatase activity: Application to enzyme immunoassay, *Biosensors, 3*: 199.
15. Cass, A. E. G., Davis, G., Francis, G. D., Hill, H. A. O., Aston, W. J., Higgins, I. J., Plotkin, E. V., Scott, L. D. L., and Turner, A. P. F. (1984). Ferrocene-mediated enzyme electrode for amperometric determination of glucose, *Anal. Chem., 56*: 667.
16. Cass, A. E. G., Davis, G., Green, M. J., and Hill, H. A. O. (1985). Ferricinium ion as an electron acceptor for oxidoreductase, *J. Electroanal. Chem., 190*: 117.
17. Digleria, K., Hill, H. A. O., McNeil, C. J., and Green, M. J. (1986). Homogeneous ferrocene-mediated amperometric immunoassay, *Anal. Chem., 58*: 1203.
18. Ngo, T. T., and Lenhoff, H. M. (1983). Antibody-induced conformational restriction as basis for new separation-free enzyme immunoassay, *Biochem. Biophys. Res. Commun., 114*: 1097.
19. Eggers, H. M., Halsall, H. B., and Heineman, W. R. (1982). Enzyme immunoassay with flow-amperometric detection of NADH, *Clin. Chem., 28*: 1848.
20. Johannsson, A., Stanley, C. J., and Self, C. H. (1985). A fast, highly sensitive colorimetric enzyme immunoassay system demonstrating benefits of enzyme amplification in clinical chemistry, *Clin. Chim. Acta, 148*: 119.
21. Self, C. H. (1985). Enzyme amplification: A general method applied to provide an immunoassisted assay for placental alkaline phosphatase, *J. Immunol. Methods, 76*: 389.
22. Stanley, C. J., Johannsson, A., and Self, C. H. (1986). Enzyme amplification can enhance both the speed and the sensitivity of immunoassays, *J. Immunol. Methods, 83*: 89.
23. Johannsson, A., Ellis, D. H., Bates, D. L., Plumb, A. M., and Stanley, C. J. (1986). Enzyme amplification for immunoassays detection limit of one hundredth of an attomole, *J. Immunol. Methods, 87*: 7.
24. Stanley, C. J., Cox, R. B., Cardosi, M. F., and Turner, A. P.F. (1988). Amperometric enzyme-amplified immunoassays, *J. Immunol. Methods, 112*: 153.
25. McNeil, C. J., Spoors, J. A., Cocco, D., Cooper, J. M., and Bannister, J. V. (1989). Thermostable reduced nicotinamide adenine dinucleotide oxidase: Application to amperometric enzyme assay, *Anal. Chem., 61*: 25.
26. Cocco, D., Rinaldi, A., Savini, I., Cooper, J. M., and Bannister, J.V. (1988). NADH oxidase from the extreme thermophile, *Thermus aquaticus* YT-1, *Eur. J. Biochem., 174*: 267.
27. Cardosi, M. F., Rich, S. W., Stanley, C. J., Johannsson, A., and Turner, A. P. F. (1989). An electrochemical immunoassay for prostatic acid phosphatase incorporating enzyme amplification, *Am. Biotechnol. Lab., 7*: 50.

4

Potentiometric Enzyme Electrodes

Jean-Michel Kauffmann

Université Libre de Bruxelles
Brussels, Belgium

George G. Guilbault

University of New Orleans
New Orleans, Louisiana

INTRODUCTION

Potentiometric enzyme electrodes are analytical devices that incorporate a biological or biologically derived sensing element intimately associated with a physiochemical transducer and may be classified among the growing area of biosensors (1). Although the nature of the process involved at the biological site of potentiometric biosensors may be quite distinct, for example biocatalysis, immunoreaction, and chemoreception, the measurable signal is clearly a potential change due to the "specific" biological reaction. As is illustrated later, the notion of specificity must be considered with criticism since it is not uncommon that interfering analytes may be sensed by the biological element (enzyme or antibody) depending on the degree of purity of the latter. The first potentiometric biosensor was developed by Guilbault and Montalvo for urea in 1969 (2). Since then over 100 different potentiometric enzyme electrodes have appeared in the literature (3–6). Originally ion-selective electrodes, described as membrane sensors, served as the physical transducer. The potential that develops across an ion-selective membrane separating two solutions is measured at virtually zero current. A second impulse in the development of potentiometrically operating enzyme sensors came in the early 1980s from the works of Caras and Janata describing a penicillin-responsive device that comprised a pH-sensitive ion-selective field effect transistor (ISFET) and an enzyme-immobilized ISFET (ENFET) (7). At present a great amount of research is devoted to the construction and characterization of ENFET with the hope of solving the problems related to the marketing of "classic" potentiometric biosensors (8,9). Besides some interesting features of ISFET (microconception), the need to compensate for variations in the pH and buffering capacity of the sample still exist (9). For a detailed description of the performance of ENFET, the reader is referred to Chapter 7.

A rapid survey of the literature on electrochemical biosensors indicates a general trend toward the development and commercialization of amperometric biosensors (10–14). This tendency may be explained by the interesting characteristics of these devices, such as high sensitivity, versatility, and fast response (see Chap. 3).

Potentiometric electrode systems involving immobilized enzymes as biocatalysts have also been extensively investigated with respect to their performance and possible applications in clinical (10,11), environmental, and industrial analyses (3,12,13). Besides the continuous improvements of classic enzyme-based probes, further recent progress has been concerned with the development of new sensor concepts, for example incorporating immunological reagents (15).

ELECTRODE CHARACTERISTICS AND PERFORMANCE

In designing potentiometric enzyme electrodes the different steps that may govern the sensor response must be optimized: (1) mass transport of the reactant to the sensor surface, (2) reactant diffusion through the immobilized biological reagent, (3) reaction within the biological reagent, (4) product diffusion through the biological layer to the surface of the electrode, and (5) potential variation at the electrode surface. The first step, transport of the substrate, is most critically dependent on the stirring rate of the solution. Rapid stirring brings the substrate rapidly to the electrode surface. If the membrane is kept very thin using highly active enzyme, then steps 2 and 4 are eliminated or minimized. Since step 3 is very fast, the response of an enzyme electrode should theoretically approach the response time of the base sensor. These ideal conditions can be approached by using a thin membrane and rapid stirring.

pH-Glass and Ion-Selective Electrodes

The pH-glass electrode is universally available, and that many enzymatic reactions produce pH changes may explain the vast amount of effort devoted to the construction of enzyme-based pH sensors. Among some drawbacks encountered with the use of pH-sensing devices, and as already mentioned, the sample buffer capacity may be regarded as the most serious (9). This is particularly true when operating in real samples, such as biological fluids and fermentation broths, in which the buffer capacity is generally high and fluctuations in pH with time or from sample to sample may occur. Variations in pH can be dealt with by differential measurements using a second identical electrode but without immobilized enzyme. The construction of a differential pH meter enabling a change in the pH medium up to 1×10^{-4} units to be recorded has been reported and successfully applied to biological samples (16). Actually, to detect slight pH changes within the biological layer low buffering power is generally required. This may be achieved by sample dilution with concomitant risks of contamination and loss of sensitivity. Other crystal membrane and cation-selective glass ion-selective electrodes, such as iodide, fluoride, chloride, cyanide, and ammonium sensors, have been successfully applied to biosensor construction. Although interesting results have been reported, the use of the electrodes suffers from the fact that the measurements are not reagentless (I^-, F^-, CN^-, and Cl^-), surface film formation may occur (F^-), and poor selectivity may be encountered, for example a selectivity ratio of NH_4^+ over K^+ with ammonium sensors and urea, uric acid, and acetaminophen interference with the fluoride-selective electrode (17).

Solid-State pH and Redox Electrodes

In addition to pH-glass sensors various metal-metal oxide electrodes have been advantageously employed in combination with enzymes. The pH variation that occurs within the biological layer due to the enzymatic reaction is supposed to affect the structure of the oxide layer with concomitant modification of the electrode potential (18). Antimony oxide (18–20), palladium oxide (21), and titanium electrodes coated with oxides (IrO_2 and RuO_2) (22) have been proposed and have been shown to offer interesting advantages over the classic pH-glass electrode in terms of response time, miniaturization, and robustness.

Similarly, noble metals (platinum and gold) and carbon electrodes may be functionalized by oxidation procedures leaving oxidized surfaces. Actually, the potentiometric response of solid electrodes is highly governed by the surface state (23). Enzymes have been attached (chemically or physically) to these pretreated electrodes, and the biocatalytic reaction that occurs at the sensor tip may create potential shifts related to the amount of reactant. Products of the enzymatic reaction that may modify the redox state of the surface, such as hydrogen peroxide and protons, have been suspected to play a major role in the observed potential shifts (23). Compared to membrane electrodes the responses are rapid, but limitations in the use of solid-state electrodes come from the practical difficulties of controlling the surface state in a reproducible manner and the risk of interference by reactive redox species.

Recently an original all-solid-state electrode for NH_4^+ was reported and successfully combined with urease for the assay of urea (24). This electrode conception, which consisted of a conductive resin (epoxy + graphite) covered by a nonactin-polyvinyl chloride (PVC) matrix, may offer interesting commercial advantage over membrane electrodes with internal solution and a reference electrode.

Gas Electrodes

Improved selectivity may be achieved by casting the ion-selective electrodes with synthetic membranes, such as microporous Teflon, polypropylene, or homogeneous plastic films of Teflon, silicone rubber, or Mylar. Potentiometric gas-sensing electrodes consist generally of a combination of pH-glass and reference electrodes and a gas-permeable membrane. Between the glass electrode and the membrane is a layer of sodium bicarbonate solution (CO_2 sensor) or ammonium chloride (NH_3 sensor; Fig. 1). When the whole electrode is exposed to a solution of the substrate to be analyzed, the enzymatic reaction produces a pH-changing gas (CO_2 and NH_3). The gas diffuses both back to the sample and toward the pH sensor, passing through the enzyme and the gas-permeable membrane and then dissolving in the internal filling solution. This changes the pH and gives rise to the potentiometric response of the electrode system (25). Linear Nernstian responses are generally observed in the range of substrate concentration from 10 μM to 0.1 M, with a response time ranging from 1 to 10 min and a baseline recovery time of the same order of magnitude. Improvements in the response of gas-sensing electrodes may be observed by reducing the time of baseline recovery, for example by changing the internal filling solution after measurement. A thorough examination of the different steps controlling the functioning of gas-sensing electrodes, along with a detailed description of a new electrode design that allows easy change of the internal filling solution, has been reported (26). Despite their unique conception, gas-sensing electrodes may respond to gases other than those liberated by the enzymatic reaction. Indeed, interfering volatile gases (organic amines and CO_2) may be present in the analyzed sample and produce erroneous and abnormally high responses. Several theoretical analyses and experimental verification of the responses of gas sensors to interfering gases have been reported (27–30). The selectivity of a gas CO_2

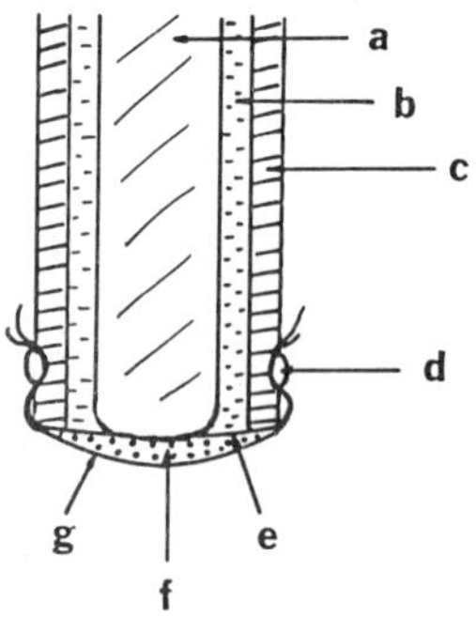

Figure 1 Cross-sectional view of a potentiometric gas-sensing enzyme electrode: (a) glass pH electrode; (b) internal filling solution; (c) electrode outer jacket; (d) O-ring; (e) gas-selective membrane; (f) enzyme layer; (g) dialysis membrane.

probe has been shown to be time and concentration dependent (27). In the case of NH_3 sensors the selectivity was shown to be determined by the acidity and basicity of the interferents rather than by the volatility (28,29). For homogeneous plastic films the solubility of the gas in the membrane material must be considered (30).

POTENTIOMETRIC IMMUNOBIOSENSORS

As a result of the progressive marketing of monoclonal antibodies that exhibit better selectivity during molecular recognition, new immunosensors have been developed in the last few years (31). Actually, the binding of antigens or antibodies (labeled or not) to a solid carrier, such as glass, collagen, and graphite, can be effected in a manner similar to that for an enzyme (3). Various enzymes, for example glucose oxidase, catalase, urease, cholinesterase, asparaginase, and adenosine deaminase (33), may be successfully linked to antigens for labeling provided that the enzyme activity as well as the immunological properties of the antigen are retained. Amperometric detection modes are generally preferred for immunobiosensor construction (33), but several interesting applications with potentiometric membrane electrodes have been reported, as listed in Table 1. Potentiometric immunoassays may be performed (1) by recording the change in potential (mixed surface potentials), at either an antibody- or antigen-immobilized electrode, when the specific binding occurs, and (2) by measuring the product (H^+, CO_2, and NH_3) generated by the enzyme-labeled antigen. The enzyme-labeled antigen and the unlabeled antigen to be determined compete for the binding sites of antibodies retained on a membrane. After completion the active membrane may be cast over a potentiometric electrode. The unknown concentration of unlabeled antigens is inversely proportional to the activity of the marker enzymes measured by the electrode. This assay is generally heterogeneous; that is, the antibody-bound antigen must be separated from the antigen at some point in the procedure. In contrast, homogeneous assays require no separation step and are simpler and faster but may suffer from poor detection limits and a higher susceptibility to interference and fouling problems. Original innovations in potentiometric membrane electrode design, in which the antibody-antigen reaction is limited to the sensor tip region, have recently been reported (34,35). The electrode contains an antigen coupled to an ionophore, and the reaction of the antigen with its antibody modulates the ion carrier properties of the ionophore and produces a potential change (Fig. 2). This configuration has been found to have highly attractive features since the entire immunoreaction occurs within the tip of the sensor (antibody conservation) and since reversible antigen responses are exhibited. Examples have been reported that show the ability of this probe to distinguish selectivity of monoclonal antibody over polyclonal antibody for 2,4-DNP (dinitrophenol) (36). In addition, such a concept may allow determination of the specificity of the antibody for the homolo-

Table 1 Antibody-Immobilized Potentiometric Sensors[a]

Antigen (Ag)	Antibody (Ab)	Electrode	Antibody immobilization	Remarks	Reference
2, 4-DNP	Anti-DNP	PVC membrane + 2,4-DNP	Ab solution + collagen	Reversible; DL = 1 μM; OS = 17 days	34
HCG	Anti-HCG	pH-sensitive TiO_2	Cyanogen bromide	Reversible; DL = 0.1 μg/ml	36
HCG	Anti-HCG	pH-glass	Polyethylene + GA + BSA	Acetylcholine esterase as label enzyme; DL = 1 nM; OS = 1 week;	37
Riboflavin	Aporiboflavin	Membrane	Ab + FAD bound membrane	DL = 0.1–2 μM	38
Protein MRP 14	Anti-MRP 14	Sodium-glass	—	Reversible; OS = 3 months	39
HBsAg	Anti-HBsAg	I^-	Gelatin	HRP as label; DL = 0.1 μg/L	40 41
17β-Estradiol	Anti-17β-estradiol	I^-	Gelatin	HRP as label; DL = 57 pmol/L	42

[a]HCG, human chorionic gonadotropin; HBsAg, hepatitis B surface antigen; GA, glutaraldehyde; DL, detection limit; OS, operational stability; HRP, horseradish peroxidase.

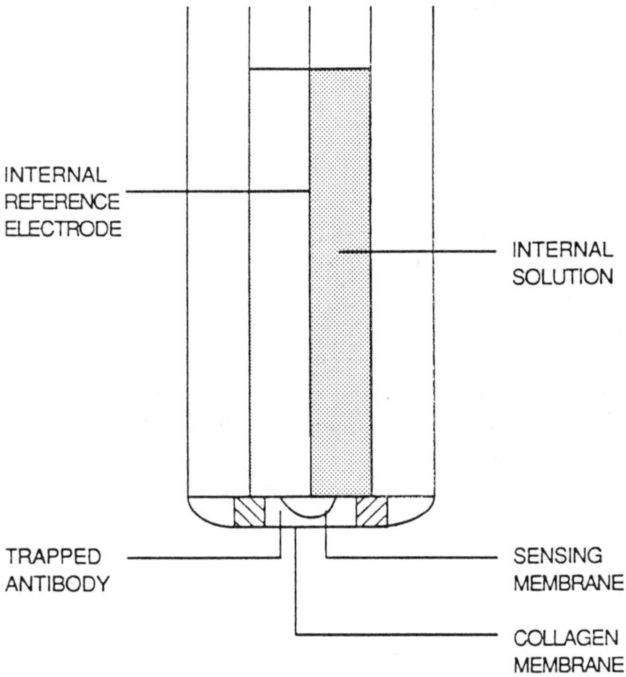

Figure 2 Monoclonal antibody biosensor for antigen monitoring. (From ref. 34, with permission of publisher.)

gous hapten over the heterologous hapten (36). Further information regarding immunobiosensors may be obtained by consulting Chapter 14.

IMMOBILIZATION METHODS

A great diversity of procedures has been developed to fix enzymes or proteins on electrode supports. The "soluble" enzyme electrode originally described consisted of maintaining the biological element close to the sensor by an appropriate membrane (43). Further strategies were conducted to improve the stability of the biological layer by physical or chemical insolubilization and optimization of the intimate contact between enzyme and electrode (44). Physical entrapment in an inert matrix, such as polyacrylamide, starch, or gelatin, is often applied because of the ease of preparation over many chemical procedures. The resulting enzymatic gel may be cast over the potentiometric electrode and maintained by a dialysis membrane. Physically entrapped enzymes present generally lower stabilities than chemically immobilized enzymes (3). Under ideal conditions a physically entrapped enzyme may be stable for about 3 weeks or 100 assays and a chemically bound enzyme for over a year and up to 1000 assays. Chemical immobilization may be performed by cross-linking the enzyme to itself or, better, to an inert protein (bovine serum albumin, BSA, and gelatin) or to a solid support by a suitable bifunctional reagent, such as glutaraldehyde, and the resulting membrane assembled on the sensor tip. Better performances in terms of response time can be achieved by chemically attaching the enzyme to the electrode surface (45) or directly to the polymer membrane, which is the integral part of gas membrane electrodes (74,75). In Tables 2 through 4 are listed representative examples of immobilization procedures generally applied with potentiometric sensors. Detailed information regarding enzyme immobilization techniques have been published elsewhere (46–48). Actually there is no universal ''ideal'' immobilization mode. Each enzyme is unique in terms of structure and activity, and a rule of thumb in enzyme immobilization procedures is to ensure that the treatment does not inactivate those functional groups of the enzyme that are essential for its catalytic activity. Trends in enzyme immobilization are directed toward the synthesis (chemical and electrochemical) of improved polymers in terms of substrate permeation (49,50), along with the investigation of biocompatible biopolymers (saccharides, polypeptides, and polyalcohols) as immobilizing agents for in vivo sensors (51).

ANALYTICAL APPLICATIONS

Potentiometric enzyme probes have been applied to the assay of a great variety of substances of clinical, biotechnological, and environmental interest. Once the operating parameters are optimized, the analysis may be performed as simply as pH measurements, and by referring the responses to precalibration curves, that is, concentration-activity versus potential in a linear logarithmic plot. A survey of Table 2 indicates that the determination of amino acids has been widely investigated with potentiometric enzyme electrodes. Actually, the quantitation of amino acids is of prime importance in clinical practice since their presence and their activity level in biological fluids and tissues (such as phenylalanine and tyrosine) may be a useful indicator of certain diseases or disorders. Similarly, control of the evolution of biotechnological processes (including bioreactors and fermentation), as well as the determination of specific amino acids (including lysine and arginine) in foods, requires reliable measuring devices that are easy to use and handle. In this respect, and as already suggested, potentiometric enzyme sensors exhibit attractive features. Potentiometric gas electrodes are mostly utilized because of their inherent selectivity, the reliability of the results being generally governed by the purity of the attached enzyme. Lysine, arginine, and histidine biosensors have been shown to be highly selective probes and have been successfully applied in clinical and food controls (Table 2). Lysine decarboxylase enzymes from different sources and of different degrees of purity, for example *Bacterium cadaveris* (66), *Escherichia coli* (68), and *Klebsiella pneumoniae* (67), have been combined with the pCO_2 probe. Operational stabilities for longer than 6 weeks have been reported (68).

Table 2 Amino Acid Potentiometric Biosensors

Substrate	Enzyme	Electrode	Immobilization	Reference
N-acetylmethionine	Amino acylase + amino acid oxidase	pNH_3	Soluble enzyme GA + BSA Glass beads	52
Alanine	Alanine dehydrogenase + lactate dehydrogenase	pNH_3	Coimmobilized with dextran-bound NAD^+	53
Arginine	Arginase and urease	pNH_3	Controlled pore glass beads	54
	Arginase and urease	NH_4^+	GA + BSA	55
Asparagine	Asparaginase	pNH_3	Soluble enzyme	56
	Asparaginase	pNH_3	GA or agar gel	65
Aspartate	Aspartase	pNH_3	GA	57
Aspartame	Aspartase	pNH_3	GA + BSA	58
	Carboxypeptidase + aspartase	pNH_3	GA + BSA	59
Glutamine	Glutaminase	pNH_3	Soluble enzyme	60
	Glutaminase	Cation-glass	Soluble enzyme	61
Histidine	Histidine decarboxylase	pCO_2	GA + BSA	63
	Histidine amonialyase	pNH_3	Soluble enzyme or GA + BSA	64
Lysine	Lysine decarboxylase	pCO_2	GA + BSA + Teflon	66
	Lysine decarboxylase	pCO_2	Agglutination on Teflon	67
	Lysine decarboxylase	pCO_2	Gelatin + GA	68
Methionine	Methionine lyase	pNH_3	GA + BSA	69
Phenylalanine	*l*-Amino acid oxidase + HRP	I^-	Chemical immobilized in polyacrylamide gel	70
Tryptophan	Crude tryptophanase	pNH_3	Soluble enzyme or GA or agar gel	71
Tyrosine	Tyrosine decarboxylase	pCO_2	Small intestine + GA + BSA	72
	apo-Tyrosine decarboxylase	pCO_2	Small intestine + GA + BSA	73

Table 3 Urea Potentiometric Sensors[a]

Electrode	Immobilization	Reference
pNH_3	Cyanuric chloride cross-linking with polyvinylalcohol −g− butyl acrylate	49
pNH_3	GA + Teflon or polypropylene	74 75
pNH_3	Agar gel or GA + cellulose acetate	76
Glass NH_4^+	polyacrylamide gel	77
pCO_2 or glass NH_4^+	GA + BSA	78
Silicone rubber-based nonactin NH_4^+	Polyacrylic gel	79
PVC-based nonactin NH_4^+	GA + BSA + aminosilanized polyester	55
PVC-based nonactin NH_4^+	GA	80
pH-glass	Polyacrylamide hydrazide + glyoxal	50
Sb + gas membrane	Ga + BSA	20
Graphite-epoxy	Acrylamide gel	24
PVC-nonactin NH_4^+		62

[a]Substrate was urea; enzyme was urease.

Several possible enzymatic pathways may be followed for the degradation of substrates, and a judicious selection of the most appropriate enzyme in terms of purity and activity is generally preferable for biosensor construction. Aspartame, an artificial sweetener that consists of a combination of two amino acids (phenylalanine and aspartic acid), may be analyzed by different biosensor configurations (Fig. 3). Actually two distinct biosensors have been developed, and their performance has been established by analyzing the sweetener in different diet products (58,59).

Limitations to enzymatic biosensor performance may arise from the instability of the biological layer. Upon immobilization, however, less susceptibility to inactivation and, in the case of dual enzymes, faster responses (due to closer contacts) may be observed (54). Unfortunately there is no general procedure for enzyme immobilization, and each new biosensor development requires a comparative study of the different possible immobilization modes. Alternatively, comparison between different biocatalysts, for example enzyme, bacteria, and tissue (vegetable and animal) may be performed to select the most judicious biosensor, as shown for glutamine (60) and tryptophan (Table 5) (71).

The sensitivities of the measurements lie generally within the millimolar and micromolar ranges, with detection limits as low as 10 μM that is, well below the sensitivities required in fermentation and bioprocess control. Biomedical analyses often require lower detection limits, which may preclude the application

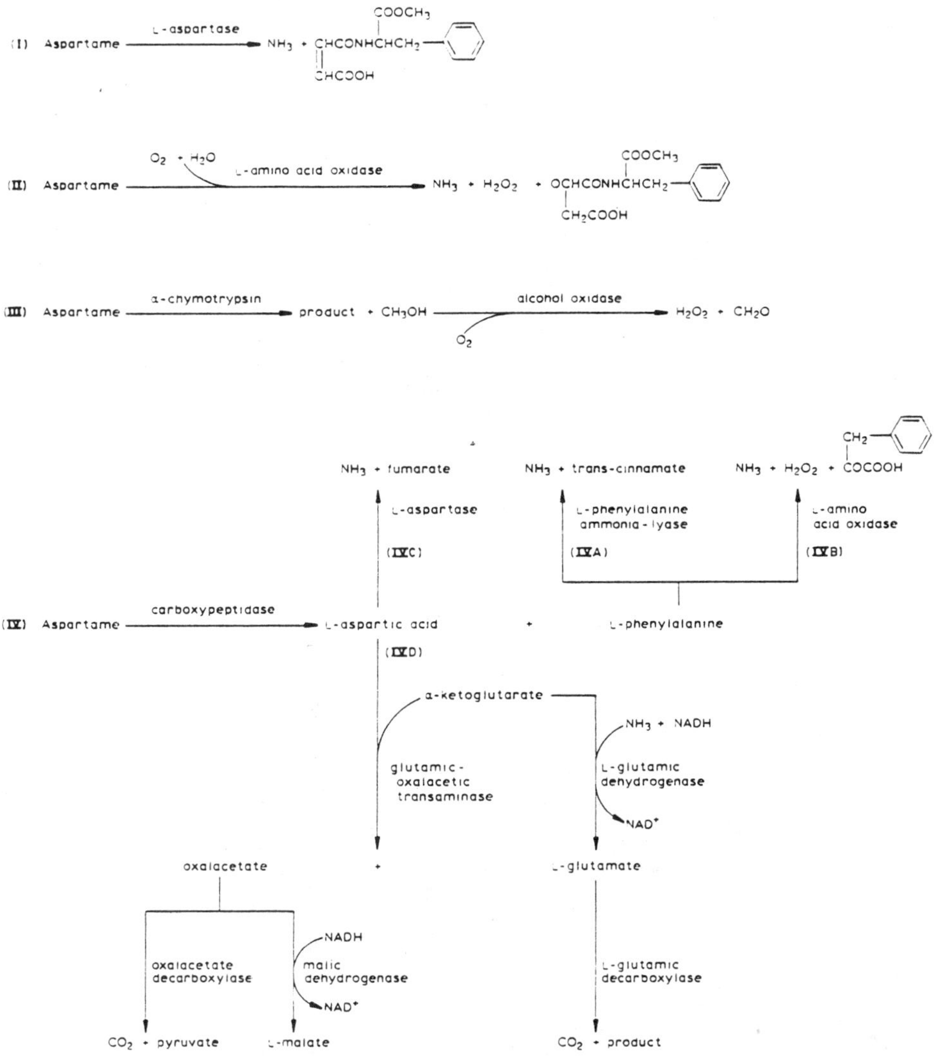

Figure 3 Possible enzymatic pathways for aspartame degradation.

of potentiometric biosensors. Urea may be ideally analyzed in biological fluids by biosensors combining the quite stable and commercially available enzyme urease to potentiometric detectors (pNH_3,NH_4^+, and pH-glass), as listed in Table 3. Over the last few years several improvements were reported to accelerate the sensor response (49,75) and to fabricate micro and solid-state devices (20,24,62).

Table 4 Examples of Potentiometric Enzyme Biosensors

Substrate	Enzyme	Electrode	Immobilization	Reference
Adenosine	Adenosine deaminase	pNH_3	Soluble enzyme	81
5′-AMP	AMP deaminase	pNH_3	Soluble enzyme	82
Chloroform	Crude extract (*Pseudomonas putida*) hydrolase	Cl^-	GA + BSA	83
Creatinine	Creatininase	pNH_3	BSA	84
	Creatininase	pNH_3	GA + BSA	85
	Creatine deaminase	pNH_3	Activated collagen or pig intenstine	86
	Glutamate dehydrogenase + creatinine deiminase	pNH_3	Adsorption on wet PVC	87
FAD	Alkaline phosphatase + adenosine deaminase	pNH_3	Soluble enzyme BSA + GA	88
Fluoride	Urease	pNH_3	GA + polypropylene	89
Glucose	GOD	pH-glass	Polyacrylamide gel	90
	GOD + HRP	I^-	PVC matrix	91
	GOD + catalase	Pt	Polyacrylamide gel	92
	Glucose dehydrogenase	Pt	Polyacrylamide gel	93
	GOD	Pt/graphite/polymer	Soluble enzyme	94

Lactate	Lactate dehydrogenase	Redox membrane (PVC + ferrocene)	Gelatin gel	95
	Lactate oxidase + HRP	F^-	Gelatin + GA	96
Maltose	Glucoamylase + GOD + HRP	F^-	Gelatin + GA	96
Oxalate	Oxalate oxidase	PCO_2	Soluble enzyme	97
	Oxalate decarboxylase	pCO_2	Soluble enzyme or glutamine + BSA	98
Oxaloacetate	Artificial enzyme	pCO_2	Soluble enzyme	99
Penicillin	Penicillinase	Sb/Sb_2O_3	Carbodiimide	100
			Polyacrylamide gel BSA + GA	
Pesticides	Butyrylcholine esterase	pH-glass	GA + HSA[a]	101
Salicylate	Salicylate hydroxylase	pCO_2	Soluble enzyme	102
Uric acid	Uricase	pCO_2	Soluble enzyme BSA + GA	103

[a]Human serum albumin.

Table 5 Comparison of the Characteristics of Enzyme and Bacterial Electrodes for Tryptophan Determination

Criteria	Bacterial electrode	Enzyme electrode (solubilized)
Linearity range, mol/L	1×10^{-2}–3×10^{-4}	1×10^{-2}–5×10^{-4}
Slope, mV/decade	22	18
Response time, min	10–15	10–15
Stability	3 weeks	4 days
Cofactor concentration	2.5 mg per 100 ml	20 mg per 100 ml
Interferences	Asparagine, glutamine, aspartic acid	Asparagine, glutamine, aspartic acid
Stability of stock material	Several years on HIA slants at 4°C	Less than 2 months (lyophilized form)

Source: From Reference 65.

Numerous potentiometric biosensors measure the ammonia generated during an enzymatic reaction; when applied in real sample analysis, however, interference by endogenous ammonia is likely to occur. Several strategies may be developed to circumvent this difficulty, for example by performing differential measurements using two probes, one of which contains no enzyme. Alternatively a dual configuration may be realized, as shown recently for creatinine measurements (87). In this device inactivation of endogenous ammonia occurs within the first enzymatic layer as follows:

$$NH_4^+ + \alpha\text{-ketoglutarate} + NADH \xrightarrow{\text{GlDH}} \text{glutamate} + NAD^+ + H_2O$$

Creatinine does not react in the first layer and reaches the immobilized creatinine deiminase with production of ammonia, which is detected at the electrode.

Recent developments in potentiometric biosensors use the advantage of flow injection analysis (automation, speed, minor fouling problems, and others) (80,101). In pollution control a pH-glass probe with immobilized butyrylcholinesterase (BuChE) has been successfully applied to pesticide detection (101). The assay consisted of monitoring the inhibition of the activity of the immobilized enzyme due to the presence of organophosphorus pesticides, as follows:

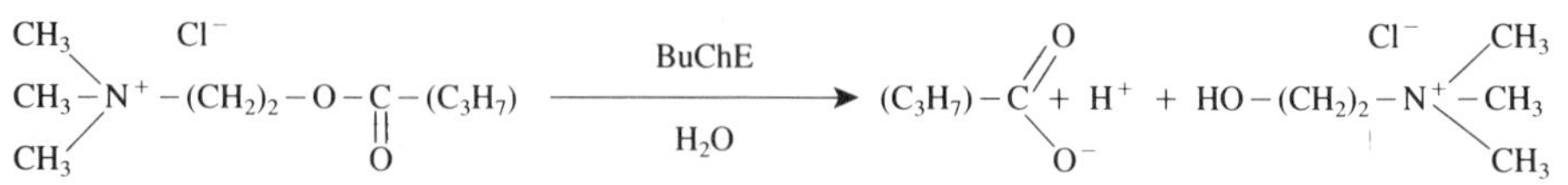

The presence of pesticides in river waters induced a proportional pH increase; the probe was reactivated by 2-pyridine aldoxime. Detections of pesticides as low as 10 nM have been reported. Other automated systems have advantageously combined multienzyme layers over a fluoride-selective electrode for the assay of maltose, glucose, and lactose (96). Coimmobilized peroxidase catalyzes the liberation of fluoride ions from a suitable reagent present in the analyzed solution (hydrogen peroxide being liberated by the substrate-enzyme reaction):

$$R-F_n + H_2O_2 \xrightarrow{\text{HRP}} R-F(n-1) + F^- + H_20$$

Finally, new probes may arise from the development of synthetic enzymes (synzymes), which have been shown to exhibit improved stability over natural enzymes in a potentiometric biosensor configuration (99).

COMMERCIALLY AVAILABLE DEVICES

Universal Sensors, Inc. (New Orleans) markets self-contained enzyme potentiometric electrodes for urea, amino acids (lysine, methionine, tyrosine, and phenylalanine), creatinine, and others (3,12,104).

REFERENCES

1. Turner, A. P. F., Karube, I., and Wilson, G. S. (Eds) (1987). *Biosensors: Fundamentals and Applications*. Oxford University Press, Oxford.
2. Guilbault, G. G. and Montalvo, J. (1969). A urea specific enzyme electrode, *J. Am. Chem. Soc., 91*: 2164.
3. Guilbault, G. G. (1984). *Analytical Uses of Immobilized Enzymes*, Marcel Dekker, New York.
4. Guilbault, G. G., and Kauffmann, J.-M. (1987). Enzyme based electrodes as analytical tools, *Biotechnol. Appl. Biochem., 9*: 95.
5. Arnold, M. A., and Meyerhoff, M. E. (1984). Ion-selective electrodes, *Anal. Chem., 56*: 20R.
6. Kuan, S. S. and Guilbault, G. G. (1987). Ion selective electrodes and biosensors based on ISEs. In: *Biosensors: Fundamentals and Applications*, Oxford University Press, Oxford, p. 135.
7. Caras, S., and Janata, J. (1980). Field effect transistor sensitive to penicillin, *Anal. Chem., 52*: 1935.
8. Karube, I., and Kubo, I. (1987). Micro-biosensor. In: *Analytical Uses of Immobilized Biological Compounds for Detection, Medical and Industrial Uses* (G. G. Guilbault and M. Mascini, Eds), NATO ASI Series, D. Reidel Publishing, New York, Vol. 226, p. 207.
9. van der Schoot, B. H., and Bergveld, P. (1987/88). ISFET based enzyme sensors, *Biosensors, 3*: 161.
10. Janata, J., and Bezegh, A. (1988). Chemical sensors, *Anal. Chem., 12*: 62R.

11. Mascini, M., and Guilbault, G. G. (1986). Clinical uses of enzyme electrode probes, *Biosensors, 2*: 147.
12. Scheller, F., Schubert, F., Pfeiffer, D., Hintsche, R., Dransfeld, I., Renneberg, R., Wollenberger, U., Riedel, K., Pavlova, M., Kuhn, M., Muller, H.- G., Tan, P., Hoffmann, W., and Moritz, W. (1989). Research and development of biosensors, *Analyst, 44*: 653.
13. Tran-Minh, C., and El Yamani, M. (1988). Enzyme sensors for the determination of toxic chemicals in environmental samples. In: *Electrochemical Detection Techniques in the Applied Biosciences* (G. A. Junter, Ed), Ellis Horwood Ltd., Chichester, Vol. 2, p. 131.
14. Turner, A. P.F. (1989). Current trends in biosensor research and development, *Sensors Actuators, 17*: 433.
15. Rechnitz, G. A. (1988). Biosensors, *Chem. Eng. News, 9*: 24.
16. Ripamonti, M., Mosca, A., Rovida, E., Luzzana, M., Luzi, L., Ceriotti, F., Cottini, F., and Bernardi, L. R. (1984). Urea, creatine and glucose determined in plasma and whole blood by a differential pH technique, *Clin. Chem., 30*: 556.
17. Cowell, D. C., and Ford, P. A. E. (1987). Interference in an electrochemical detection system for peroxidase linked reaction based on fluoride ion-selective electrode, *Clin. Chem., 33*: 1458.
18. Flanagan, M. T., and Carrol, N. J. (1984). Thin-film antimony-antimony oxide enzyme electrode for urea determination, *Biotechnol. Bioeng., 26*: 642.
19. Kulys, J. J., Gureviciene, V. V., Laurinavicius, V. A., and Jonuska, A. V. (1986). Urease sensors based on differential antimony electrodes, *Biosensors, 2*: 35.
20. Joseph, J. P. (1985). An enzyme microsensor for urea based on ammonia gas electrode, *Anal. Chim. Acta, 169*: 249.
21. Szuminski, N. J., Chen, A. K., and Liu, C. C. (1984). A miniature palladium-palladium oxide enzyme electrode for urea determination, *Biotechnol. Bioeng., 26*: 642.
22. Roberts, D. C., Osborn, J. A., and Yacynych, A. M. (1986). Proteolytic enzyme modified metal oxide electrodes as potentiometric sensors, *Anal. Chem., 58*: 140.
23. Wingard Jr., L. B., and Castner, J. (1987). Potentiometric biosensors based on redox electrodes. In: *Biosensors: Fundamentals and Applications*, Oxford University Press, Oxford, p. 153.
24. Alegret, S., and Martinez-Fabregas, E. (1989). Biosensors based on conducting filled polymer all-solid state PVC matrix membrane electrodes, *Biosensors, 4*: 287.
25. Ross, J. W., Riseman, J. M., and Krueger, J. A. (1973). Potentiometric gas electrodes, *Pure Appl. Chem., 36*: 473.
26. Guilbault, G. G., Czarnecki, J. P., and Nabi Rahni, M. A. (1985). Performance improvements of gas diffusion ion-selective and enzyme electrodes, *Anal. Chem., 57*: 2110.
27. Morf, W. E., Mostert, I. A., and Simon, W. (1985). Time response of potentiometric gas sensors to primary and interfering species, *Anal. Chem., 57*: 1122.
28. Lopez, M. E., and Rechnitz, G. A. (1982). Selectivity of the potentiometric ammonia gas sensing electrode, *Anal. Chem., 54*: 2085.
29. Viejo, V., and Vallon, J. J. (1989). Selectivité de l'électrode à ammoniac vis à vis d'amines volatiles, *Anal. Lett., 22*: 25.

30. Kobos, R. K., Parks, S. J., and Meyerhoff, M. E. (1982). Selectivity characteristics of potentiometric carbon dioxide sensors with various gas membrane material, *Anal. Chem., 54*: 1976.
31. Ngo, T. T. (Ed.). (1987). *Electrochemical Sensors in Immunological Analysis*, Plenum Press, New York.
32. Gebauer, C. R., and Rechnitz, G. A. (1982). Deaminating enzyme labels for potentiometric enzyme immunoassay, *Anal. Biochem., 124*: 338.
33. Green, M. (1987). New approaches to electrochemical immunoassays. In: *Biosensors: Fundamentals and Applications* (A. P. F. Turner, I. Karube, and G. Wilson, Eds), Oxford University Press, Oxford, p. 60.
34. Bush, D. L., and Rechnitz, G. A. (1987). Monoclonal antibody biosensor for antigen monitoring, *Anal. Lett., 20*: 1781.
35. Bush, D. L., and Rechnitz, G. A. (1988). Comparison of antibodies as molecular recognition elements for biosensor design, *Anal. Lett., 21*: 1947.
36. Yamamoto, N., Nagasawa, Y., Shuto, S., Tsuburoma, H., Sawai, M., and Okurama, H. (1980). Antigen-antibody reaction investigated with the use of a chemically modified electrode, *Clin. Chem., 26*: 1569.
37. Mascini, M., Zolesi, F., and Paleschi, G. (1981). pH electrode based enzyme immunoassay for the determination of human chorionic gonadotropin, *Anal. Lett., 15*: 101.
38. Yao, T., and Rechnitz, G. A. (1987). Potentiometric biosensor for riboflavin based on the use of aporiboflavin-binding protein, *Anal. Chem., 59*: 2115.
39. Camman, K., Jahn, K., Rohen, A., Weiss, T., Wilken, H., and Winter, B. (1989). Comparison of electrochemical with optical biosensors. In: *Biosensors in Medicine, Environmental Protection and Process Control* (R. D. Schmid and F. Scheller, Eds), GBF Monographs, VCH Publishers, New York, p. 201.
40. Boitieux, J. L., Desmet, G., and Thomas, D. (1979). Antibody electrode: Preliminary report on a new approach in enzyme immunoassay, *Clin. Chem., 25*: 318.
41. Boitieux, J. L., Desmet, G., and Thomas, D. (1978). Determination potentiométrique de l'antigène de surface de l'hépatite B dans les liquides biologiques, *Clin. Chim. Acta, 88*: 329.
42. Boitieux, J. L., Le May, C., Desmet, G., and Thomas, D. (1981). Use of a solid phase biochemistry for potentiometric enzyme immunoassay of oestradiol-17— preliminary report, *Clin. Chim. Acta, 113*: 175.
43. Clark, L., and Lyons, C. (1962). Electrode systems for continuous monitoring in cardiovasculary surgery, *Ann. NY Acad. Sci., 102*: 29.
44. Barker, S. A. (1987). Immobilization of the biological component of biosensors. In: *Biosensors: Fundamental and Applications* (A. P. F. Turner, I. Karube, and G. S. Wilson, Eds), Oxford University Press, Oxford, p. 85.
45. Ianniello, R. M. and Yacynych, A. M. (1981). Chemically modified graphite electrode with immobilized enzyme as a potentiometric sensor for some l-amino acids, *Anal. Chim. Acta, 13*: 123.
46. Attiyat, A. S., and Christian, G. D. (1984). Immobilization of enzymes, *Am. Biotechnol. Lab., 2*: 8.
47. Gekas, V. C. (1986). Artificial membranes as carriers for the immobilization of biocatalysts, *Enzyme Microb. Technol., 8*: 450.

48. Guilbault, G. G., and de Olivera Neto, G. (1985). Immobilized enzyme electrodes. In: *Immobilized Cells and Enzymes: A Practical Approach* (J. Woodword, Ed), IRL Press Oxford, p. 55.
49. Hsiue, G. H., Chou, Z. S., Yu, N., and Ksiung, K. P. (1987). Urease immobilized in polyvinylalcohol-g-butyl acrylate for urea sensor, *J. Appl. Polym. Sci., 34*: 319.
50. Tor, R., and Freeman, A. (1986). New enzyme membrane for enzyme electrode, *Anal. Chem., 58*: 1042.
51. Campanella, L., Mazzei, F., Samartino, M. P., and Tomasetti, M. (1989). Biopolymers as biocompatible immobilizing agents. In: *Strategies for In Vivo Sensing* (M. Mascini, Ed), 2nd Workshop of Biomedical Engineering Action of the European Community on Chemical Sensors for in vivo monitoring, Firenze, p. 3.
52. Nikolelis, D. P., and Hadjiioannou, T. P. (1983). Construction of a *N*-acetyl-L-methionine electrode and determination of acylase with an ammonia gas sensor, *Anal. Lett., 16*: 401.
53. Pau, C. P. and Rechnitz, G. A. (1984). Bound cofactor/dual enzyme electrode system for l-alanine, *Anal. Chim. Acta, 160*: 141.
54. Valle-Vega, P., Young, C. T., and Swaisgood, H. E. (1980). Arginase-urease electrode for determination of arginine and peanut maturity, *J. Food. Sci., 45*: 1027.
55. Yasuda, K., Niyagi, H., Hamada, Y., and Takata, Y. (1984). Determination of urea in whole blood using a urea electrode with an immobilized urease membrane, *Analyst, 109*: 61.
56. Nikolelis, D. P. (1984). Construction of an immobilized asparaginase sensor and determination of asparagine in human blood, *Anal. Chim. Acta, 161*: 343.
57. Fatibello-Filho, O., Suleiman, A., and Guilbault, G. G. (1989). Enzyme electrode for the determination of aspartate, *Biosensors, 4*: 313.
58. Guilbault, G. G., Lubrano, G. J., Kauffmann, J. -M., and Patriarche, G. J. (1988). Enzyme electrode for the sugar substitute aspartame, *Anal. Chim. Acta, 206*: 369.
59. Fatibello-Filho, O., Suleiman, A. A., Guilbault, G. G., and Lubrano, G. J. (1988). Bienzymatic electrode for the determination of aspartame in dietary products, *Anal. Chem., 60*: 2397.
60. Arnold, M. A., and Rechnitz, G. A. (1980). Comparison of bacterial, mitochondrial, tissue and enzyme biocatalysts for glutamine selective membrane electrodes, *Anal. Chem., 52*: 1170.
61. Guilbault, G. G., and Shu, F. (1971). An electrode for the determination of glutamine, *Anal. Chim. Acta, 56*: 333.
62. Palleschi, G., Mascini, M., Martinez-Fabregas, E., and Alegret, S. (1988). Urea determination in human sera with an ammonium ion selective electrode made with solid inner electric contact and immobilized urease, *Anal. Lett., 21*: 1115.
63. Kovach, P. M., and Meyerhoff, M. E. (1982). Development and application of a histidine-selective biomembrane electrode, *Anal. Chem., 54*: 217.
64. Walters, R. R., Johnson, P. A., and Buck, R. P. (1980). Histidine ammonia-lyase enzyme electrode for determination of l-histidine, *Anal. Chem., 52*: 1684.
65. Vincke, B. J., Vire, J.-C. and Patriarche, G. J. (1986). Potentiometric determinations of amino acids using enzyme and bacterial electrodes. In: *Electrochemistry, Sensors and Analysis* (M. R. Smyth and J. G. Vos, Eds), Elsevier, Amsterdam, p. 147.

66. White, W. C., and Guilbault, G. G. (1978). Lysine specific enzyme electrode for determination of lysine in grains and food stuffs, *Anal. Chem., 50*: 1481.
67. Weissbach, F., Schindler, W., Huber, J., and Kreibich, G. (1987). Enzymelektrode zur bestimmung von l-lysin in Aminosäuregemischen, *Acta Biotechnol, 7*: 377.
68. Romette, J. L. (1988). Electrochemical biosensors in clinical analysis. In: *Electrochemical Detection Techniques in the Applied Biosciences* (G. A. Junter, Ed), Ellis Horwood Ltd., Chichester, Vol. 1, p. 103.
69. Fung, K. W., Kuan, S. S., Sung, H. Y., and Guilbault, G. G. (1979). Methionine selective enzyme electrode, *Anal. Chem., 51*: 2319.
70. Guilbault, G. G., and Nagy, G. (1973). Enzyme electrode for the determination of l-phenylalanine, *Anal. Lett., 6*: 301.
71. Linders, C. R., Vincke, B. J., Devleeschouwer, M. J. and Patriarche, G. J. (1985). Determination du tryptophane à l'aide d'électrodes bactériennes et enzymatiques, *J. Pharm. Belg., 40*: 19.
72. Havas, J., and Guilbault, G. G. (1982). Tyrosine selective enzyme probe and its application, *Anal. Chem., 54*: 1991.
73. Guilbault, G. G., and Shu, F. (1972). Enzyme electrode based on the use of carbon dioxide sensor urea and l-tyrosine electrode, *Anal. Chem., 34*: 1437.
74. Anfalt, T., Graneli, A., and Jagner, D. (1973). A urea electrode based on the ammonia probe, *Anal. Lett., 6*: 969.
75. Guilbault, G. G., and Mascini, M. (1977). Urease coupled ammonia electrode for urea determination in blood serum, *Anal. Chem., 49*: 795.
76. Fischer, M., Laudet, A., Vire, J.-C., Patriarche, G. J., and Dufrane, S. P. (1982). Determination de l'urée sanguine à l'aide d'une électrode à diffusion gazeuse d'ammoniac, *J. Pharm. Belg., 37*: 202.
77. Guilbault, G. G., and Montalvo, J. G. (1970). An enzyme electrode for the substrate urea, *J. Am. Chem. Soc., 92*: 2533.
78. Tran-Minh, C., and Brown, G. (1975). Construction and study of electrode using cross-linked enzymes, *Anal. Chem., 47*: 1359.
79. Guilbault, G. G., and Nagy, G. (1973). Improved urea electrode, *Anal. Chem., 45*: 417.
80. Petersson, B. A. (1988). Enzymatic determination of urea in undiluted whole blood by flow-injection analysis using an ammonium ion-selective electrode, *Anal. Chim. Acta, 209*: 239.
81. Deng, I., and Enke, C. (1980). Adenosine-selective electrode, *Anal. Chem., 52*: 1937.
82. Papastathopoulos, D. S., and Rechnitz, G. A. (1976). Highly selective enzyme electrode for 5'-adenosine monophosphate, *Anal. Chem., 48*: 862.
83. Rahni, M. A. N., Kuan, S. S., and Guilbault, G. G. (1986). Aerobic microbial degradation of chloroform: Construction of an immobilized enzyme electrode for chloroform assay, *Enzyme Microb. Technol., 8*: 300.
84. Meyerhoff, M., and Rechnitz, G. A. (1976). An activated enzyme electrode for creatinine, *Anal. Chim. Acta, 85*: 277.
85. Guilbault, G. G., Chen, S. P., and Kuan, S. S. (1980). A creatinine specific enzyme electrode, *Anal. Lett., 13*: 1607.

86. Guilbault, G. G., and Coulet, P. R. (1983). Creatinine selective enzyme electrode, *Anal. Chim. Acta, 152*: 223.
87. Kihara, K., and Yasukawa, E. (1986). Determination of creatinine with a sensor based on immobilized glutamate dehydrogenase and creatinine deiminase. *Anal. Chim. Acta, 183*: 75.
88. Ayatin, A. R. M., and Riedel, T. L. (1984). A dual enzyme electrode for the determination of flavin adenine dinucleotide in the presence of riboflavin and flavin mononucleotide, *Anal. Lett., 17*: 835.
89. Tran-Minh, C., and Beaux, J. (1979). Enzyme electrode for inhibitors determination: Urease-fluoride system, *Anal. Chem., 51*: 91.
90. Nilsson, H., Ackerlund, A. C., and Mosbach, K. (1973). Determination of glucose, urea and penicillin using enzyme pH electrodes, *Biochim. Biophys. Acta, 320*: 529.
91. Al-Hitti, I. K., Moody, G. J., and Thomas, J. D. R. (1984). Glucose oxidase membrane system based on poly(vinyl chloride) matrices for glucose determination with an iodide ion-selective electrode, *Analyst*, 109:1205.
92. Liu, C. C., Wingard Jr., L. B., Wolfson Jr., S. K., Yao, S. J., Drash, A. L., and Schiller, J. G. (1979). Quantitation of glucose concentration using a glucose oxidase-catalase electrode by potentiometric measurements, *Bioelectrochem. Bioenerg., 6*: 210.
93. Liu, C. C., Weaver, J. P., and Chen, A. K. (1981). Potentiometric measurement of glucose concentration using a gel-immobilized glucose dehydrogenase electrode, *Bioelectrochem. Bioenerg., 8*: 379.
94. Xie, S. L., and Schmidt, H.-L. Mechanistic studies on a potentiometric glucose electrode, *Biosensors*, submitted 1990.
95. Shinbo, T., Sugiora, M., and Kamo, N. (1979). Potentiometric enzyme electrode for lactate, *Anal. Chem., 51*: 100.
96. Dransfeld, I., Hintsche, R., and Scheller, F. (1989). Enzymelektroden für lactat und maltose auf der basis einer fluoridionensensitiven electrode, *Fresenius'Z. Anal. Chem., 333*: 23.
97. Bradley, C. R., and Rechnitz, G. A. (1986). Comparison of oxalate oxidase enzyme electrodes for urinary oxalate determination, *Anal. Lett., 19*: 151.
98. Kobos, R. K., and Ramsey, T. A. (1980). Enzyme electrode system for oxalate determination utilizing oxalate decarboxylase immobilized on a carbon dioxyde sensor, *Anal. Chim. Acta, 121*: 111.
99. Ho, M. Y. K., and Rechnitz, G. A. (1987). Highly stable biosensor using artificial enzyme, *Anal. Chem., 59*: 536.
100. Flanagan, M. T., and Carroll, N. J. (1986). Thin-film antimony-oxide enzyme electrode for penicillin determination, *Biotechnol. Bioeng., 28*: 1093.
101. El Yammani, H., Tran-Minh, C., Abdul, M. A., and Chavanne, D. (1988). Automated system for pesticide detection, *Sensors Actuators, 15*: 193.
102. Fonong, T., and Rechnitz, G. A. (1984). Enzyme electrode for the determination of salicylate, *Anal. Chim. Acta, 158*: 357.
103. Kawashima, T., and Rechnitz, G. A. (1976). Potentiometric enzyme electrode for uric acid, *Anal. Chim. Acta, 83*: 9.
104. Turner, A. P. F., and Swain, A. (1988). Commercial perspectives for diagnostics using biosensor technologies, *Am. Biotechnol. Lab., 6*: 10.

5

Enzyme Thermistor Devices

Bengt Danielsson

University of Lund
Lund, Sweden

INTRODUCTION

The many attractive features of bioanalytical calorimetry were recognized early. Most biological reactions are exothermic, which makes calorimetry an almost universal detection principle insensitive to the optical or electrochemical properties of the sample. Since enzymic reactions are usually associated with rather high enthalpy changes, in the range of 20–100 kJ/mol, it is often possible to construct very simple metabolite assays based on measurements of only one enzymic step. On the other hand, the combined heat development of sequential or cycling enzymic reactions can also be utilized to increase the overall sensitivity.

In earlier studies a wider application of calorimetry in bioanalysis, especially in routine analysis, was prevented by high cost, sophisticated equipment, and the relatively slow response of available microcalorimeters (1). In the middle of the 1970s, however, several simple calorimetric devices based on immobilized enzymes were introduced that combined the general detection principle of calorimetry with the high specificity of enzymes (2). The advantages of these instruments included reusability of the biocatalyst, possibility for continuous-flow operation, insensitivity to the optical and electrochemical properties of the sample, and simple procedures. Over the years several of the concepts developed, including the enzyme thermistor designed in our laboratory, have been used in numerous applications. The general interest in the technique has increased as a consequence of the intense attention paid to biosensors in recent years. Initial applications focused on the determination of glucose and urea; subsequently the enzyme thermistor devices have been applied in the determination of a wide variety of molecules (3).

Several research groups attempted to develop simple, low-cost calorimeters for routine analysis (thermal enzyme probes, TEP) by attaching the enzyme directly onto the temperature transducer, usually a thermistor (4,5). In this straightforward arrangement, however, the majority of the heat evolved in the enzymic reaction was lost to the surroundings, which resulted in low sensitivity. The TEP concept was primarily intended for batch operation and was a "small-volume calorimeter" in which the enzyme was attached to thin aluminum foil placed on the surface of a Peltier element acting as temperature sensor (6).

Even if the sensitivity was improved in later TEP designs (7,8), a considerably more efficient detection of the reaction heat was achieved in systems employing a small column with the enzyme immobilized on a suitable support. In this case the heat is transported by the liquid passing through the column to or along a temperature sensor mounted at the top of the column or at its outlet. Several constructions of this type were presented around 1975, including the "enzyme thermistor" (3,4) and the "immobilized enzyme flow-enthalpimetric analyzer" (9), as well as a similar concept also intended for glucose determination (10). Furthermore, a combination of a commercial flow-enthalpimeter with an immobilized enzyme column has been described (11).

Recently, several miniaturized prototypes have been constructed, for instance a thermal biosensor probe designed as an integrated circuit biocalorimetric sensor for glucose with total dimensions of only 1 × 1 × 0.3 mm (12). Another recent report describes a small thermoelectric glucose sensor that employs a thin-film thermopile to measure the evolved heat. This device is reported to be less affected by external thermal effects than calorimetric sensors based on thermistors and can be operated without environmental temperature control (13). Work is in progress in the author's laboratory to construct a miniaturized biothermal flow injection system suitable for portable monitoring. An instrument with 0.1–0.2 mm (ID) flow channels and a flow rate of 25–50 μl/min with sample volumes of 1–10 μl is being evaluated at present. A 1 × 3 mm enzyme column permits determination of concentrations down to 0.1 mM. Recently a device equipped with thin-film temperature sensors of thermistor type (0.1 × 0.1 mm or smaller) for glucose measurements has been presented (14).

PRINCIPLE OF MEASUREMENT

The total heat evolved in a reaction is proportional to the molar enthalpy:

$$Q = -n_p \, \Delta H \tag{1}$$

where Q = total heat, n_p = moles product, and ΔH = molar enthalpy change. It is also dependent on the heat capacity C_s of the system including the solvent:

$$Q = C_s \, \Delta T \tag{2}$$

The temperature change ΔT recorded by the enzyme thermistor is thus dependent on the molar enthalpy change and on the heat capacity:

$$\Delta T = - \frac{\Delta H n_p}{C_s} \tag{3}$$

Since the heat capacity of most organic solvents is two to three times lower than that of water, a considerably higher sensitivity is expected when using organic solvent, provided the enthalpy change remains unaltered. This field of investigation is further treated in a later section.

Table 1 lists the molar enthalpy changes of some enzyme-catalyzed reactions (2). A thermometric measurement is based on the sum of all enthalpy changes in the reaction mixture. Thus it is common practice to coimmobilize oxidases with catalase, which more than doubles the sensitivity at the same time as the oxygen consumption is reduced and the deleterious effects of hydrogen peroxide are minimized. As indicated in Table 1, the high-protonation enthalpy of buffer ions like Tris can be utilized to increase the total enthalpy change of proton-producing reactions. Drastic increases in the sensitivity can be obtained in substrate or coenzyme recycling enzyme systems, in which the net enthalpy change of each turn of the cycle adds to the overall enthalpy change (15). A later section provides further details on chemical and enzymic amplification.

An inherent disadvantage of calorimetry is the lack of specificity: all enthalpy changes in the reaction mixture contribute to the result of the measurement. It is therefore important to avoid nonspecific enthalpy changes from dilution or solvation effects, for example. In the majority of cases this is not a serious problem. An efficient way of coping with nonspecific and matrix effects is to

Table 1 Molar Enthalpy Changes for Some Enzyme-Catalyzed Reactions

Enzyme	Substrate	$-\Delta H$ (kJ/mol)
Catalase	Hydrogen peroxide	100
Cholesterol oxidase	Cholesterol	53
Glucose oxidase	Glucose	80
Hexokinase	Glucose	28 (75)[a]
Lactate dehydrogenase	Sodium pyruvate	62
NADH dehydrogenase	NADH	225
β-Lactamase	Penicillin G	67 (115)[a]
Trypsin	Benzoyl-L-argininamide	29
Urease	Urea (phosphate buffer, pH 7.5)	61
Uricase	Urate	49

[a]The ΔH values in parentheses were obtained in Tris buffer (protonation enthalpy −47.5 kJ/mol).

apply a reference column containing an inactive filling in a differential determination (16).

The enzyme thermistor assay is normally performed by a flow injection technique. The sample pulses employed are too small to give a thermal steady state but result in a temperature peak that can be registered by a recorder. The height of the peak is proportional to the enthalpy change and therefore to the substrate concentration. The area under the peak and the ascending slope of the peak have also been found to be linearly related to the substrate concentration (17). A sample introduction of sufficient duration (several minutes) leads to a thermal steady state resulting in a new temperature level, which is also proportional to the substrate concentration.

THE TRANSDUCER

In this type of instrumentation thermistors are normally used as temperature transducers. Thermistors are resistors with a very high negative temperature coefficient of resistance. They are ceramic semiconductors made by sintering mixtures of metal oxides, such as manganese, nickel, cobalt, copper, iron, and uranium. They can be obtained in many different configurations, sizes (down to 0.1–0.3 mm beads), and resistance values from such manufacturers, as Fenwal Electronics (Framingham, MA), Victory Engineering Co. (Springfield, NJ), and Siemens AG (Munich, West Germany). The hitherto best empirical expression to describe the resistance-temperature relationship is the Steinhart-Hart equation:

$$\frac{1}{T} = A + B(\ln R) + C(\ln R)^3 \qquad (4)$$

where T = temperature (K); $\ln R$ = the natural logarithm of the resistance, and A, B, and C are derived coefficients. For narrow temperature ranges the relationship can be approximated by the equation:

$$R_T = R_{T0}e^{\beta(1/T - 1/T0)} \qquad (5)$$

where R_T and R_{T_0} are the zero-power resistances at the absolute temperatures T and T_0, respectively, and β is a material constant that ranges between 4000 and 5000 K for most thermistor materials. This results in a temperature coefficient of resistance between -3 and -5.7% per Celsius degree. In our enzyme thermistor devices resistances of 3–100 kΩ have been used (25°C).

Other temperature transducers employed in enzyme calorimetric analyzers include Peltier elements (6), Darlington transistors (12), and thermopiles (13). The thermistor is the most sensitive of the common temperature transducers, however.

EXPERIMENTAL TECHNIQUES

Equipment

In our early studies on calorimetric biosensors the immobilized enzyme column was housed in Plexiglas and thermostated using accurate water baths. The temperature at the top of the column was monitored with a small thermistor probe connected to a Wheatstone bridge constructed for temperature measurements (Knauer Wissenschaftlicher Geraetebau, West Berlin, FRG). A temperature change of 0.01°C resulted in a potential change of up to 100 mV. These simple devices were adequate for determinations of concentrations down to 0.01 mM and could be used to determine, for instance, serum urea concentrations (18).

Construction of a two-channel instrument increased the sensitivity and the stability. The water bath was replaced by a carefully thermostated metal block (Fig. 1). The calorimeter part of the instrument, which is placed in a container insulated with polyurethane foam, consists of an outer aluminum cylinder (80 × 250 mm) thermostated to 25, 30, or 37°C and a second inner aluminum cylinder with two column ports and a pocket for a reference thermistor. The solution to be analyzed is introduced via thin-walled acid-proof steel tubing (0.8 mm ID). Two-thirds of the tubing's length is in close contact with the thermostated jacket; the last third is in close contact with the inner cylinder, which acts as a heat sink. This arrangement results in exceedingly small short-term temperature variations in the column. The column is inserted by a long Delrin tube into the calorimeter. The tube also contains the temperature sensor (for example Type A395 isocurve thermistors from Victory Engineering Corp.; 16 kΩ at 25°C, temperature coefficient −3.9%/°C), which is mounted with heat-conducting epoxy on a short piece of gold capillary through which the column effluent passes.

The present system has a Wheatstone bridge of the dc-type, using precision resistors with low temperature coefficients and a chopper-stabilized low-drift operational amplifier. The maximum sensitivity of this bridge is 100 mV per 0.001°C. Temperatures in the range of 0.01–0.05°C are commonly measured. A reaction enthalpy change of 80 kJ/mol should give a temperature change of 0.01°C for 1 ml of a 1 mM sample at a flow rate of 1 ml/min.

Procedure

A typical experimental arrangement is shown in Figure 2. An essentially pulse-free pump, usually a high-quality peristaltic pump, is employed to produce a continuous flow of buffer (0.5–5 ml/min). Since water has a very high heat of vaporization, it is important that the apparatus is handled to prevent spillage inside the calorimeter and to avoid the entry of air bubbles into the system. A pulse damper capable of trapping incoming gas bubbles is recommended, and solutions should be partially degassed. Samples are usually introduced with an

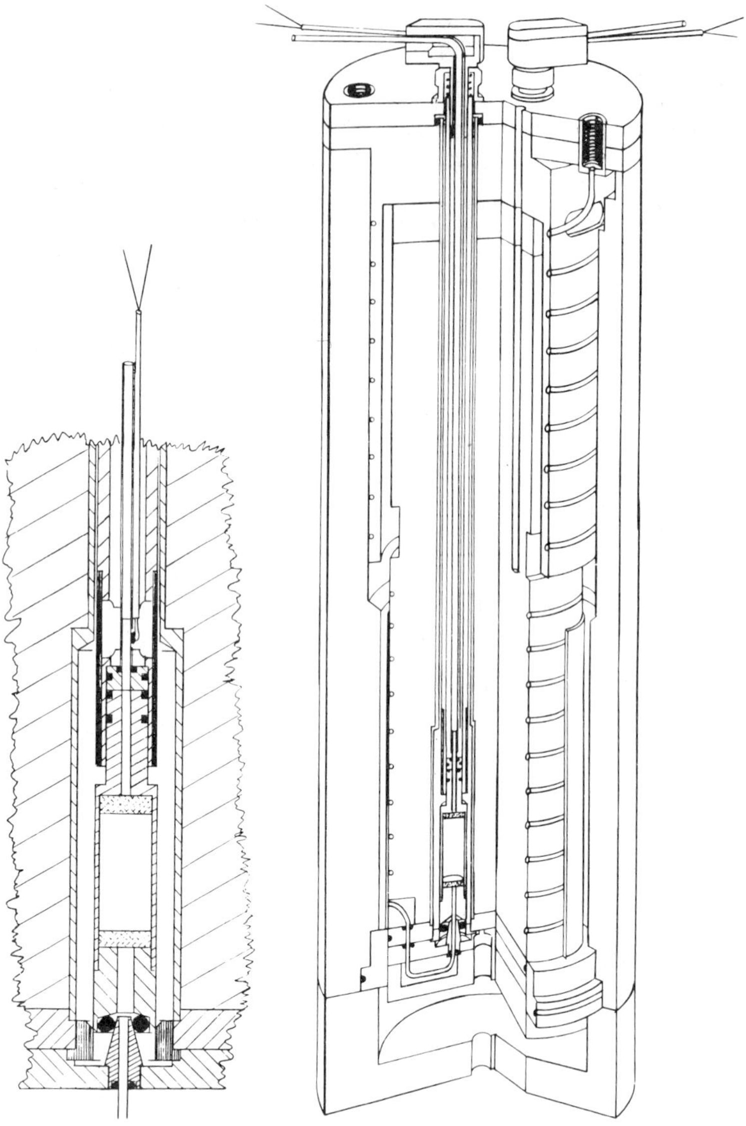

Figure 1 Cross section of the calorimeter of an enzyme thermistor.

injection valve (Type 50 from Rheodyne, Cotati, CA) using 0.1–0.5 ml sample loops. Small sample volumes (<25μl) can be injected undiluted; larger samples should be diluted at least 5- to 10-fold in buffer to avoid nonspecific thermal effects due to mixing.

The most common enzyme support used in our investigations has been

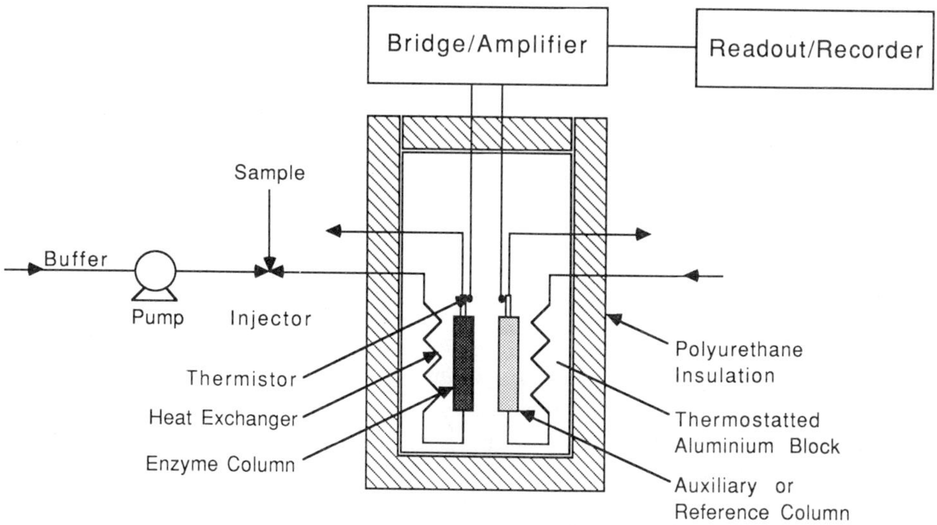

Figure 2 Experimental arrangement of an enzyme thermistor system.

propylamino derivatized controlled pore glass (CPG) with a pore size in the range of 500–2000 Å and a particle size of the order of 80 mesh (0.18 mm). Usually a large excess of enzyme is applied, often 100 units or more, and immobilized with glutardialdehyde to the support. CPG offers sufficient binding capacity and good mechanical and chemical as well as microbial stability and relatively simple coupling procedures (3). Eupergit C (oxirane acrylic beads from Röhm Pharma, Weiterstadt, FRG) (3,19) and VA-Epoxy BIOSYNTH, a copolymer of vinyl acetate and divinylethylene-urea modified with oxirane groups (Riedel-de Haën, Seelze, FRG) are, however, suitable alternative support materials. Sepharose-CL-6B (Pharmacia-LKB, Uppsala, Sweden) has a very good binding capacity with simple coupling procedures but can only be used at flow rates < 0.5 ml/min since it is easily compressible. It has mostly been used in the antibody columns in the TELISA (thermometric ELISA) technique.

Columns up to 1 ml (at 7 mm ID) can be used in the present apparatus. The major factor limiting column life appears to be mechanical obstruction. If the solutions and samples are filtered through a 1–5 μm filter and if microbial growth is prevented, good operational stability and reproducibility for large series of samples (thousands) can be achieved. The column lifetime can be several months.

Normally the thermograms are evaluated by peak height determination, which is simple to perform either manually from the recorder diagram or automatically by an integrator or computer. The area under the peak and the ascending slope of the peak have also been found to be linearly related to the substrate concentration (17).

TYPE OF BIOLOGICAL ELEMENT

Because of the general applicability of calorimetric sensors, very satisfactory results have been obtained with enzyme thermistor devices in many bioanalytical fields, including clinical chemistry, process control, fermention monitoring and other analyses within biotechnology, and environmental control. Table 2 gives an overview of the different application fields and operating principles studied in our laboratory.

Immobilized Enzymes

For increased clarity this subsection is divided into four parts of which substrate determination is the largest. In addition, inhibitor determination is mentioned as well as chemical and enzymic amplification of the sensitivity. The last part deals with determinations in organic solvents, which is a recent field of investigation.

Substrate Determinations

Enzyme thermistor assays for a large number of different metabolites have been proposed. The lower limit of detection in most of these assays is 1–10 μM. In some cases recycling techniques give about 1000 times greater sensitivity. If enzyme columns with high enzyme loadings are used, the linear range can extend to 500 mM or more. In the case of oxidases the concentration of the electron donor, oxygen, usually sets the upper limit of the linear range to 0.5 mM, or about 0.7–0.8 mM if the oxidase is coimmobilized with catalase. With other electron donors the linear range may be extended considerably. Thus, using benzoquinone the linear range can be expanded to 75 mM (11). In the following, assays for a number of metabolites are described in more detail.

Lower alcohols have been measured with alcohol oxidase (EC 1.1.3.13) from *Candida boidinii* (20). In this case coimmobilization with catalase was found to profoundly increase the stability of the enzyme column, rendering it stable for several months with an operating range of 0.01–1 mM using 0.1 M sodium

Table 2 Application Fields of Enzyme Thermistor Devices

Mode	Column	Sample type
Metabolite determination	Immobilized enzymes or cells	Physiological fluids; fermentation broth
Inhibition; toxic monitoring	Immobilized enzymes or cells	Environmental control and water samples
Enzymic activity	Inactive; mix with substrate solution	Enzyme solutions, e.g. at enzyme purification
Cell metabolism	Empty; mix with substrate solution	Microbial suspensions
TELISA	Immunosorbent	Proteins, hormones

phosphate, pH 7.0, as buffer. This assay is useful for the determination of ethanol in samples from beverages and blood and for monitoring fermentation. Recently an alternative alcohol oxidase preparation has become available from *Pichia pastoris* (Sigma Chemical Co., A 2404), which has a higher specific activity and a somewhat altered substrate specificity (Fig. 3).

Ascorbic acid was measured with ascorbate oxidase (EC 1.10.3.3) in 0.1 M sodium acetate buffer, pH 5.5, in the 0.01–0.6 mM range (21). In this experiment the enzyme was bound to concanavalin A-conjugated Sepharose by a reversible immobilization method that permits rapid change or renewal of the enzyme.

ATP/ADP can be measured with relatively high sensitivity using immobilized apyrase, which is a mixture of ATPase and ADPase. An assay scheme with nanomolar sensitivity is discussed in a later section.

The heat produced by the hydrolysis of cellobiose with β-glucosidase is too low to give reasonable sensitivity in a direct determination. Instead a procedure was designed in which the glucose formed in a precolumn containing β-glucos-

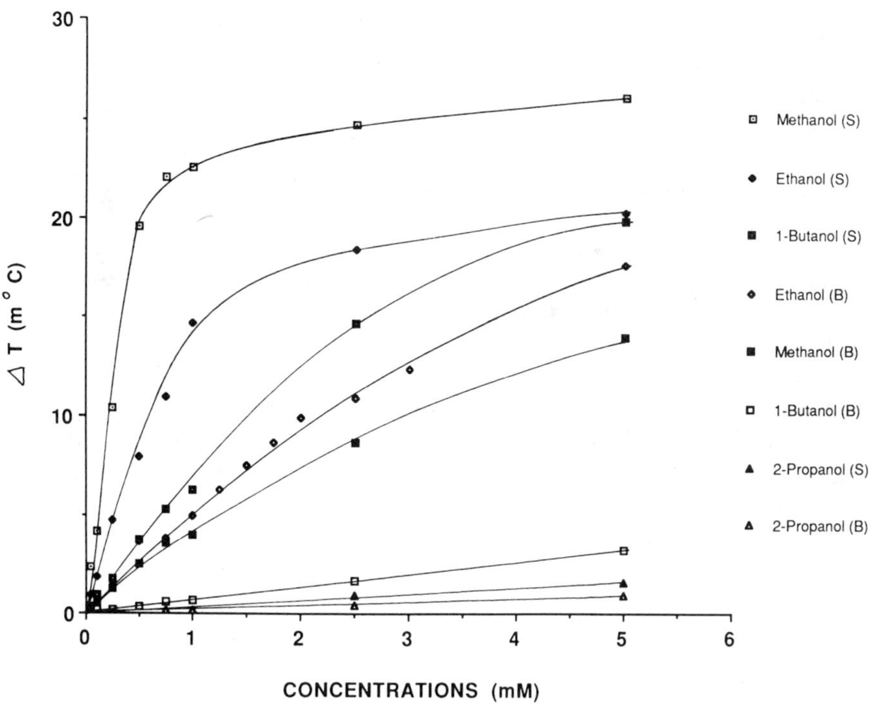

Figure 3 Calibration curves for different alcohols with alcohol oxidase from *Candida boidinii* (=B) and from *Pichia pastoris* (=S), respectively.

idase, which could be switched in and out of the flow through the enzyme thermistor, was measured by glucose oxidase and catalase (22). The operating range in this study was between 0.05 and 5 mM. Good results were obtained with samples from fermentation broths taken by on-line dialysis.

A group-specific determination of cephalosporins can be made with special types of β-lactamases (EC 3.5.2.6) with enhanced cephalosporinase activity. An enzyme from *Enterobacter cloacae* P99 (Miles Laboratories, Elkhart, IN) resulted in values that correlated well with high-performance liquid chromatographic (HPLC) analysis (23).

Cholesterol was dissolved in 0.16 M phosphate buffer, pH 6.5, containing 12% (vol/vol) ethanol and 8% (vol/vol) Triton X-100 and measured either with cholesterol oxidase (EC 1.1.3.6) immobilized on CPG in a differential ET unit or with a catalase-loaded ET using the cholesterol oxidase column as a precolumn. By treating the samples with cholesterol esterase (EC 3.1.1.13) cholesterol esters could also be measured (24). Unfortunately, the cholesterol oxidase from *Nocardia erythropolis* was found to give rather unstable enzyme columns. More recently a number of cholesterol oxidases from several different sources have become commercially available. These may provide better operational stability.

Several suppliers, including Sigma Chemical Co., offer a microbial creatinine deiminase (EC 3.5.4.21) that converts creatinine to *N*-methylhydantoin and ammonia. This enzyme works well in the enzyme thermistor, resulting in a linear range from 0.01 mM to at least 10 mM. Unfortunately the determination of creatinine in human serum requires about 10 times greater sensitivity. A split-flow arrangement may improve the situation somewhat, but it is likely that this assay must await a more sensitive instrument before it can be realized in a practical way.

Galactose was measured with D-galactose oxidase (EC 1.1.3.9) from *Dactylium dendroides* in 0.1 M Tris-HCl, pH 7.4. Storage of the enzyme column in buffer supplemented with 1 mM copper sulfate has been shown to prolong column life.

Glucose has been the subject of many studies (3,5,9–13). In most cases glucose oxidase (EC 1.1.3.4) has been used, except in one case (9) in which hexokinase (EC 2.7.1.1) was used. This enzyme has the disadvantage of requiring ATP as a cofactor. On the other hand, linear ranges of up to 25 mM can be expected.

Glucose oxidase is usually coimmobilized with catalase (3). The catalase reaction more than doubles the total enthalpy change, eliminates the deleterious effects of the hydrogen peroxide formed in the glucose oxidase reaction, and restores half the oxygen consumed by this reaction. The glucose oxidase/catalase system provides several other advantages, including good sensitivity and high specificity, and has no external cofactor requirement. The useful concentration range is limited, however, to <0.7–1.0 mM by the supply of oxygen, which is soluble only to a small extent in water, about 0.25 mM at 25°C. The limit of detection can be as low as 1 μM. By using a more soluble electron acceptor than oxygen, such as benzoquinone (11), the useful working range can be increased considerably to at least 75 mM. We found a precision (relative standard deviation) of 0.6% for measurements made within a day using glucose oxidase/catalase. Columns can be used for several months and for several hundred serum glucose determinations (2). Glucose concentrations measured by the enzyme thermistor correlated well with conventional enzymic techniques and are very useful in routine hospital diagnoses and bioprocess analysis.

Hydrogen peroxide can be sensitively measured from 1 μM to 10 mM using catalase or peroxidase. The determination of concentrations above 10 mM is disturbed by bubble formation in the column due to oxygen production by the reaction.

L-lactate has been successfully determined in two different enzyme systems. With lactate-2-monooxygenase (EC 1.13.12.4; from *Mycobacterium smegmatis*) applied to a CPG column lactate could be determined with a linear response in the range 0.005–2 mM in 0.2 M sodium phosphate buffer, pH 7.0. Alternatively the assay can be run with lactate oxidase from *Pediococcus pseudomonas* (EC 1.1.3.2) together with catalase in 0.1 M sodium phosphate, pH 7.0. In this case a linear response in the range of 0.002–1 mM was obtained (3). Both enzyme preparations are stable for several months. A highly sensitive assay for lactate using an enzymic recycling system is described in a later section.

Since the heat of hydrolysis of lactose is low, an assay with reasonable sensitivity must be based on the oxidation of the glucose formed, much in the same way the cellobiose assay was designed. Thus a coimmobilized column of β-galactosidase (EC 3.2.1.23), glucose oxidase, and catalase gives a useful range of 0.1–10 mM (24).

L-malate has been determined with the "malic enzyme" (EC 1.1.1.40), which converts malate and $NADP^+$ to pyruvate, CO_2, and NADPH. The pyruvate formed is subsequently determined by the lactate dehydrogenase/lactate oxidase recycling system described in a later section.

There are two different enzymes available for the determination of oxalate: oxalate decarboxylase (EC 4.1.1.2) has a working range of 0.1–3 mM, although the enzyme column has a rather poor stability. Increased enzyme stability is obtained with an oxalate oxidase (EC 1.2.3.4) from barley seedlings (25). This system is stable for 3–4 weeks and has a useful range of 0.005–0.5 mM in 0.1 M sodium citrate buffer, pH 3.5, containing 2 mM EDTA and 0.8 mM 8-hydroxyquinoline. The assay is suitable for the determination of oxalate in urine, beverages, and food samples. Samples should be diluted 10-fold and passed through a C_{18} cartridge to remove interfering substances.

Penicillins G and V and other β-lactams have been very successfully assayed using β-lactamases, such as penicillinase type I from *Bacillus cereus* (EC 3.5.2.6) (23). The useful linear range is about 0.01–100 mM. The enzyme columns are very stable and can be used for several months or for thousands of samples. The values obtained correlate very well with those from other techniques, such as HPLC. Consequently many enzyme thermistors are used in industrial laboratories for penicillin analysis.

Phospholipids have been determined with the use of three consecutively acting enzymes (26). Phospholipase D (EC 3.1.4.4), which was found to be rather unstable, was added directly to a 0.05 ml sample, which was then injected into the buffer stream. The enzyme column contained choline oxidase (EC 1.1.3.17) and catalase coimmobilized on CPG and gave a linear range of 0.03–0.19 mM. This was sufficient for the direct determination of phospholipids in 10-fold diluted serum samples and correlated well with conventional methods. The system performed well during an 8 week period and for at least 1600 analyses per column.

In contrast to most other disaccharide splitting enzymes, invertase (EC 3.2.1.26) produces enough heat to allow direct determinations. Thus sucrose can be directly determined in the range of 0.05–100 mM in samples, even in the presence of glucose (27). Moreover, invertase columns are very stable.

Practically useful routine methods for the determination of the main blood lipid classes, cholesterol and cholesterol esters, phospholipids, and triglycerides, have been developed (26). Triglycerides can be determined with lipoprotein lipase (EC 3.1.1.34) immobilized on CPG with a pore size of 2000 Å. A suitable assay buffer is 0.1 M Tris buffer, pH 8.0, containing 0.5% Triton X-100. A split-flow system gave a linear response between 0.05 and 10 mM tributyrin and 0.1 and 5 mM triolein. Twofold diluted serum samples could be directly analyzed for triglyceride concentrations up to 3 mM. The values correlated well with conventional spectrophotometric enzymic methods.

Another common metabolite assay that has attracted a great deal of interest is biosensor analysis of urea (8,18). With urease (EC 3.5.1.5), which gives a linear range of at least 0.01–200 mM, a clinically useful assay has been designed with a precision of greater than 1%. This method has the advantage of being independent of the ammonium concentration in the sample. Addition of 1 mM EDTA and 1 mM reduced glutathione to the buffer protects the urease against heavy metals and renders a very stable enzyme column.

Uric acid can be determined using the enzyme uricase (EC 1.7.3.3) in the range of 0.05–4 mM (24). That an oxidase could give such a large linear range suggests that the bound enzymic activity is rather low. Unfortunately, the high cost of the commercial enzyme has prevented more extensive studies of this assay.

Xanthine and hypoxanthine have been measured in the range of 0.1–0.5 mM using xanthine oxidase (EC 1.1.3.22):

$$\text{Hypoxanthine} \xrightarrow{\text{XOD}} \text{xanthine} + H_2O_2 + H_2O + O_2 \quad (6)$$

$$\text{Xanthine} \xrightarrow{\text{XOD}} \text{uric acid} + H_2O_2 + H_2O + O_2 \quad (7)$$

This technique was found to be suitable for continuous monitoring of xanthine in flow streams, although the stability of the enzyme column was comparatively low. It could, however, be used during 1 week for at least 150 samples (28).

Inhibitor or Effector Determination

This approach has mainly been studied in environmental control analysis. These systems have exploited the calorimetric detection principle in two different ways. The first measures the heat of conversion of a pollutant by a cell metabolic route or a specific enzyme directly. Thus the enzyme rhodanese (EC 2.8.1.1) has been used for determining cyanide concentrations in sewage (3). Furthermore, calorimetric analysis could be developed for a number of enzymes using pesticides and other hazardous chemicals as substrates.

The second alternative, which is much more sensitive, measures the inhibitory effect of a pollutant on an enzyme or on the metabolism of suitable cells (3). To detect an environmental pollutant by quantitating its effect on a biological system is ideal. For example, a technique was developed for highly sensitive determination of heavy metals (Hg^{2+}, Cu^{2+}, and Ag^{+}) based on the inhibition of urease (3). A small amount of urease was applied to a CPG column. The response obtained for a 0.5 ml standard pulse of urea in phosphate buffer at a flow rate of 1 ml/min was noted, after which 0.5 ml of the sample was injected. A new 0.5 ml standard pulse of urea was injected 30 sec after the sample pulse (timing is important) and the response compared with that of the noninhibited peak. After a sample had been run the initial response could be restored by washing the column with 0.1–0.3 M NaI plus 50 mM EDTA for 3 min. A calibration curve was made with standard samples, and the heavy metal concentration in an unknown sample was determined. For the conditions given here a 50% inhibition (50% of the initial response) was obtained for a 0.5 ml pulse of 0.04–0.05 mM Hg^{2+} or Ag^{+} or 0.3 mM Cu^{2+}. Longer sample pulses resulted in a considerably higher sensitivity (sub-ppb levels).

In some cases the inhibition of the enzyme is irreversible. In this situation a reversible enzyme immobilization technique that allows simple replacement of spent enzyme is required. In one case this has been accomplished by reversibly binding cholinesterase (EC 3.1.1.7) to a concanavalin A-Sepharose column (3). A relatively sensitive determination of cholinesterase inhibitors has been made using butyrylcholine as substrate (17). The old enzyme is removed by injecting a pulse of 0.2 M glycine-HCl, pH 2.2. Fresh enzyme can then be immobilized on the column simply by injecting the enzyme preparation while the column remains in the enzyme thermistor apparatus.

Many enzymes require metal ions in their active sites to be catalytically active. In many cases the metal ion can be removed using strong chelating agents. The resulting apoenzyme is inactive. The apoenzyme can be reactivated upon exposure to a sample containing the appropriate metal ion. This measuring principle has been used to develop calorimetric detection of a variety of metal ions (29). Zinc(II) was determined in the range 25–250 nmol/ml via its activation of immobilized, metal-free carbonic anhydrase. The enzyme reactor was regenerated by pumping a 2.5 ml portion of 2,6-pyridine dicarboxylic acid, pH 5.0, through it. With the same system Co(II) was determined in the range 25–100 nmol/ml. Nanomolar amounts of Cu(II) were assayed via its activation of ascorbate oxidase. *N,N*-diethyldithiocarbamate solution, pH 8, was used for very efficient regeneration of the apoenzyme.

Chemical and Enzymatic Amplification

When protons are produced (or taken up) by an enzymic reaction, as is the case with proteolytic enzymes, a considerably greater total heat production can be obtained by using a buffer with a high protonation enthalpy, such as a Tris buffer (−47.5 kJ/mol compared to −4.7 kJ/mol for phosphate buffer) (1). Several examples are shown in Table 1.

In addition, the total temperature signal can be increased using sequentially acting enzymes. As already mentioned, this is commonly done with hydrolytic enzymes, such as disaccharide splitting enzymes, which can be combined with an enzyme acting on the monosaccharide formed. A typical combination is β-galactosidase and glucose oxidase/catalase for the determination of lactose. Oxidases are almost always combined with catalase or occasionally with peroxidase for two reasons. First, the destruction of hydrogen peroxide more than doubles the total enthalpy change since the molar enthalpy change for the catalase reaction is −100 kJ/mol. Second, the damaging effect of hydrogen peroxide on the enzymes is greatly decreased.

Amplification of the sensitivity by substrate or coenzyme recycling is especially efficient in thermometric analysis, since in many cases all the reactions employed contribute to increase the total temperature change. A case in point is the determination of lactate or pyruvate by substrate recycling using coimmobilized lactate oxidase and lactate dehydrogenase (15). L-lactate is oxidized by lactate oxidase to pyruvate, which is reduced to lactate again by LDH. The total enthalpy change of this system is further increased by the addition of catalase, which makes the overall enthalpy change as high as −225 kJ/mol. Signal enhancements of greater than 1000-fold have been obtained. Using this arrangement lactate (or pyruvate) concentrations as low as 10 nM can readily be determined.

A coenzyme recycling system for the determination of ATP/ADP was recently reported (30). Concentrations as low as 10 nM could be determined using the enzymes pyruvate kinase and hexokinase with phosphoenolpyruvate and glucose as substrates, respectively. This cycle produces pyruvate, which is then used in the pyruvate recycling system presented previously. This combination results in an enormous amplification and a corresponding increase in sensitivity.

An example of NAD(H) recycling is described in Reference 3. NAD(H) was recycled by the two enzymes lactate dehydrogenase and glucose-6-phosphate dehydrogenase with pyruvate and glucose-6-phosphate as substrates. This resulted in an 80-fold increase in sensitivity.

Operation in Organic Solvents

Thermometric measurements in organic solvents should be up to three times more sensitive than in water because organic solvents have a lower heat capacity than water [see Eq. (3)]. That enzymes are usually active in organic solvents under a variety of conditions has attracted increasing interest in biotechnology, since this may open new enzymic routes and offers increased solubility of the reactants involved. It would also be advantageous to be able to directly measure reactants in processes that are carried out in organic solvents. Furthermore, it would be of great interest to develop assays that use organic solvents, especially for substrates with poor solubility in aqueous solutions. For example, triglycerides have been determined directly in cyclohexane. The enzyme lipoprotein lipase functions well in this solvent, and the response is about 2.5 times higher than in water, which corresponds to the difference in heat capacity (31). It is, however, often difficult to foresee the effect of an organic solvent on the thermal response. Thus, even small amounts (a few percentage) of lower alcohols were found to considerably increase the temperature response of the glucose oxidase/catalase and the β-lactamase reactions, presumably by increasing the enthalpy change. Higher alcohol concentrations, however, reduced the temperature response. Interesting effects were also observed with peroxidase, which produced an as much as 45 times higher temperature signal in toluene than in buffer. Additional increases in sensitivity were obtained in the presence of low concentrations of toluene (Fig. 4).

Since the calorimetric measuring principle is equally useful with endothermic as with exothermic reactions, it was found that the enzyme thermistor could follow synthetic as well as hydrolytic reactions catalyzed by α-chymotrypsin (32). With an α-chymotrypsin column mounted in the enzyme thermistor, peptide concentrations to below 0.1 mM could be measured in an assay buffer with pH slightly above neutrality, and the concentration of the peptide-forming components (esters and amides) could be followed at pH 10 in 50% dimethylformamide and 50% 0.1 M sodium borate.

Immobilized Cells

Immobilized whole cells offer several advantages over pure enzyme systems (3). First, the purified enzyme may be unstable. Second, the enzyme may be difficult to purify. The cost savings will be especially pronounced in reactions involving sequential multienzyme systems. Third, cells provide built-in systems for coenzyme regeneration, eliminating the need to supplement the buffer stream with expensive coenzymes. Finally, immobilized cells provide a better model for

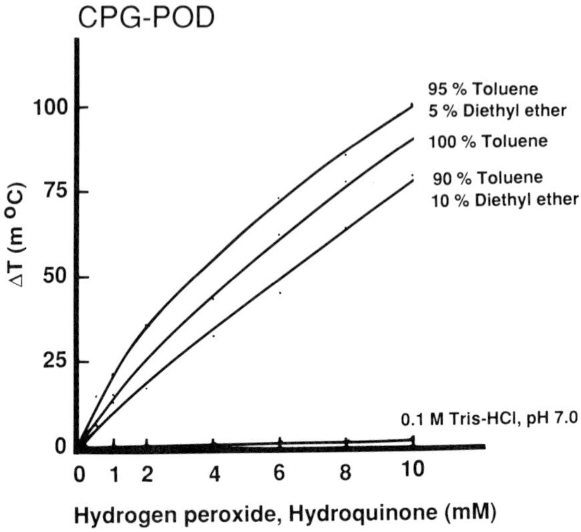

Figure 4 Calibration curves for a peroxidase column with 0.5 ml samples of hydrogen peroxide and hydroquinone (1:1) in toluene and toluene-ether mixtures and in 0.1 Tris-HCl buffer, pH 7.0, at a flow rate of 1.0 ml/min.

studying the effects of agents, such as vitamins or poisons, on living systems. The general applicability of calorimetry is vast, especially when compared with systems based on oxygen consumption or ammonia production.

Substrate Determination

A wide variety of cell immobilization techniques are available (33). It is important to maintain a sterile environment and to make the beads sufficiently small, preferably <1–2 mm, to guarantee a short response time. For example, *Gluconobacter oxydans* was immobilized in calcium alginate for calorimetric studies (3). Determination of glycerol was carried out in 0.1 M sodium succinate, pH 5.0, containing 10 mM $CaCl_2$. The response curve was linear up to 2 mM for 1 ml sample pulses with a sensitivity of 0.003°C/mM at a flow rate of 0.9 ml/min. Although the response was slower than with an enzyme column, 15–20 samples per hr could be analyzed. Continuous sample introduction resulted in a sensitivity of 0.008°C/mM and linearity up to 1 mM.

Effector Determination

The use of immobilized cell columns in calorimetric studies of metabolic effectors has great potential as a result of its general applicability. The metabolic heat is very likely to be affected by the presence of a metabolic poison, such as a nerve gas, or substances with an activating effect, such as vitamins. Model studies in this area were carried out early in enzyme thermistor development (17,34). Unfortunately not much published material is available on this topic.

The determination of the vitamin nicotinic acid using *Lactobacillus plantarum* entrapped in gelatin beads has been briefly described (3). Although the sensitivity was high, the assay system had a relatively slow response time and the column had a short lifetime (it could often be used for only one experiment). If large amounts of the cells are available, however, the assay could be designed to run in a continuous-flow fashion with a continuous response.

TELISA: Reversible Immobilization

Rapid, sensitive determination of larger molecules, such as hormones and antibodies present, for instance, in fermentation broth, can be accomplished by an automated, flow-through thermometric enzyme-linked immunosorbent assay (TELISA). The use of this assay was demonstrated in the determination and monitoring of the production and release of human proinsulin by genetically engineered *Escherichia coli* cells (35). Antiinsulin antibodies were affinity purified against beef insulin. The fraction that was eluted with 0.2 M glycine-HCl, pH 2.2, was coupled to Sepharose-4B (0.2 mg/g wet gel) activated with tresyl chloride (6 μl/g wet gel). A beef insulin-peroxidase conjugate was prepared, and the fraction with a weight ratio of insulin to peroxidase of 1:2 was isolated by chromatography on Sephacryl S-200.

In this form of TELISA, which is a competitive assay, the unlabeled antigen in the sample (or standard) is mixed with a fixed amount of enzyme-labeled antigen and the mixture is then applied to the immunosorbent column mounted in the enzyme thermistor apparatus. When the amount of unlabeled antigen in the sample increases, the amount of enzyme-labeled antigen bound to the column decreases. The amount of peroxidase bound to the column is determined by injecting a substrate pulse containing 2 mM H_2O_2, 14 mM phenol, and 0.8 mM 4-aminoantipyrine. Finally, bound antigen is removed from the immunosorbent by washing with 0.2 M glycine-HCl, pH 2.2, thereby regenerating the column for the next assay. The automated assay can be controlled by a programmable controller (Hizac D28, Hitachi, Japan) that operates the valves for different eluants, substrates, and the sample changer. It takes 7 min after sample introduction to obtain a reading, and a single assay cycle is completed after 13 min. Insulin concentrations in the range of 0.1–50 μg/ml were determined. These values correlated well with conventional radioimmunoassay. Standard curves were reproducible over a period of several days, even when the immunosorbent column was stored inside the ET (35). This method is very suitable for monitoring fermentations with respect to time, sensitivity, and stability. It is also useful when rapid determinations on a limited number of samples are required, for instance to establish a proper dosage of drugs in critical medical therapy.

To increase the sensitivity, use of the LOD/LDH recycling system (described in an earlier section) in combination with TELISA has recently been investigated (36). Alkaline phosphatase is employed as the enzyme label. The immunosorbent is packed in a separate column and placed in the second enzyme thermistor channel. The substrate, phosphoenolpyruvate, is converted into pyruvate by alkaline phosphatase. The pyruvate is then injected into the recycling column, which increases the sensitivity by 10- to 100-fold. Since separate columns are used for the detection and for the immunosorbent, the sample introduced into the recycling column is very clean, with no risk of inhibition or contamination of the recycling column. Another advantage is that it is simple to operate several immunosorbent columns in combination with one detection column, which has an at least five times shorter cycle time. Alkaline phosphatase is a common enzyme label, and it has recently become possible to construct conjugates by gene fusion, such as the in-frame fusion between human proinsulin and *E. coli* alkaline phosphatase (37) that was used in the study described earlier.

Determinations of Enzyme Activities in Solution

The ET unit can be used for the determination of the activity of soluble enzymes by modifying the flow system slightly and using an empty or inactive column, preferably made of Teflon, as a reaction chamber. The sample solution and buffer containing an appropriate substrate in excess are passed through heat exchangers and thoroughly mixed. The mixture is then passed through a short heat exchanger to eliminate mixing and solvation heats (17). In the present enzyme thermistor construction each flow channel has one long heat exchanger tube in contact with the outer aluminum jacket and one short tube in contact with the inner heat sink (see Fig. 1). These facilitate adaptation to enzyme activity determination. Even when the residence time is <1 min the sensitivity is high enough to determine enzyme activities down to 0.01 U/ml. The technique is quite general. It can be used with both sample pulses and continuous sample introduction. This system is of particular interest for monitoring enzyme purification processes. It has been demonstrated as a specific chromatographic detector with about the same sensitivity as A_{280} monitoring (38). It can also be employed as a direct, continuous-flow detection method on crude samples using cheap substrates for controlling downstream processing, for example in affinity precipitation (Fig. 5) and in coupled chromatographic procedures (39).

An alternative technique for the determination of enzyme activity, which is of particular importance at low enzyme concentrations, involves enrichment of the enzyme by affinity binding (preferably reversible) to an affinity column in the enzyme thermistor unit. The activity is then determined by introducing a pulse of substrate in excess (17). This technique could also be used for the reversible immobilization of labile enzymes when the enzyme column must be renewed frequently.

Measurements on Cell Suspensions

By mixing a flow stream of a cell suspension with a substrate stream in a manner similar to that described for enzyme activity determination, or simply by letting a

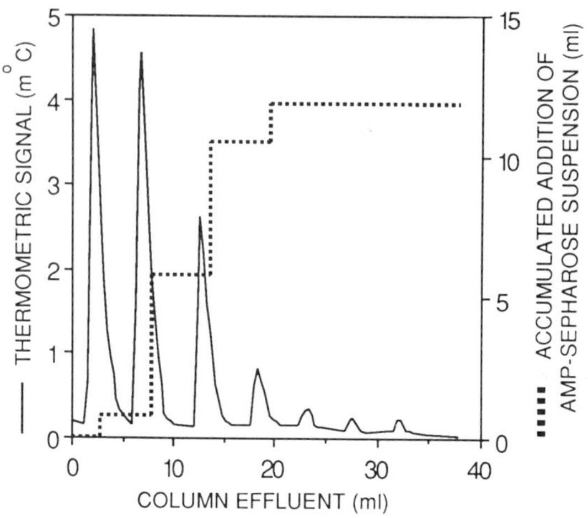

Figure 5 Computer control of the isolation of lactate dehydrogenase by affinity adsorption to AMP-Sepharose. The accumulated amount of adsorbent added and the thermometric signal representing the unbound LDH activity as measured by repeated injections of 30 sec pulses of LDH solution. The control system was set to give a few percentage rest activity of LDH.

flow with cells, for instance from a fermentor, pass through an empty column in the ET unit, the enzyme thermistor can be used for studies of the total metabolic heat generated by the cells (17). However, aerobic cells must be run at a sufficiently low concentration not to deplete the available oxygen while the cells are passed through the flow system. This technique should be useful in studies of metabolic effectors and in toxicological studies and as a general method of following the overall metabolic state of a fermentation. The heat or the temperature level recorded represents the total metabolism of the microorganisms, and changes in the metabolism are directly detectable as changes in the temperature level. This has been exemplified by studies on the ampicillin susceptibility of *E. coli* (40). A bacterial suspension from a 3 L fermentor was pumped at a low flow rate of 17 ml/hr through a 0.9 ml Teflon column in the enzyme thermistor apparatus. Changes in metabolic behavior during cultivation under standardized conditions were clearly observable, even in the presence of ampicillin at a concentration 10 times lower than the minimal inhibitory concentration (MIC).

UPDATED LITERATURE AVAILABLE

Table 3 summarizes important analyses providing the concentration ranges studied and updated literature references. In addition, Reference 3 gives rather detailed information on assays for some common metabolites, including ethanol, glucose, L-lactate, oxalate, sucrose, and urea. Reference 17 is an overview of a

Table 3 Updated Literature References and Concentration Ranges for Enzyme Thermistor Analyses

Analysis	Enzyme(s) used	Concentration range (mM)	References
Ascorbid acid	Ascorbate oxidase	0.01–0.6	21
ATP (or ADP)	Pyruvate kinase + hexokinase	10 nM	30
Cellobiose	β-Glucosidase + glucose oxidase/ catalase	0.05–5	22
Cephalosporins	Cephalosporinase (β-lactamase)	0.005–10	23
Creatinine	Creatinine iminohydrolase	0.01–10	17
Glucose	Hexokinase	0.5–25	9
Glucose	Glucose oxidase/catalase	0.001–0.8	3, 11
L-lactate	Lactate-2-mono-oxygenase	0.005–2	3
L-lactate	Lactate oxidase/catalase	0.002–1	3
L-lactate	Lactate oxidase/catalase + lactate dehydrogenase	10 nM	15
Oxalate	Oxalate oxidase	0.005–0.5	25
Penicillin	β-Lactamase	0.01–100	23
Sucrose	Invertase	0.05–100	27
Urea	Urease	0.01–200	3, 8

variety of subjects, such as evaluation of thermograms, experiments with cell suspensions, and reversible immobilization. Assays carried out in organic solvents are discussed in Reference 31, and Reference 39 summarizes measurements of soluble enzyme activities in downstream processes.

COMMERCIALLY AVAILABLE DEVICES

Despite many attractive features only a few biosensor-based instruments have so far had more than a very limited success on the market. Of the thermal biosensors only the enzyme thermistor is commercially available to our knowledge. ThermoMetric Co. (Järfälla, Sweden) is marketing a thermal assay probe, which is based on the enzyme thermistor concept developed in Lund.

Several temperature-measuring instruments using thermistors with adequate sensitivity for calorimetric experiments are commercially available, however, for instance the Knauer Temperature Measuring Instrument from Knauer Wissenschaftlicher Gerätebau (W. Berlin, FRG) and the instrument from Japan Electron Optics Laboratory (Tokyo, Japan) used in Reference 11.

TRENDS AND PROSPECTS

As a result of the universal detection principle thermal flow analyzers have a multitude of application possibilities, as indicated by Table 2. In clinical chemistry there is an increasing interest in biosensors for their use in decentralized health care. Biosensors should also be very suitable for home monitoring of certain metabolites, such as glucose in diabetes control. In these contexts thermal biosensors should be of special interest because they are easy to operate and simple to maintain and have a general applicability. The thermistor is virtually driftless, which increases the operational stability of the system. A very promising development of miniaturized thermal biosensor devices is in progress, which further increases the potential in biomedical applications.

In the rapidly growing and important field of biotechnology, there will be an increasing demand for flow stream analyzers in process control, fermentation monitoring, and downstream analyses. We are currently developing special versions of the enzyme thermistor for bioprocess monitoring with special attention to robustness and sterile sampling. Thermal flow analyzers appear to be particularly attractive for bioprocess and downstream analyses because of their ability to analyze turbid, particulate, or colored samples using highly specific continuous-flow techniques, not only for enzyme substrates but also for enzymes, cells (metabolic activity), and proteins (TELISA). In environmental control the possibility of developing very sensitive assays that can be directly related to the biological effect of the pollutants should be of special importance. This could also be an interesting field for portable monitoring using miniaturized devices.

REFERENCES

1. Spink, C., and Wadsö, I. (1976). Calorimetry as an analytical tool in biochemistry and biology, *Methods Biochem. Anal., 23*: 1.
2. Danielsson, B., and Mosbach, K. (1986). Theory and applications of calorimetric sensor. In: *Biosensors: Fundamentals and Applications* (A. P. F. Turner, I. Karube, and G. S. Wilson, Eds), Oxford University Press, Oxford, p. 575.
3. Danielsson, B., and Mosbach, K. (1988). Enzyme thermistors, *Methods Enzymol., 137*: 181.
4. Mosbach, K., and Danielsson, B. (1974). An enzyme thermistor, *Biochim. Biophys. Acta, 364*: 140.
5. Weaver, J. C., Cooney, C. L., Fulton, S. P., Schuler, D., and Tannenbaum, S. R., (1976). Experiments and calculation concerning a thermal enzyme probe, *Biochim. Biophys. Acta, 452*: 285.

6. Pennington, S. N. (1976). A small volume microcalorimeter for analytical applications, *Anal. Biochem., 72*: 230.
7. Tran-Minh, C., and Vallin, D. (1978). Enzyme-bound thermistor as an enthalpimetric sensor, *Anal. Chem., 50*: 1874.
8. Rich, S., Ianielli, R. M., and Jespersen, R. D. (1979). Development and application of a thermistor enzyme probe in the urea-urease system, *Anal. Chem., 51*: 204.
9. Bowers, L. D., and Carr, P. W. (1976). Immobilized-enzyme flow-enthalpimetric analyzer: Application to glucose determination by direct phosphorylation catalyzed by hexokinase, *Clin. Chem., 22*: 1427.
10. Schmidt, H. -L., Krisam, G., and Grenner, G. (1976). Microcalorimetric methods for substrate determinations in flow streams with immobilized enzymes, *Biochim. Biophys. Acta, 429*: 283.
11. Kiba, N., Tomiyasu, T., and Furusawa, M. (1984). Flow enthalpimetric determination of glucose based on oxidation of 1,4-benzoquinone with use of immobilized glucose oxidase column, *Talanta, 31*: 131.
12. Muramatsu, H., Dicks, J. M., and Karube, I. (1987). Integrated-circuit biocalorimetric sensor for glucose, *Anal. Chim. Acta, 197*: 347.
13. Muehlbauer, M. J., Guilbeau, E. J., and Towe, B. C. (1989). Model for a thermoelectric enzyme glucose sensor, *Anal. Chem., 61*: 77.
14. Urban, G., Jachimowicz, A., Kohl, F., Kuttner, H., Olcaytug, F., Goiser, P., and Prohaska, O. (1990). High-resolution thin-film temperature sensor arrays for medical applications. *Sensors and Actuators, A21–A23:* 650.
15. Scheller, F., Siegbahn, N., Danielsson, B., and Mosbach, K. (1985). High-sensitivity enzyme thermistor assay of L-lactate by substrate recycling, *Anal. Chem., 57*: 1740.
16. Mattiasson, B., Danielsson, B., and Mosbach, K. (1976). A split-flow enzyme thermistor, *Anal. Lett., 9*: 867.
17. Danielsson, B., Mattiasson, B., and Mosbach, K. (1981). Enzyme thermistor devices and their analytical applications, *Appl. Biochem. Bioeng., 3*: 97.
18. Danielsson, B., Gadd, K., Mattiasson, B., and Mosbach, K. (1976). Determination of urea with an enzyme thermistor using immobilized urease, *Anal. Lett., 9*: 987.
19. Wehnert, G., Sauerbrei, A., Bayer, T., Scheper, T., and Schügerl, K. (1987). Application of an enzyme thermistor for the determination of glucose in complex fermentation media, *Anal. Chim. Acta, 200*: 73.
20. Guilbault, G. G., Danielsson, B., Mandenius, C. F., and Mosbach, K. (1983). A comparison of enzyme electrode and thermistor probes for assay of alcohols using alcohol oxidase, *Anal. Chem., 55*: 1582.
21. Mattiasson, B., and Danielsson, B. (1982). Calorimetric analysis of sugars and sugar derivatives with aid of an enzyme thermistor, *Carbohydr. Res., 102*: 273.
22. Danielsson, B., Rieke, E., Mattiasson, B., Winquist, F., and Mosbach, K. (1981). Determination by the enzyme thermistor of cellobiose formed on degradation of cellulose, *Appl. Biochem. Biotechnol., 6*: 207.
23. Decristoforo, G. and Danielsson, B. (1984). Flow injection analysis with enzyme thermistor detector for automated determination of β-lactams, *Anal. Chem., 56*: 263.
24. Mattiasson, B., Danielsson, B., and Mosbach, K. (1976). Enzyme thermistor assay of cholesterol, glucose, lactose, and uric acid in standard solutions as well as in biological samples, *Anal. Lett., 9*: 217.

25. Winquist, F., Danielsson, B., Malpote, J. -Y., Persson, L., and Larsson, M. -B. (1985). Enzyme thermistor determination of oxalate with immobilized oxalate oxidase, *Anal. Lett., 18*: 573.
26. Satoh, I. (1988). Biomedical applications of the enzyme thermistor in lipid determination, *Methods Enzymol., 137*: 217.
27. Mandenius, C. F., Bülow, L., Danielsson, B., and Mosbach, K. (1985). Monitoring and control of enzymic sucrose hydrolysis using on-line biosensors, *Appl. Microbiol. Biotechnol., 21*: 135.
28. Satoh, I., Inoue, K., and Arakawa, S. (1988). Continuous determination of xanthine using enzyme sensor and thermistor probes. In: *Technical Digest of the 7th Sensor Symposium*, Tokyo, p. 229.
29. Satoh, I. (1989). Continuous biosensing of heavy metal ions with use of immobilized enzyme-reactors as recognition elements. In: *Materials Research Society International Meeting on Advanced Materials*, Tokyo, Vol. 14, p. 45.
30. Kirstein, D., Danielsson, B., Scheller, F., and Mosbach, K. (1989). Highly sensitive enzyme thermistor determination of ADP and ATP by multiple recycling enzyme systems, *Biosensors, 4*: 231.
31. Danielsson, B., Flygare, L., and Velev, T. (1989). Biothermal analysis performed in organic solvents, *Anal. Lett., 22*: 1417.
32. Stasinska, B., Danielsson, B., and Mosbach, K. (1989). The use of biosensors in bioorganic synthesis: Peptide synthesis by immobilized α-chymotrypsin assessed with an enzyme thermistor, *Biotechnol. Technol., 3*: 281.
33. Nilsson, K., Scheier, W., Katinger, H. W. D., and Mosbach, K. (1987). Entrapment of animal cells, *Methods Enzymol., 135*: 399.
34. Mattiasson, B., Larsson, P. -O., and Mosbach, K. (1977). The microbe thermistor, *Nature, 268*: 519.
35. Birnbaum, S., Bülow, L., Hardy, K., Danielsson, B., and Mosbach, K. (1986). Rapid automated analysis of human proinsulin produced by *Escherichia coli, Anal. Biochem., 158*: 12.
36. Li, H., Johansson, C., Lindbladh, C., Mecklenburg, M., Mosbach, K., and Danielsson, B. (1990). To be published.
37. Lindbladh, C., Persson, M., Bülow, L., Stahl, S., and Mosbach, K. (1987). The design of a simple competitive ELISA using human proinsulin-alkaline phosphatase conjugates prepared by gene fusion, *Biochem. Biophys. Res. Comm., 149*: 607.
38. Danielsson, B., Bülow, L., Lowe, C. R., Satoh, I., and Mosbach, K. (1981). Evaluation of the enzyme thermistor as a specific detector for chromatographic procedures, *Anal. Biochem., 117*: 84.
39. Danielsson, B., and Larsson, P. -O. (1990). Specific monitoring of chromatographic procedures. *Trends Anal. Chem., 9:* 223.
40. Hörnsten, E. G., Nilsson, L. E., and Danielsson, B. (1990). The production of molecular hydrogen by *Escherichia coli* during ampicillin-induced spheroplast formation, *Appl. Microbiol. Biotechnol., 32*: 455.

6

Analytical Applications of Piezoelectric Crystal Biosensors

John H. T. Luong

Biotechnology Research Institute
National Research Council of Canada
Montreal, Quebec, Canada

George G. Guilbault

University of New Orleans
New Orleans, Louisiana

INTRODUCTION

The discovery of a linear relationship between the change in the oscillating frequency of a piezoelectric crystal and the mass variation on its metallic surface has produced diverse studies and applications of piezoelectric sensors in many areas of analytical chemistry. The discovery of piezoelectricity itself by the Curies dates back to 1880 (1); however, the first analytical paper on this subject was published by King only in 1964 (2).

The general approach to exploiting the piezoelectric effect is to coat a piezoelectric crystal with a material that interacts highly selectively with the target molecule. When the crystal is exposed to the particular substance of interest, adsorption occurs, causing a frequency change that can be used to determine the amount of material adsorbed. The coating material determines the selectivity and specificity of the piezoelectric sensor. For the past two decades extensive research has been performed on the use of various organic and inorganic active materials for the detection and determination of various agents in the environment and in the workplace (3–5).

Biologically active materials, such as antibodies, enzymes, and antigens, are much more specific and selective than organic and inorganic coating materials. The specificity and selectivity provided by the biological component was first exploited by Clark and Lyons in 1962 to build an enzyme-based biosensor for glucose (6). Since then, biocatalysts or other suitable biological materials, including enzymes, antigens, antibodies, membrane components, and whole cells or plant slices, have been placed onto transducers, such as electrodes (amperometric, potentiometric, and conductometric) and optical sensors, to form the biosensor. In 1972 Shons et al. (7) first demonstrated the use of piezoelectric crystals for studying immunochemical reactions. Antigen-coated crystals were proven applicable for the determination of the corresponding antibody concentration in solution. After this development the use of antigens, antibodies, and enzymes as specific coatings in piezoelectric biosensors has been widespread in liquid-phase analysis and, more recently, in gas-phase analysis.

This chapter focuses on the potential use of piezoelectric crystals as analytical sensors in which biological components serve as coating materials. Both liquid-phase and gas-phase applications are presented and the future of this technology discussed.

PRINCIPLE OF MEASUREMENT

Piezoelectric Effect

The piezoelectric effect is due to the fact that certain crystals contain positively and negatively charged ions, which separate when the crystals are subjected to stress. One of the simplest crystals, zinc sulfide (ZnS), can be used to illustrate this phenomenon. Figure 1 shows a unit cell of cubic zinc sulfide consisting of two types of atoms: zinc atoms (closed circles), with a positive charge, and sulfur atom (open circles), with a negative charge. Shearing stress, indicated by the long arrows, produces motion of the sulfur atom, indicated by the short arrows. This motion results in separation of the charge centers, that is, polarization.

Any given crystal has a natural vibrational frequency, called its resonant (or fundamental) frequency, that depends upon its chemical nature, its size and shape, and its mass. Each vibration of a crystal involves its passing from a deformed configuration through its equilibrium configuration to an opposite deformed configuration, and then back through the equilibrium configuration to the original deformed configuration. This cycle occurs repeatedly as long as the crystal is vibrating.

When the vibrating crystal is piezoelectric, this cycle of oscillating deformity produces an oscillating electrical field; the frequency of the electrical oscillation is identical to the vibration frequency of the crystal. At the same time, placing a piezoelectric crystal in an oscillating electrical field causes it to vibrate at the frequency of the oscillating field. This transfer of energy from the electrical field to the crystal is very inefficient except when the frequency of the oscillating electrical field is the same as the resonant frequency of the crystal. This is exploited by the incorporation of quartz crystals into oscillator circuits, the frequency of the entire circuit becoming the resonant frequency of the quartz

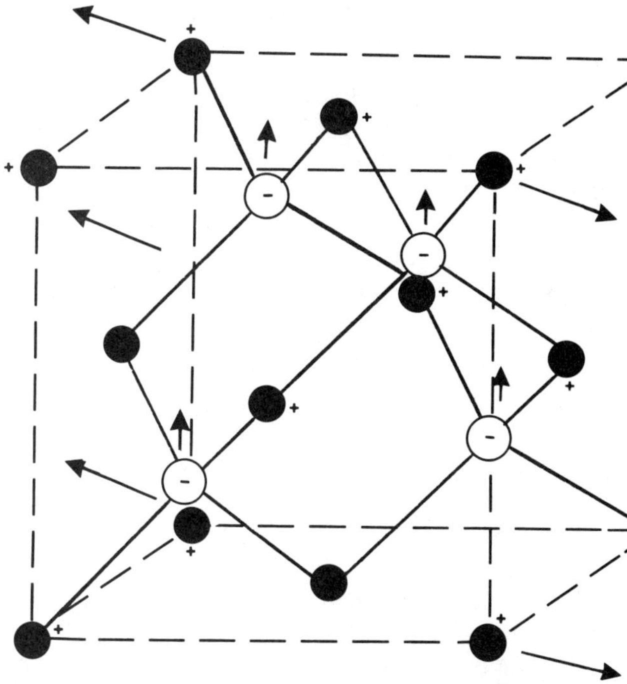

Figure 1 Unit cell of cubic zinc-sulfide crystal with positive zinc and negative sulfur atoms. Long arrows indicate applied shearing stress, short arrows the associated displacement.

crystal. From the same principle, the resonant frequency at a given crystal, for instance a crystal used as a biosensor, can be determined from the frequency of an electrical oscillator circuit in which the crystal is a component.

The principle piezoelectric materials used commercially are crystalline quartz and Rochelle salt, despite the existence of a wide range of other materials, such as tourmaline, ammonium dihydrogen phosphate, ethylenediamine tartrate, and barium titanate. Rochelle salt possesses a large piezoelectric effect, and this material is useful in acoustical and vibrational devices in which sensitivity is desired. However, it decomposes at only 55°C and requires protection against moisture. Barium titanate has lower sensitivity but greater immunity to temperature and humidity effects. Quartz has the important quality of being a completely oxidized compound (silicon dioxide). It is insoluble in water and resists temperatures of up to 579°C with no loss of piezoelectric properties. Industrially grown rather than natural quartz crystals are used almost exclusively for electrogravimetric sensors because of their higher purity, and α quartz is most often used.

Relationship Between Added Mass and Frequency Shift

The resonant frequency of the quartz crystal depends on the physical dimensions of the quartz plate and the thickness of the electrode deposited. For use as a piezoelectric detector only AT- or BT-cut quartz plates are useful. AT and BT refer to the orientation of the plate with respect to the crystal structure (35°15′ and −49°00′, respectively). The AT-cut crystal is more stable than most other piezoelectric cuts and has a temperature coefficient of about 1 ppm/°C over a temperature range of 10–50°C (4).

The thickness of the quartz plate θ is related to the fundamental frequency by

$$\theta = \frac{N}{f} \tag{1}$$

where N is the frequency constant. For an AT-cut quartz crystal N is 1.67×10^5 cm Hz.

The thickness of the quartz plate θ is also related to its mass by the equation

$$\theta = \frac{m}{A\rho_q} \tag{2}$$

where ρ_q is the density of quartz (2.65 g/cm^3), m is the mass (g) of oscillating quartz, and A is the area (cm^2) of the quartz plate undergoing oscillation (between the two electrodes on opposite faces).

Combining Equations (1) and (2) yields

$$f = \frac{\rho_q NA}{m} \tag{3}$$

If an added mass Δm is attached uniformly to one of the electrodes of area A, then

$$\Delta f + f = \frac{\rho_q NA}{m + \Delta m} \tag{4}$$

Rearrangement of Equation (4) provides

$$\Delta f = -\frac{f^2}{\rho_q NA} \frac{\Delta m}{1 + \Delta m/m} \tag{5}$$

Substituting numerical values for N and ρ_q in Equation (5) yields

$$\Delta f = -2.26 \times 10^{-6} f^2 \frac{\Delta m}{A(1 + \Delta m/m)} \tag{6}$$

When Δm is very small compared to m, interfacial mass changes Δm are related in a simple manner to changes in oscillation frequency changes Δf:

$$\Delta f = -2.26 \times 10^{-6} f^2 \frac{\Delta m}{A} \tag{7}$$

Equation (7) was originally derived by Sauerbrey (8) to describe the effect of vacuum deposition of metals on the quartz crystal. He suggested the use of the piezoelectric crystal as a sensing device for the measurement of the thickness of thin films. Sauerbrey (8) demonstrated that the shift in the resonant frequency of the crystal was proportional to the mass deposited to within $\pm 2\%$.

It is obvious from Equation (7) that when Δf is very small in relation to f (0.17% or less) Δf is a linear function of added mass Δm. The Sauerbrey equation is appropriate for many situations; however, its applicability is only good for thin and rigid layers, that is, the mass added should not experience any shear deformation during oscillation. This equation predicts that a commercially available 9 MHz piezoelectric crystal would have a mass sensitivity of about 400 Hz/μg or a 15 MHz crystal a sensitivity of 2600 Hz/μg (9). From this it can be seen that a vibrating quartz can be an extremely sensitive weight indicator. The limit of detection is determined to be about 10^{-10}–10^{-11} g if the frequency can be measured within 0.1 Hz. There are two types of oscillator sensors available, SAW (surface acoustic wave) and piezoelectric crystals. The SAW devices can operate either on the Rayleigh wave propagation principle at solid/thin-film boundaries, or as a bulk wave PZ device. The bulk wave PZ crystal devices operate according to the Sauerbrey principle: that small thin films on AT-cut crystals can be treated as equivalent mass changes of the crystal. At their simplest level PZ crystal devices are used to determine the amount of mass adsorbed on the active surface in a nonspecific manner. For more specific applications biologically active materials are applied to the surface of the crystal through simple adsorption or chemical immobilization. This immobilization process is crucial to the performance of the piezocrystal biosensor.

GENERAL EXPERIMENTAL APPARATUS AND INSTRUMENTATION

The Piezoelectric Crystal

The crystals most commonly used are 5, 9, or 10 MHz quartz of 10–16 mm disks, squares, or rectangles that are approximately 0.15 mm thick. The metal electrodes are 3000–10,000 Å thick and 3–8 mm in diameter and can be made of gold, silver, aluminum, or nickel. For most applications the gold electrode is used because of its inert property. Silver electrodes do not possess long-term stability in an aqueous environment; however, they can be treated with palladium to overcome this problem. In addition, the anodically oxidized palladium electrode (1.4 V versus Ag/AgCl) has a more highly hydroxylated surface than untreated palladium. The silver electrode can also be sputtered with silicon dioxide; however, the resulting surface is not stable during oscillation (10).

The quartz wafer is sandwiched between two electrodes bonded to the wafer surface (Fig. 2). These electrodes are used to induce an oscillating electrical field

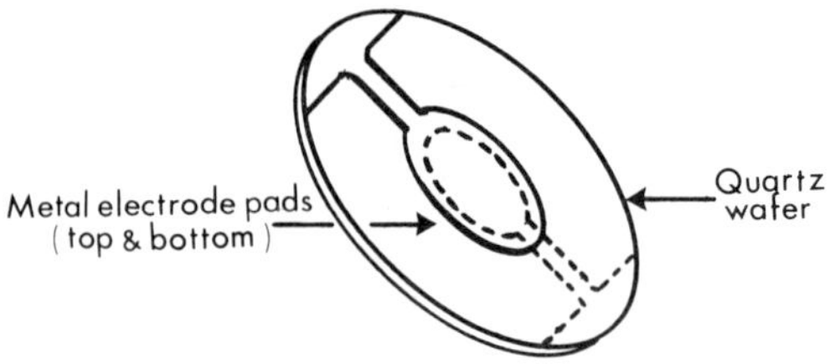

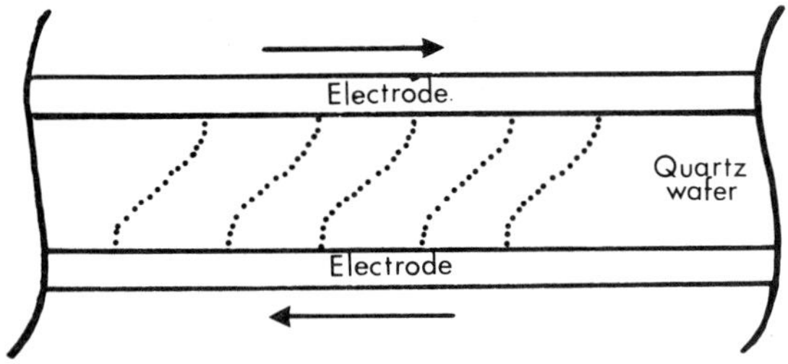

Figure 2 A typical piezoelectric crystal: orientation of electrodes and shear deformation during oscillation at the fundamental frequency.

perpendicular to the surface of the water. The electrical field then produces a mechanical oscillation, a standing wave, in the bulk of the quartz wafer. Mechanical oscillation of the crystal is maximum where the electrode pads overlap and diminishes rapidly in areas where the oscillating electrodes do not overlap.

In the manufacture of piezoelectric crystals metals are often evaporated directly on the quartz plate to serve as electrodes. The amount of metal deposited is used to adjust the crystal to a desired frequency level. Metals and many other solids do not significantly affect the crystal's ability to vibrate.

Measurement of the Resonant Frequency

When placed in an electronic oscillator circuit the portion of the quartz wafer located between the electrodes vibrates with its precise fundamental frequency. The frequency output from the oscillator, which is identical to the resonant frequency of the crystal, can be measured by a frequency counter. To improve the stability and reproducibility of the resonant frequency it is essential to minimize the distance from the oscillator to the piezoelectric crystal. Specific circuits on the oscillators are not given here because of the wide variety that are commercially available.

In practice a piezoelectric detection system consists of two oscillator circuits: a detector crystal oscillator and a reference crystal oscillator. The reference crystal oscillator is identical to the detector oscillator, except that this crystal is not coated with a biological material. The reference crystal is used to correct for temperature, humidity fluctuations, and other interference. The two oscillator frequencies are fed to a mixer, where the difference in frequency between the two crystals is obtained (Fig. 3).

The frequency of the piezoelectric crystal can be measured using either commercial or custom-designed frequency-counting equipment. With high-resolution counters, at 5 MHz, the frequency can be measured to within 0.1 Hz in 0.06 sec.

Commercially Available Instrumentation

A coated piezoelectric crystal analyzer sensitive to water vapor as low as 0.1 ppm was developed by King (2). This sorption hygrometer has been commer-

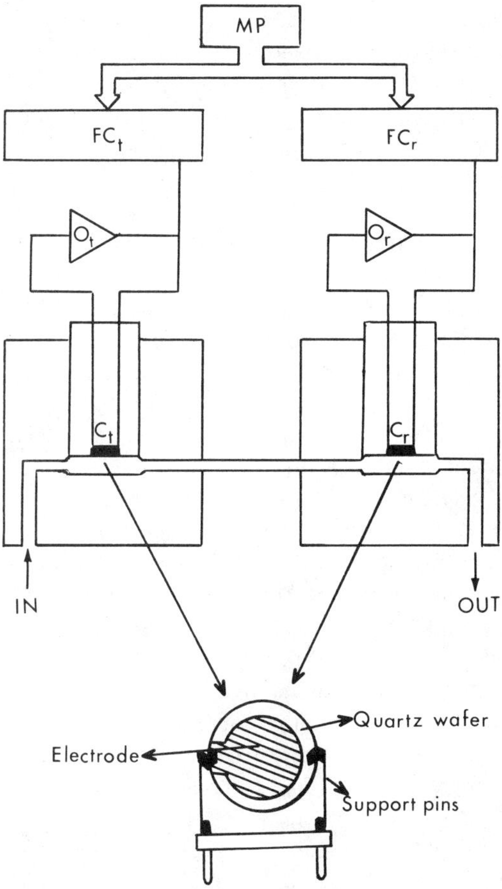

Figure 3 A typical system for piezoelectric crystal detector incorporating reference (C_r) and test (C_t) crystal sensors individually held in oscillating circuits (O_r and O_t, respectively) serviced with separate frequency counters (FC_r and FC_t, respectively) interfaced to a common microprocessor or other readout devices.

cially available since 1964 (DuPont Instruments, Wilmington, DE, Model 560). The instrument has a very high sensitivity, fast response, and long lifetime, with an accuracy of about 5%.

A gas chromatographic apparatus using a piezoelectric crystal detector was designed and developed for the determination of effluents vaporizable at up to 200°C. The qualitative and quantitative performance of this chromatographic detector was investigated by Karasek et al. (11). Janghorbani and Freund (12) pointed out that piezoelectric detectors in chromatography can be interfaced to a large, general-purpose, time-sharing computer for on-line data acquisition and processing with simple and inexpensive equipment. A prototype piezoelectric crystal mass detector for GPC was compared with the refractometer detector. The experience gained on the new type of sensor indicated that more measurements are needed to develop a generally useful detector.

Two commercially available aerosol mass concentration analyzers were described by Daley and Lundgren (13) for the determination of aerosols and particulates. The instruments are both portable and relatively rugged and operate on ordinary commercial power. Sem and Tsurubayashi (14), using an electrostatic precipitator, developed a portable system that is commercially available for mass concentration measurements in the 100 $\mu g/m^3$ range of airborne dust particles smaller than 10 μm. Chuan (15) also reported using a typical commercial instrument based on the piezoelectric quartz crystal microbalance for the rapid assessment of particulate mass concentrations in the atmosphere. The adhesive coating used in the device in nonhygroscopic and nonreactive to the usual concentrations of pollutant gases in the atmosphere, such as CO, SO_2, NO_x, and hydrocarbons.

Detection of various gases in ambient air (including sulfur dioxide, ammonia, hydrogen sulfide, hydrogen chloride, and phosgene), organophosphorus compounds (such as pesticides) and aromatic hydrocarbons has already been demonstrated with piezoelectric crystals. Portable instruments for monitoring gaseous components in air are commercially available in a unit that has rechargeable batteries, a variable-speed pump, digital display of Δf and concentration, alarm signals that can be preset for certain levels, and interfacing to both a computer and a recorder (Universal Sensors, Inc., PZ Model 105, New Orleans, LA).

A hand-calculator–sized instrument with a digital readout for measuring mercury vapor concentrations in air and/or personal exposure to mercury vapor was developed and evaluated by Scheide and Taylor (16). Measurements of concentrations below 15 ppb have a precision of 10%, with uncertainties of 5% above this level.

Fog and Rietz (17) developed a portable instrument for use as a dosimeter or a personal monitor for ozone detection in parts per billion concentrations. The accuracy of the instrument is comparable to the results obtained with the portable AID ozone analyzer and Drager detector tubes.

Turnham and coworkers (18) designed and developed a working monitor for propylene glycol dinitrate (PGDN). The dual-crystal design, containing a trap for

PGDN on the reference crystal, eliminates frequency stability and selectivity problems utilizing an auto-zero function. A detection limit of better than 0.05 ppm is possible with this device. It is highly selective for PGDN over all other normal atmospheric contaminants and is insensitive to fluctuations in humidity.

APPLICATIONS OF PIEZOELECTRIC CRYSTALS FOR DETECTING A TARGET COMPONENT IN SOLUTION

Immobilization of a Biocatalyst on the Piezoelectric Crystal

The first step to exploiting the piezoelectric effect for detecting a target analyte in solution is to immobilize a suitable biological material onto the crystal. The biological material, of course, must interact highly selectively with the target analyte molecule. Several immobilization methods have been used to construct piezobiosensors. Such methods fall into the following categories:

Immobilization of a biocatalyst on the crystal precoated with a suitable material: this is the most popular method used to construct piezobiosensors. In this procedure the piezoelectric crystal is precoated with a suitable material to create a thin layer capable of forming hydrophobic and/or covalent bonds with the biocatalyst. To date several materials have been used to precoat the piezoelectric crystal (Table 1).

Immobilization via entrapment to deposit a solution containing acrylamide, *N,N'*-methylene-bisacrylamide, potassium persulfate, and the biocatalyst onto the electrode of the piezoelectric crystal (24): The polymerization is initiated by addition of 3-dimethylaminopropionitrile. This method is limited to assaying small substrates, which have greater ease of access than large substrates.

Immobilization via cross-linking: The biocatalyst is cross-linked with bovine serum albumin via glutaraldehyde. The resulting mixture is then deposited on the crystal's surface. The biocatalyst can also be cross-linked by itself via glutaraldehyde and then deposited onto the piezocrystal (19).

Immobilization by irradiation: DNA can be grafted to the piezoelectric crystal precoated with a solution of poly (butyl methacrylate) in ethyl acetate and dimethoxyphenylacetophenone in acetone by irradiation at 365 nm ultraviolet (UV) light (21).

In biosensors a prerequisite for the electrode surface is that it should be chemically stable during the measurement process and that a high density of the biological material be immobilized onto it. However, overloading of the piezoelectric crystal with the biological material should be avoided since this may restrict the free interaction between the immobilized material and the macromolecular analyte. This feature is particularly important in piezosensors, where interaction is often between antibody and antigen macromolecule. The leakage of the immobilized materials must not occur to any extent during the use of the

Table 1 Materials Used to Precoat the Piezoelectric Crystal

Materials	Procedure	Reference
Nyebar C	Dip coating the crystal with a low surface energy plastic (Nyebar C 30% solution in 13-di[trifluoromethyl]benzene); this coating layer is capable of forming hydrophobic bonds with enzymes, proteins, antibiotics, antigens, and so on	7
Polyethyleneimine	Coating the crystal with a methanol solution containing 2–3% polyethyleneimine; after treating the precoated crystal with glutaraldehyde biocatalysts can be immobilized on the precoated crystal via the surface aldehyde groups	19
APTES	Coating the crystal with γ-aminopropyltriethoxysilane (APTES) solution (2–5% in acetone, dry benzene); biocatalysts are immobilized on the precoated crystal via glutaraldehyde activation	10
Protein A	After treatment of the crystal with concentrated hydrochloride protein A is deposited onto the surface of the crystal; the protein A-coated crystal is capable of reacting with antibodies and other biological materials	20
Polyethyleneimine and avidin	The crystal is first coated with a polyethyleneimine solution (2–3%); avidin is then immobilized on the coated crystal via glutaraldehyde activation; the biotinylated biocatalyst is immobilized on the coated crystal via a biotin avidin interaction since 1 molecule of avidin binds 4 molecules of biotin	19
PBMA	Coating the crystal with a poly(butyl methacrylate) (PBMA) solution (0.1% in ethyl acetate) and dimethoxyphenylacetophenone (0.1% in acetone); DNA is grafted to the precoated crystal by irradiation with 365 nm UV light	21
PHDMAB	Coating the crystal with 0.06% aqueous solution of poly[2-hydroxy-3-dimethylamino-1,4-butane] (PHDMAB); this coated crystal is capable of reacting with various antigens, including human gammaglobulin (IgG), L-thyroxine, human serum albumin (HSA), and human α globulin	22
Polyacrylamide	Coating the crystal with a thin layer of polyacrylamide gel; antigens and antibodies are immobilized on the polyacrylamide via glutaraldehyde activation	23

biosensor, and this layer must also be stable during oscillation. Although the literature provides a plethora of information on the immobilization and the stability of an immobilized layer on the crystal surface, the choice of an appropriate immobilization procedure for a particular application cannot yet be predicted.

Measurement Procedure

The frequency difference between the coated crystal and the reference crystal is first determined (ΔF_1). The coated crystal is then dipped in a solution containing the target analyte to allow reaction to occur between the analyte and the coated crystal. After thorough washing and air drying the coated crystal is placed back in the test cell and a new frequency is measured (ΔF_2). A shift in the frequency difference ($\Delta F = \Delta F_2 - \Delta F_1$) is related to the amount of the analyte adsorbed onto the coated crystal. As mentioned previously, the AT-cut crystal has a low temperature coefficient; however, temperature control of the detection cell is necessary to obtain maximum accuracy. In addition, the biological materials coated on the crystal, such as enzymes, proteins, and antibodies, absorb more or less water, depending on the humidity. Therefore the humidity of the detection cell must also be controlled when measuring mass changes on the surface of the coated crystal. This method is referred to as the "end-point" measurement.

Bioapplications with Coated Piezoelectric Crystals

Biological materials as piezoelectric crystal coatings were first demonstrated by Shons et al. (7). Bovine serum albium (BSA) was immobilized on the crystal precoated with a 30% solution of Nyebar C, a low surface energy plastic. The rationale of using this solution as the coating material is that proteins adsorbed on low-energy surfaces retain their antigenic characteristics. When the BSA-coated crystal was exposed to a solution containing anti-bovine serum albumin, the frequency shift of the coated crystal was proportional to the concentration of anti-bovine serum albumin in solution. The sensitivity of this technique was equal to or better than that of the passive agglutination method, a conventional procedure used for the assay of antibody. The passive agglutination method required hours, but the result using the piezoelectric crystal could be obtained in minutes.

After the demonstration of Shons et al. (7), the use of antibodies or antigens as specific coatings in PZ biosensors has been widespread (Table 2). In all these studies the reaction between the coated crystal and the target analyte is carried out in the liquid phase and the method is referred to as direct immunoassay. Although the direct immunoassay is simple and straightforward, this technique suffers from two major drawbacks. First, it is applicable only for assay of large antigens or antibodies. Second, the technique cannot be applied to situations in which significant amounts of substances other than antibody attach to the antigen-coated crystal along with the antibody. In such cases measuring the change in the resonant frequency of a crystal having an antigen-antibody layer bound to its surface would not provide an accurate determination of the amount of antibody in the measured sample, since an unknown portion of the frequency shift could be due to the nonspecific binding of extraneous substances. An indirect method for determining the amount of an antigen in a liquid sample was disclosed by Oliveira and Silver (27). The piezoelectric crystal having an antigen

Table 2 Detection of a Target Component in Solution by the Piezoelectric Crystal (Direct Assay)

Target analyte	Biocatalyst	Detailed information on the piezobiosensor	Reference
Anti-bovine serum albumin antigody	BSA	BSA was immobilized on the crystal precoated with a 30% solution of Nyebar C, a low surface energy plastic; the resonant frequency decreased linearly to the concentration of antibodies	7
Candida albicans	Anti-*Candida* antibody	The anodically oxidized palladium electrode was treated with γ-aminopropyltriethoxysilane (APTES); anti-*Candida* antibody was immobilized on the electrode via glutaraldehyde activation; the frequency shift was correlated with *C. albicans* in the range 10^6–5×10^8 cells per cm^3; the piezobiosensor showed no response to *Saccharomyces cerevisiae*	10
S. typhimurium	Anti-*Salmonella*	Among various immobilization procedures tested the best result was obtained when the antibody was immobilized on the polyethylene-imine-precoated crystal; the frequency shift was correlated with *S. typhimurium* in the range of 10^5–10^9 cells per cm^3; the piezoelectric crystal could be used for six to eight consecutive assays	19
Human transferrin	Anti-human transferrin	Anti-human transferrin antibody was immobilized on protein A and polyethyleneimine-coated crystals, respectively; human transferrin concentrations in the range of 10^{-4}–10^{-1} mg/cm^3 were detected by	25

		the piezobiosensor; the protein A-coated crystal was 10-fold more sensitive than the polyethyleneimine-coated crystal; attempts to re-use the immunosensor were not successful since dissociation of the complex protein A-antibody transferrin was not complete	
Human albumin	Anti-human albumin antibody	The crystal was coated with protein A and then with antibody; human albumin in the range 10^{-4}–10^{-1} mg/ml was detected; this paper suggests an alternative for the reuse of the crystal immunobiosensor: after each measurement the crystal was submerged in a solution of albumin to saturate all the binding sites available for the antibody; the crystal was then reacted with antibody to form another layer on the crystal surface; following this procedure, the crystal could be reused at least five times without losing its sensitivity (200 Hz; albumin concentration is 0.1 mg/ml)	26
S. typhimurium DNA	*S. typhimurium* DNA	The crystal was coated with a solution of poly(butyl methacrylate) and dimethoxyphenylacetophenone; *S. typhimurium* DNA was then immobilized on this coating layer by irradiation (365 nm for 2 hrs); the coated crystal was incubated with melted *S. typhimurium* DNA and melted calf thymus DNA as control; the hybridization produced a frequency change as high as 780–1010 Hz	21

adsorbed on the surface is exposed to a sample containing the same antigen and a predetermined amount of antibody specific for the antibody. The antigen attached to the crystal competes with the antigen (free) in the sample for the supply of the antibody. The amount of antigen in the sample, if any, can be determined by measuring the change in frequency of the piezoelectric crystal and referring to a standard curve. A wide range of proteins (antigens) were determined: human gammaglobulin (IgG), human serum albumin, thyroxine, and α-antitrypsin. The main advantage of the competitive binding assay over the direct assay is its capability of measuring both large and small molecules. However, like the direct immunoassay this technique is not very useful in any situation involving nonspecific binding of extraneous substances to the antigen-coated piezoelectric crystal.

A sandwich immunoassay using the piezoelectric crystal, patented by Rice (28), can be used to overcome this nonspecific binding. In this approach the crystal is first coated with the antigen specific for the antibody being assayed. The coated crystal is then exposed to the sample containing the antibody and its resonant frequency is measured. A solution containing an excess amount of a substance specific for the antibody, such as an anti-antibody, is applied to the crystal surface and its resonant frequency is measured again. From this reading and the previous frequency a change in reading due to the added mass of the newly attached antibody-specific substance is calculated. By comparison to a standard curve using predetermined amounts of given antibody, the actual amount of antibody in the sample can be ascertained. The sandwich immunoassay using the piezoelectric crystal is also useful in competitive assays of small molecules of clinical interest present in human sera or urine, such as vitamins, cofactors, hormones, polypeptides, prostaglandins, cyclic nucleotides, antibiotics and other drugs, sugars, and many other metabolites or host of infectious organisms.

A method for the determination of a particular subclass of antibody, such as IgA, IgE, and IgM, was also patented by Rice (22). Similarly to the sandwich immunoassay, after an antigen is bound to the piezocrystal surface the antigen-coated crystal is allowed to react with an antiserum to determine the total antigen-specific antibody content of that antiserum. A reactive material is then applied to the crystal, and this material or "sandwich" layer is specifically reactive with the class or subclass of antibody that is first bound to the antigen on the crystal. This second or sandwich layer has specificity for the type of antibody to be quantitated. The class of antibody bound to the antigen is determined by the specificity of the sandwich layer toward a particular antibody class or subclass, and the amount of a particular antibody class or subclass is determined by comparison with a standard curve. This technique was applied to the assay of human IgG and IgE antibodies against honeybee venom, human IgG antibodies against phospholipase A, and the different classes of specific antibodies to keyhole limpet hemocyanin.

The application of the piezoelectric crystal to the detection of polynucleotide hybridization was developed by Fawcett (29). In general a polynucleotide can be immobilized on a surface of a piezoelectric crystal and its resonant frequency is measured. The polynucleotide-coated piezeolectric crystal is then exposed to a separate source of polynucleotide for a sufficient length of time, under conditions suitable for hybridization. The resonant frequency of the crystal is then again measured, and the difference between the resonant frequency before and after the incubation period indicates the extent of hybridization. This approach has particular applications to medical diagnosis, bacteriology, virology, pathology, and biochemical and molecular biological research. For instance, it is useful for detecting polynucleotide hybridizations, such as DNA-RNA, DNA-DNA, and RNA-RNA hybridizations. An interesting example of this application is the detection of *Salmonella typhimurium* DNA hybridization to *S. typhimurium* DNA attached to a crystal coated with a 0.1% solution of poly(butyl methacrylate) in ethyl acetate and a 0.1% solution of dimethoxyphenylacetophenone in acetone. One strand of DNA (denatured by boiling and then quick cooling at 4°C) was grafted to the coated crystal by irradiation at 365 nm for 2 h. Hybridization was carried out by dipping the crystal in a solution containing *S. typhimurium* DNA (heated at 100°C and then cooled to 65°C). The frequency shift associated with the hybridization was reported as high as -833 Hz. Although this technique lacks the sensitivity of radioassays, it is sufficiently sensitive to permit routine screening of some clinical and food samples for bacterial contamination.

Reusability of the Coated Crystal

It is still problematical to dissociate the bound analyte from the coated crystal since the antibody-antigen interaction is very strong and irreversible. In some cases antigens, such as human IgG, could be removed from the crystal coated with antihuman IgG by washing the crystal with high ionic strength solutions. However, the immobilized antibody is expected to gradually break away from the surface with repeated use so that the sensitivity of a given crystal decreases with age.

Guilbault et al. (30) hypothesized that an immobilization procedure that would create a stable protein layer would surmount this problem to allow multiple assays. To illustrate this point, four different immobilization procedures were used to attach anti-*S. typhimurium* antibody to a PZ crystal: (1) glutaraldehyde cross-linking of the antibody directly on to the PZ crystal, (2) precoating the crystal with a thin silane layer and antibody was attached via glutaraldehyde activation, (3) anti-*Salmonella* antibody was biotinylated and reacted with an avidin-coated crystal, and (4) precoating the crystal with a thin layer of polyethyleneimine and antibody was attached via glutaraldehyde activation. It was observed that the poor stability of some coatings is due to their inability to withstand repeated submersion into a solution of 8 M urea, a necessary condition to remove adsorbed bacterial cells of *S. typhimurium*. Dissociating agents, such as 10 mM HCl, 0.2 M glycine-HCl, pH 2.8, and 0.2 M ethanolamine, pH 8, did not remove the adsorbed cells effectively. Immobilization of antibody onto a crystal surface precoated with a thin layer of polyethyleimine gave the optimal combination of sensitivity and stability. The coated crystal was stable for six to seven assays and gave very good sensitivity (ΔF = 330 Hz; analyte is 10^8 cells per ml *Salmonella*). Only after eight assays was the immobilized antibody possibly denatured and/or desorbed from the crystal surface. The silane-coated crystal gave poor sensitivity (ΔF = 80 Hz) and lasted for only one or two assays. The avidin-coated crystal gave reasonable sensitivity (ΔF = 280 Hz) but allowed only one or two assays. Finally, the glutaraldehyde–cross-linked antibody-coated crystal was limited to a single assay, which is unfortunate considering its excellent sensitivity (ΔF = 530 Hz).

A new approach to the development of a reusable piezoelectric crystal biosensor for human albumin was also reported (26). The crystal was coated with protein A and then reacted with anti-human albumin antibody. Human albumin in the range 10^{-4}–10^{-1} mg/ml was detected by the piezobiosensor system. After the immunosensor was used for determining albumin the three-protein layer (protein A-antibody-albumin) crystal was allowed to react with albumin again to saturate all the binding sites available from the antibody. The crystal was then reacted with antialbumin antibody; that is, an additional layer was formed onto the crystal surface, and the resulting crystal was ready for the next measurement. Following this manner the piezoelectric crystal was reused at least five times without losing its sensitivity. Frequency changes caused by subsequent antibody and albumin binding were reproducible (frequency change is about 200 Hz corresponding to 0.1 mg/ml of albumin concentration). In this study attempts to dissociate human albumin from the antibody-coated crystal were not successful even at pH 2. Dissociation of the crystal with phosphate-citric acid buffer (pH 3) for 3 min resulted in only a partial removal of albumin. A longer incubation (>20 min) at pH 3 resulted in a significant desorption of antibody (40%) from the protein A-coated crystal. The regeneration of the crystal in this conventional manner failed to provide sensitive and reproducible results.

Undoubtedly the challenges in the piezoelectric crystal are more understanding its behavior than instrumental. The reproducible immobilization of the biological material on the crystal surface must be further developed. Whether the reusability of the piezoelectric crystal remains an important issue, this open question depends on the production cost of the crystal. At the present time disposable transducers are not yet feasible since each crystal still costs a few dollars. It is still hoped that the production cost will become so low that very small and disposable crystals could be used for single analyses.

Critique of the End-Point Measurement Procedure

As described, determinations of components in solution are carried out by reaction of components in solution with the crystal, after which the latter is washed and dried before the frequency change is measured. This procedure requires several steps and is time consuming. In some cases the procedure requires several hours before the result can be obtained.

Despite these shortcomings the end-point measurement still provides the most reliable result compared to the measurement in which the crystal is immersed in liquid. As discussed later, a crystal that has the electrode on one side immersed in a solution also oscillates; however, the frequency change is nonspecific and depends on several unknown parameters. The other drawback of the direct liquid-phase assay is its poor sensitivity and limits of detection due primarily to unspecific adsorption on the crystal surface. It is well known that some antibodies show severe cross-reactivity with some metabolic intermediates or by-products structurally similar to their haptens. This varies from one antibody preparation to another and is usually accompanied by differences in binding rate constants. In the end-point measurement the problem of unspecific bindings can be alleviated by washing the crystal with high ionic strength solutions.

APPLICATIONS OF THE PIEZOELECTRIC CRYSTAL IN THE LIQUID PHASE

Behavior of the Piezoelectric Crystal in Liquid Phase

Determinations of components in solution reported thus far have been carried out by reaction of the component in solution with the crystal, after which the surface is washed and dried before the frequency change is measured. Nomura and Minemura (31) reported that when immersed in liquid the piezocrystal also oscillates, and the oscillation frequency is dependent on the viscosity, density, and specific conductivity of the solution. Konash and Bastiaans (32) also reported that a crystal with one electrode in contact with an inorganic solvent also oscillates. The frequency shift again depends on the density, and they proposed that this arrangement may be used as a detector for liquid chromatography. For such measurements the crystal must be fixed with a liquid-tight seal to avoid leakage of liquid to the electrode on the other side. Nomura and Okuhara (33) later demonstrated that the frequency of a crystal immersed in an organic solvent containing no electrolyte is influenced only by the density and viscosity:

$$\Delta F = a\eta^{1/2} + bd^{1/2} - c \qquad (8)$$

where η = viscosity (cP) = density (g/ml), and a, b, and c are constants that depend on the properties of the crystal.

In liquids of high density and/or viscosity, such as carbon tetrachloride (d = 1.5848 g/ml, η = 0.903 cP), 1=butanol (d = 0.857 g/ml, η = 2.46 cP), and nitrobenzene (d = 1.1986 g/ml, η = 1.811 cP), the crystal does not oscillate.

Bruckenstein and Shay (34) conducted an in situ measurement of the resonant frequency of an oscillating quartz having an electrode face in contact with an electrolytic solution. As experimentally demonstrated, the height of liquid above the crystal has no significant effect on the resonant frequency, and the resonant frequency varies as the square root of the solution viscosity. In situ measurements of the crystal frequency show marked temperature sensitivity as a result of the temperature effect on the viscosity of liquids.

In a recent study Yao and Zhou (35) reported that the frequency shift is dependent mainly on density and viscosity for 47 organic liquids tested including alcohols, ketones, ether, esters, and aliphatic and aromatic hydrocarbons. The influence of dielectric constant is not significant, and the influence of density is more pronounced than that of viscosity. The authors also reported that for a crystal to oscillate in the liquid phase it is necessary to maintain the working temperature at a level higher than the relevant temperature at which the crystal ceases to oscillate. The temperature at which oscillation ceases increases with the concentration of electrolyte in an aqueous solution. For a homologous series of organic liquids this temperature increases with the length of the carbon chain, that is, with the carbon atom number. The temperature at which the crystal ceases to oscillate differs for different crystals, even with the same rated frequency. Kanazawa and Gordon (36) recently presented a theoretical treatment of the properties of liquids on a piezoelectric crystal's oscillation frequency and proposed the equation

$$\Delta f = f_0^{3/2} \left(\frac{\eta_L d_L}{\pi \eta_Q d_Q} \right)^{1/2} \tag{9}$$

where f_0 is the oscillation frequency of the free (dry) crystal, η_L and d_L are the absolute viscosity and density of the liquid, respectively, and η_Q and d_Q are the elastic modulus and density of the quartz. This model was confirmed by using aqueous solutions of glucose and ethanol at various concentrations.

Analytical Application of the Piezoelectric Crystal in Liquid Phase

The behavior of piezoelectric crystals in liquid has been exploited for determination of traces of ions (37–40) and the total salt content of natural waters (41). However, few liquid-phase biological applications of piezoelectric crystals have been reported.

A piezoelectric quartz surface acoustic wave (SAW) device was developed to detect IgG in solution using the ST-cut crystal with a resonant frequency of 10.3 MHz (42) The area between electrodes was etched with HF/NaF/$NH_4F \cdot HF$ solution to increase surface area. After coating the crystal with glycidoxypropyltrimethoxysilane (GOPS), the goat antibody to IgG was bound to the crystal

surface. The reference crystal contains oxidized GOPS but no antibody. Specific adsorption between the immobilized antibody and the analyte was demonstrated as reflected by the shift in the frequency of the crystal. To test the response of the piezoelectric crystal, several antigen-containing solutions were used. The detection limit of this device was found to be 13 μg antigen, and the frequency response was linearly proportional to IgG concentration, ranging from 0.0225 to 2.25 mg/ml. Antigen could be removed from the crystal surface by washing with high ionic strength solutions so that the crystal could be used for many analyses. Although this work demonstrated the applicability of immunoassay with piezoelectric crystals in the liquid phase, the greatest limitation to this technique is poor sensitivity and limits of detection due primarily to unspecific adsorption occurring at both detector and reference crystal surfaces.

Davis and Leary (20) developed a continuous liquid-phase piezoelectric biosensor for kinetic immunoassays. At AT-cut crystal with a fundamental frequency of 10 MHz was mounted in a Plexiglas holder, and one side was directly coated with protein A. A special oscillator circuit was used to compensate for the degraded quality of the crystal's efficiency in an energy-adsorbing medium. When the protein A-coated crystal was exposed to solutions containing rabbit or human IgG, a decrease in the resonant frequency was observed. Subsequent addition of sheep anti-human IgG to the crystal caused a specific threefold further decrease in the resonant frequency. A frequency change of approximately 1 Hz for each 10 ng added immunoglobulin was observed. Decreasing the pH to 3 released the bound IgG but not the protein A and permitted reuse of the crystal for further IgG binding.

It is thus obvious that the behavior of piezoelectric crystals in the liquid phase also obeys the Sauerbrey equation; that is, a frequency decrease is associated with the adsorption of small amounts of material on the surface of the crystal. However, Thompson and coworkers (23) revealed that it is feasible to oscillate piezoelectric crystals in water with a reasonably high degree of reproducibility but that frequency shifts are not necessarily associated with the classic microgravimetric signal. Such behavior was observed when they attempted to detect human IgG in the liquid phase with goat anti-human IgG immobilized on an AT-cut crystal. As IgG was introduced to the reaction chamber housing the antigen-coated crystal, immunoprecipitation produced a frequency decrease. Although good frequency shift correlations were obtained between antisera-water and precipitin-water at different concentrations, the authors insisted that there was no relationship between the amount of antibody in solution or the bound antibody and the frequency decrease was possibly caused by the effect of bulk viscosity changes.

With time, as more anti–IgG-IgG (antibody-antigen) complexes formed, there was a frequency increase. This apparent contradiction of the classic microgravimetric behavior, they explain, represents the interfacial immunochemical reaction, which is different from the bulk antigen-antibody interaction. Based on the experimental data obtained, it was proposed that the frequency response was associated with changes in surface interactions during the immunochemical reaction. Precipitin formation generally leads to increased hydrophilicity, and the viscoelastic properties of proteins under shear stress at interfaces are strongly influenced by cross-binding and intermolecular interactions. Further theoretical and experimental work on nonequilibrium interfacial viscosity effects is required to substantiate these arguments.

The theory of kinetics involved in antigen-antibody complexation at the crystal surface was examined along with the interfacial contributions by Thompson et al. (43). These authors addressed the questions of antigen-antibody interactions on the crystal in the liquid phase and concluded that careful immobilization of antibodies yields surfaces with kinetics close to those in homogeneous environments with complete coverage in a short time. The treatment omitted precipitin formation, which should not really come into play if the reactions are over in a few seconds. Further study is needed to investigate the variables involved in the immunoreactions.

Unlike the Thompson group's findings (23), Muramatsu et al. (44) reported a correlation between the amount of antibody in solution and the frequency shift. In this study protein A was immobilized onto the surface of an AT-cut, 9 MHz crystal precoated with (γ-aminopropyl)triethoxysilane. The crystal was positioned inside a thermostated flow cell, and measurement of the resonant frequency was performed in constantly flowing deionized and distilled water. The resonant frequency shift resulting from the affinity interaction of protein A and human IgG was linearly proportional to that of human IgG in the range 10^{-6}–10^{-2} mg/ml. The analysis pattern of mouse IgG subclasses obtained by using a stepped gradient of buffer solution identified three distinctive peaks of IgG_1, IgG_{2a}, and IgG_{2b}.

A flow-through piezobiosensor system was recently developed for the detection of glucose (24). Hexokinase was entrapped within a poly (acrylamide) matrix and placed onto the gold electrode of a 5 MHz AT-cut quartz crystal. The response of this biosensor system was roughly linear, with the glucose concentration in the range found in blood (approximately 7 mM). Also, the response time was quite short; that is, real-time detection is possible. Although the frequency changes were quite reproducible, they appeared to be too large to arise solely from the mass change due to the binding reaction between glucose and hexokinase. A possible explanation for this behavior is that the piezoelectric crystal senses changes in the viscoelastic properties of the poly (acrylamide)/enzyme film induced by the binding reaction. Since the frequency change is related to the viscoelastic properties of any medium present at this surface, the conformational changes in hexokinase induced by glucose binding may well be changing the viscoelastic properties of the poly (acrylamide) matrix, that is, causing the large decreases in Δf. Undoubtedly further experimentation is necessary to elucidate such a phenomenon.

Muramatsu and coworkers (45) proposed an interesting approach to monitor the concentration of endotoxin, a fever-inducing pyrogen. Endotoxin is a lipopolysaccharide containing lipid A in the center of the molecule, and it is widely found in water samples. The determination of endotoxin is very important for examination of medicinal products. The gelation of *Limulus* amebocyte lysate by endotoxin involves several enzyme reactions, which amplify the scale of the reaction at each stage and induce rheological change. Maramatsu et al. (45) used an AT-cut 9 MHz quartz crystal and an impedance analyzer to measure the electrical resistance of quartz crystals in contact with liquids at their resonant frequency:

$$R = (2\pi f d_L \eta_L)^{1/2} \frac{A}{k^2} \tag{10}$$

where R is the resistance, which is included as one of the parameters in the electrical equivalent circuit, A is the area of the quartz crystal, and k is the electromechanical coupling factor of the quartz crystal. The gelation time was computed by using an approximation to a polynomial equation from the resistance [Eq. (10)] or the resonant frequency change [Eq. (9)]. Both the gelation time and the R and f changes exhibited a linear correlation with endotoxin. Although the gelation time gives a lower detection limit (1 pg/ml), the R and f indices are considered better for determination of high concentrations of endotoxin. The detection limits obtained by measurement of R and f changes were 10 and 100 pg/ml, respectively.

In another study Muramatsu et al. (46) used a quartz crystal resonator to monitor the gelation time of fibrinogen on mixing with thrombin solution and to determine fibrinogen concentration. The effect of temperature was eliminated by addition of aluminum oxide particles to the thrombin solution. When an aluminum particle suspension was injected, the particles sedimented and accumulated on the bottom of the reaction chamber containing the piezocrystal, that is, on the quartz crystal surface, and induced a decrease in the resonant frequency. The frequency decrease was stopped when aluminum oxide particles become trapped in the gel network. The gelation time obtained showed a good linearity with fibrinogen concentration over the range 5–50 mg/ml.

A new apparatus consisting of two piezoelectric membranes was developed for determination of microbial populations of *Saccharomyces cerevisiae*, *Bacillus subtilis*, and *Klebsiella* sp. (47). The apparatus was placed in a medium containing microbial cells, and an arbitrary potential (5.1 V) was charged to the piezoelectric membrane by an oscillator. The ultrasonic waves generated were transmitted in the medium and vibrate the other piezoelectric membrane. The output voltage generated was measured with the ac voltmeter. The surface of the two piezoelectric membranes was coated by an epoxy resin adhesive (thickness $\simeq$ 1 mm) for electrical insulation. A linear relationship between the output voltage and cell population of *S. cerevisiae* was observed in the range of 10^6–10^8

cells per ml. The output voltage was reproducible, with an average relative error of 6%, when a medium containing 1.5×10^7 *S. cerevisiae* cells per ml was used for the experiments. Similarly, this system was applicable to the determination of bacterial cells of *B. subtillis* and *Klebsiella* sp., respectively. The continuous monitoring of the *S. cerevisiae* cell population during cultivation in the medium was also attempted by this piezoelectric membrane system. The values determined by the system, however, were slightly higher than those measured by hemacytometric and turbidimetric methods. The deviations were especially large in the exponential phase of microbial growth. The budding of *S. cerevisiae* and the increase in the conductivity of the medium were mentioned as the main causes of such deviations. The apparatus is sterilizable with an autoclave at 120°C for 10 min, and it can be reused five times. However, on the sixth reuse, the output voltage disappeared, probably because the electrical leakage caused by the small cracks on the surface of the adhesive layer stopped the vibration of the piezoelectric membrane. The stability or reusability of this apparatus is therefore not yet satisfactory and requires further improvements.

In a study using quartz crystals in conjunction with cyclic voltametry (48) the adsorption and desorption of IgG and anti-IgG at a silver electrode was evaluated as a function of the applied potential. Reversible frequency changes were observed, with a frequency decrease in the anodic region and the reverse in the cathodic region. The frequency decrease was associated with insoluble IgG complexes formed with the silver ions, which precipitate on the electrode during silver oxidation. A frequency shift of 10 Hz was reported to correspond to 6.1×10^{-8} g/cm^2. Although this work does not fall under the definition of that of a typical biosensor, it is included here because of the use of antigens and antibodies.

The combination of the two techniques, cyclic voltammetry with microgravimetry, demonstrates the potential applications and capabilities of quartz crystal in biotechnology applications. It is possible to monitor the progress of an antigen-antibody interaction as a function of time or antigen and/or antibody concentration by monitoring the frequency versus potential profile of a quartz crystal in intimate contact with the reactants at a solid/liquid interface. As the concentration of the antibody and/or antigen changes, accompanying frequency shifts are observed at specific IgG-IgG ratios. Piezoelectric crystals can thus be used reliably to study some of the fundamental properties of large biological molecules in solution and can be used quantitatively to study the adsorption and desorption processes at liquid/solid interfaces or the viscoelastic properties of some liquids.

GAS-PHASE ANALYSIS

Coated piezoelectric crystal detectors have become of increasing interest for the detection of traces of toxic atmospheric pollutants. By coating the surface of crystal with a suitable biological component that selectively adsorbs a particular gas, it is possible to determine the concentration of that gas quantitatively. For each application there is an optimum weight of coating biological material that gives an optimum sensitivity response to the analyte to be detected, and this level must be established a priori.

In gas-phase applications the coating procedure is simply done by dropping the desired biological solution onto the crystal's surface and then air drying the crystal before measurement of its resonant frequency. The coated crystal is mounted in a ceramic holder and then placed in an oscillator circuit, and the resonant frequency is measured using conventional electronic techniques as described earlier. A reference crystal is also used to correct for temperature fluctuations and differences in pressure or gas flow rate. A typical experimental setup used for the evaluation of piezoelectric detectors is shown schematically in Figure 4. The carrier gas can be supplied from compressed cylinders or, for air, by a vibrating diaphragm pump. The carrier gas must be partially humidified before entering the detection chamber because the biologically active coatings require a certain amount of moisture to behave as they normally would in solution. Beyond a certain relative humidity, however, condensation on the crystal occurs, leading to an overloaded crystal, which ceases to vibrate. Therefore in gas-phase analysis the amount of humidity must be carefully controlled, and this level depends on both the coating and the oscillator. The optimum flow rate of the carrier gas must be determined for each coating and application and is usually about 120 ml/min (49–51).

Unlike liquid-phase analysis, analytical applications of the piezoelectric crystal gas-phase biosensor using biologically active materials are somewhat limited. Guilbault, a principal contributor in this field, identified detectors for the determination of formaldehyde (49), cocaine (50), and parathion (51) (Table 3).

The use of polymeric bilayer-forming amphiphiles as coating materials was reported for the detection of various odorants and perfumes by Okahata and Shimizu (53). AT-cut, 9 MHz crystals were prepared by casting 0.47 μm thick films of a synthetic multibilayer films of amphiphilic lipids on both sides of the crystal. Upon exposure to various odorants or perfumes in the gas phase, the frequency shift was proportional to the odorant or perfume concentration. The authors also reported preliminary data and partition coefficients of various odorants obtained by determining frequency shifts on exposure to the crystal. This work was done in closed containers in a vapor-saturated atmosphere. The frequency decrease was reversed by simple exposure to the atmosphere.

A fragrance meter was recently introduced to the market by Sogo Pharmaceutical Co. of Japan. The equipment consists of a sensor for detecting odor using a disk-shaped crystal oscillator of about 8 mm in diameter. A synthetic bilayer film functioning exactly like the bimolecular lipids overlays the quartz oscillator, producing vibrations 9 million times per second and adsorbing odor substances free-floating in the air. Human olfactory cells have membranes made up of bimolecular lipids where the "odor" ingredients are adsorbed. This human smell structure and organ is now artificially re-created by Sogo Pharmaceutical

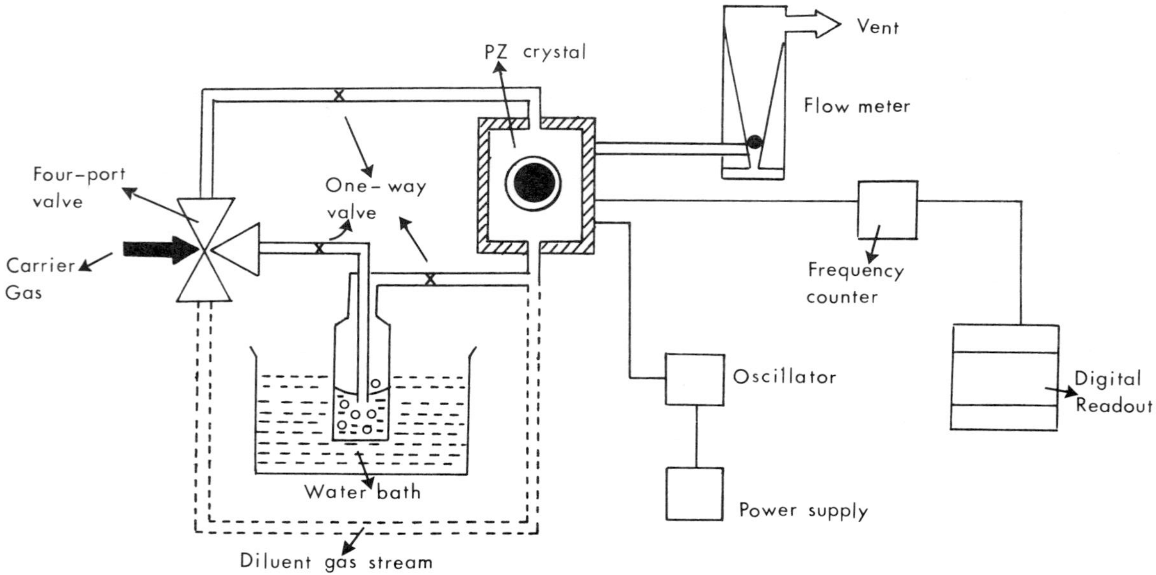

Figure 4 A typical setup for a gas-phase piezocrystal detector.

Table 3 Gas-Phase Applications of Piezoelectric Crystal Biosensor

Analyte	Coating material	Performance of the coated piezobiosensor	Reference
Formaldehyde	Formaldehyde dehydrogenase	The coated crystal is good for 3 days and/or 100 analyses; linear response from 10 ppb to 10 ppm with frequency changes from 2 to 500 Hz; it is fairly specific for formaldehyde; little response to acetaldehyde, propionaldehyde, and ethanol; not responsive to methanol and benzaldehyde at concentrations of 1–10 ppm	49
Cocaine	Anti-benzoyl ecgonine antibody	Sensitivity of the coated crystal is 50 Hz/ppb benzoyl hydrochloride; response is flat in the 10–50% relative humidity (RH) range, but overloading is observed at 52–55% RH; the response and recovery times were 30 sec; the coated crystal is good for 7 days; 84% of sensitivity is maintained after 7 days, with a gradual decline to 36% after 13 days	50
Parathion	Antiparathion antibody	The response time is 1–2 min, the recovery time is 2–5 min; the frequency response is linear in the ppb to ppm range; 36 ppb parathion is required to produce a frequency change of 400 Hz; to obtain this change, however, 106 ppb malathion or 158 ppb methylparathion is required; the reproducibility is 6%; the coated piezocrystal has a lifetime of 18 days	51
Parathion	Antiparathion antibody	Sensitivity of the crystal is 155 Hz/μg; also responsive to malathion (372.7 Hz/μg), disulfoton (443 Hz/μg), valproic acid (3.91 Hz/μg), *o*-nitrotoluene (2.69 Hz/μg), and toluene (0.02 Hz/μg); the response is not much better than that of the crystal coated with IgG (211 Hz/μg), bovine serum albumin (255 Hz/μg), or valproic acid and antiserum (224 Hz/μg)	52

with its newly introduced "fragrance meter." When the odor substances are adsorbed the weight of the bilayer-coated crystal increases and the oscillation rate decreases. By measuring the rate of this decrease the quality and volume of odor substances can be obtained with a fine degree of accuracy. Sogo Pharmaceutical further claims that the device registers figures extremely close to the accuracy of professional fragrance testers who judge complex odors by their sense of smell.

Preliminary studies utilizing biological materials, such as enzymes and antibodies, have proven successful as coatings for the piezoelectric crystal biosensor. Excellent reproducibilities, coating lifetimes, response times, and selectivities are observed. This technique is potentially useful and convenient in the area of environmental analysis and is an alternative to radioimmunoassay. However, a number of unknown parameters must be determined before such devices can be commercialized. The nature of antigen-antibody interactions in the gas phase must be investigated more carefully since the degree of avidity and sensitivity in solution appear to be significantly higher than the value obtained in the gas phase. Although much is known about antigen-antibody behavior in solution and physiological media, little or nothing is known about this behavior in the gas phase. Binding affinities, association and dissociation rates and constants, and antigen-antibody ratios needed for detectable reactions must be determined.

To exploit the piezoelectric effect and the Sauerbrey equation, both the amount of coating and the even distribution of this material on the crystal must be reproducible. Rajakovic and coworkers (52) used scanning electron microscopy to show that the coating layer possesses an extremely irregular form of densely packed fibrous solid that incorporates a large number of cavities and channels. Obviously, mass transport of gas for interaction with the protein surface is significantly hindered by this structure. The coating process by simple deposition of the biological material on the crystal surface is neither reliable nor satisfactory. The reproducible immobilizations of the desired materials on the surface must be further developed.

It is an open question whether gas-to-solid phase immunochemistry can be expected since the antibody-coated crystal exhibits no selectivity in terms of immunochemical reactions. As experimentally demonstrated (52) the performance of the parathion antibody-coated crystal in terms of frequency response and sensitivity to parathion was almost similar to that of the crystal coated with valproic acid antiserum, IgG, and even bovine serum albumin, respectively. In addition, the parathion antibody-coated crystal was responsive to parathion, malathion, disulfoton, toluene, *o*-nitrotoluene, and valproic acid. Similarly, the behavior of the valproic acid antiserum-coated crystal to valproic acid was identical to that of the crystal coated with parathion antibody, IgG, and bovine serum albumin, respectively. Such data clearly demonstrate that nonspecific chemisorption to antibody must be taken into account in any evaluation of the antibody-based piezoelectric response to antigenic species. It is indeed a difficult task to distinguish the two types of interaction.

CONCLUSIONS

The future use of biologically active molecules as a coating material on the surface of quartz crystals appears very promising. Biological materials, like antibodies and enzymes, that are very selective and specific can enhance the selectivity of piezoelectric crystals manyfold. Antigen-antibody interactions, in theory, are very selective and specific for only one substance, even in the presence of large amounts of similar molecules. Piezoelectric crystal biosensors have been suggested as possible detectors for the assay of substrates in both liquid phase and gas phase and for the study of antigen-antibody interaction at liquid/surface interfaces. The device can be easily automated or combined with a flow injection system. This gives the capability of repeated and continuous assays and raises the exciting possibility of using different crystal assays in series to assay for different contaminants in complex samples with an on-line display of the results. The design and performance of the experimentation is fairly simple, with very little time for the data to be acquired compared to any analytical technique. The total absence of sample preparation is another distinctive advantage of the piezoelectric crystal biosensor. This is probably the reason that so many potential applications of this technique have been suggested in environmental analysis to analyze hazardous materials. Last but not least, a piezoelectric sensor (crystals plus circuitry) is relatively inexpensive: a few dollars for a PZ crystal and under a few thousand dollars for electronics. Many potential applications for piezoelectric detection other than those mentioned can be envisioned. For instance, receptors for many physiologically active substances, such as hormones, drugs, and toxins, have recently been isolated and characterized. The piezobiosensor system should be an ideal technique for assaying receptor activity, since the interaction of the receptor with its ligand would be studied directly without the use of intermediate assays or labeling techniques. However, like any new area of science piezobiosensors still have a long way to go before they can truly become a routine analytical tool. Despite the very promising results reported in the literature, a number of unknown parameters and/or shortcomings must be determined and/or overcome.

1. To establish a general correlation between the frequency shift and the analyte concentration: In most of the articles cited in this review, the Sauerbrey equation is used as the basis of the frequency shift measurements, but none of the researchers reported any agreement in the frequency shifts predicted by the Sauerbrey equation and those obtained in the experimental results. This is probably a result of the inability to reproduce monolayer coatings and because the added mass on the crystal may deform during oscillation. For piezoelectric quartz crystals to function properly both the amount of coating and the distribu-

tion on the crystal must be reproducible. The effect of the orientation of coated materials on the performance of the piezobiosensor has yet to be determined. For thicker, less rigid solids a more complex theory is required. The use of Z-match theory, which considers the acoustic impedance of a layer attached to the oscillator, makes the quartz crystal microbalance more useful (54). However, the density and shear modulus of the attached layer must be known.

2. The problem of cross-reactivity and interference is particularly troubling, especially in the gas phase and in the liquid phase. This problem has been reduced significantly with the use of reference crystals but not totally eliminated. The greatest limitation to this technique for liquid-phase applications is poor sensitivity and limits of detection due primarily to nonspecific adsorption occurring at both detector and reference crystal surfaces. It will be necessary to improve detection limits of three orders of magnitude or more to demonstrate the clinical significance of piezoimmunosensors. This degree of improvement may be achievable through the deactivation of nonspecific adsorption sites, the use of more sensitive crystals, and more extensive surface coverage with immobilized material.

3. Another critical point for consideration is the molecular size of the analyte to be assayed by the PZ biosensor. The system appears to work very well for the assay of large molecules. However, attempts to detect small molecules in the liquid phase have not been successful. Although Lasky and Buttry (24) recently reported a piezoelectric biosensor for glucose detection in the liquid phase, the frequency shift is too large to arise solely from the mass change during the binding reaction between glucose and hexokinase immobilized on the crystal.

4. None of the liquid-phase biosensors were capable of multiple assays since the bound analyte cannot be easily removed from the crystal surfaces. The antigen-antibody interaction in liquid is strong and irreversible so that the bound antigen or antibody can only be removed from the crystal surface with very high ionic strength solutions. Under such circumstances the immobilized material gradually breaks away from the surface with repeated use so that the sensitivity of a given crystal decreases with age. This problem may not be important if small quartz piezoelectric crystals can be manufactured in quantity at very low cost so that disposable transducers are economically feasible. At present each crystal costs a few dollars.

5. In the gas phase the amount of humidity and temperature must be carefully controlled since the biologically active materials require a certain amount of moisture and a very narrow optimal temperature to function properly. These requirements and stringent conditions have made this technique less attractive to some practical applications. Since piezoelectric crystals can vibrate in a liquid phase, a little condensation on the crystal in the gas phase is tolerable. Unfortunately, if the crystal continues to vibrate in the presence of condensation the frequency shifts obtained are difficult to define and correlate with the analyte concentration.

6. The change in oscillation frequency on immersing crystals in solutions must be better understood. So far it is known only that the frequency shift depends principally on the density and viscosity of the solution. Further, the resonant frequency also decreases with increasing electrolyte concentration and specific conductivity, largely through their proportionality with solution density. Some researchers report a linear relationship between the frequency shift and the amount of material adsorbed on the surface of the crystal, but others insist that the frequency shift exhibits no association with the classic microgravimetric signal. Both the elasticity (shear modulus) and the viscosity of the medium adjacent to the crystal electrodes can influence the nature of the frequency response. A quantitative treatment of the influence of viscoelasticity is not yet available, but some strategies that allow an evaluation of its importance in a given system have been proposed (55). As discussed earlier, the frequency response is not very specific; therefore, the interfacial chemistry must be adjusted to give specific and reproducible measurements.

7. The determination of antibody or antigen in solution using solid-phase radioimmunoassays or enzyme immunoassays involves reaction of a solid support with an appropriate antigen so that antigen is immobilized on this solid support. This method can detect picomolar levels of antibody present and provides information about the size of the antibody and other molecular properties of interest. However, it remains to be seen if this approach is equally applicable to the detection of the target analyte in either liquid phase or gas phase by immobilizing an appropriate enzyme or antibody on the crystal surface. Of course it is somewhat difficult to make a direct comparison of piezoelectric crystal antigen-antibody responses with those of solid-phase radioimmunoassays or enzyme immunoassays because these latter techniques are fundamentally better controlled than piezoelectric crystal immunoassays. Although the binding of antibodies to antigens immobilized on a solid support is well characterized (56), the nature of the antigen-antibody interaction on the crystal is still unknown. This is a crucial research area that requires immediate attention before piezoelectric crystal biosensors can become practical.

Despite these drawbacks or obstacles to be overcome, many applications should be easily accessible with the piezoelectric biosensor since it offers relatively low cost and experimental simplicity. As piezoelectric detection systems are optimized their potential uses in both liquid phase and gas phase will be realized.

REFERENCES

1. Curie, J., and Curie, P. (1880). Développement, par pression, de l'électricité polarisée dans les cristaux hémiedriés et faces inclinées, *Comp. Rendu, 91*: 294.
2. King, W. H. Jr. (1964). Analytical uses of the piezoelectric crystal, *Anal. Chem., 36*: 1735.

3. Hlavay, J., Guilbault, G. G. (1977). Applications of the piezoelectric crystal detector in analytical chemistry, *Anal. Chem.*, *49*: 1890.
4. Guilbault, G. G., and Jordan, J. M. (1988). Analytical uses of piezoelectric crystals: A review, *CRC Crit. Rev. Anal. Chem.*, *19*: 1.
5. Guilbault, G. G. (1980) Uses of the piezoelectric crystal detector in analytical chemistry, *Ion Selective Electrode Rev.*, *2*: 4.
6. Clark, L. C., and Lyons, C. (1962). Electrode systems for continuous monitoring in cardiovascular surgery, *Ann. NY Acad. Sci.*, *102*: 29.
7. Shons, A., Dorman, F., and Najarian, J. (1972). An immunospecific microbalance, *J. Biomed. Mater. Res.*, *6*: 565.
8. Sauerbrey, G. Z. (1959). Use of quartz vibrator for weighing thin films on a microbalance, *Z. Phys.*, *155*: 206.
9. Guilbault, G. G., and Luong, J. H. T. (1988). Gas phase biosensors, *J. Biotechnol.*, *9*: 1.
10. Muramatsu, H., Kajiwara, K., Tamiya, E., and Karube, I. (1986). Piezoelectric immunosensor for the detection of *Candida albicans* microbes, *Anal. Chim. Acta*, *188*: 257.
11. Karasek, F. W., Guy, P., Hill, H. H., and Tiernay, J. M. (1976). Chromatographic design and temperature-related characteristics of the piezoelectric detector, *J. Chromatogr.*, *124*: 179.
12. Janghorbani, M., and Freund, H. (1973). Application of a piezoelectric quartz crystal as a partition detector, *Anal. Chem.*, *45*: 325.
13. Daley, P. S., and Lundgren, D. A. (1975). The performance of piezoelectric crystal sensors used to determine aerosol mass concentrations, *Am. Ind. Hyg. Assoc. J.*, *36*: 518.
14. Sem, G. J., and Tsurubayashi, K. (1975). A new mass sensor for respirable dust measurement, *Am. Ind. Hyg. Assoc. J.*, *36*: 791.
15. Chuan, R. L. (1975). Rapid assessment of particulate mass concentration in the atmosphere with a piezoelectric, *Adv. Inst.*, *30* (2): 620.
16. Scheide, E. P., and Taylor, J. K. (1974). A piezoelectric crystal detector for mercury in air, *Environ. Sci. Technol.*, *8*: 1087.
17. Fog, H. M., and Rietz, B. (1975). Piezoelectric crystal detector for the monitoring of ozone in working environments, *Anal. Chem.*, *57*: 634.
18. Turnham, B. D., Yee, L. K., and Luoma, G. A. (1985). Coated piezoelectric quartz crystal monitor for determination of propylene glycol dinitrate vapor levels, *Anal. Chem.*, *57*: 2120.
19. Prusak-Sochaczewski, E., Luong, J. H. T., and Guilbault, G. G. (1990). Development of a piezoelectric immunosensor for the detection of *Salmonella typhimurium*, *Enzyme Microb. Technol.*, *12*: 173.
20. Davis, K. A., and Leary, T. R. (1989). Continuous liquid-phase piezoelectric biosensor for kinetic immunoassays, *Anal. Chem. 61*: 1227.
21. Fawcett, N. C., Evans, J. A., Chien, L. C., and Flowers, N. (1988). Nucleic acid hybridization detected by piezoelectric resonance, *Anal. Lett.*, *21*: 1099.
22. Rice, T. K. (1980). Method for the assay of classes of antigen-specific antibodies, U.S. Patent 4,236,893.
23. Thompson, M., Arthur, C. L., and Dhaliwal, G. K. (1986). Liquid-phase piezo-

electric and acoustic transmission studies of interfacial immunochemistry, *Anal. Chem., 58*: 1206.
24. Lasky, S. J., and Buttry, D. A. (1989). Sensors based on biomolecules immobilized on the piezoelectric quartz crystal microbalance, *ACS Symp. Ser. 403*: 237–241.
25. Prusak-Sochaczewski, E., and Luong, J. H. T. (1990). Detection of human transferrin by the piezoelectric crystal, *Anal. Lett., 23* (2): 183.
26. Prusak-Sochaczewski, E., and Luong, J. H. T. (1990). A new approach to the development of a reusable piezoelectric crystal biosensor, *Anal. Lett.*, 23(3):401
27. Oliveira, R. J., and Silver, S. F. (1980). Immunoassay with coated piezoelectric crystals, U.S. Patent 4,242,096.
28. Rice, T. K. (1982). Sandwich immunoassay using piezoelectric oscillator, U.S. Patent 4,314,821.
29. Fawcett, N. C. (1987). Piezoelectric device for detection of polynucleotide hybridization, World Intellectual Property Organization, WO 87/02066.
30. Guilbault, G. G., Luong, J. H. T., and Prusak-Sochaczewski, E. (1989). Immobilization methods for piezoelectric crystal biosensors, *Bio/technology, 1*: 349.
31. Nomura, T., and Minemura, A. (1980). Behavior of a piezoelectric quartz crystal in an aqueous solution and the application to the determination of minute amount of cyanide, *Nippon Kagaku Kaishi, 1980*: 1261.
32. Konash, P. L., and Bastiaans, G. J. (1981). Piezoelectric crystal detectors in liquid chromatography, *Anal. Chem., 59*: 1921.
33. Nomura, T., and Okuhara, M. (1982). Frequency shifts of piezoelectric quartz crystals immersed in organic liquids, *Anal. Chim. Acta, 142*: 281.
34. Bruckenstein S., and Shay, M. (1985). Experimental aspects of use of the quartz crystal microbalance in solution, *Electrochim. Acta, 30*: 1295.
35. Yao, S.-Z., and Zhou, T.-A. (1988). Dependence of the oscillation frequency of a piezoelectric crystal on the physical parameters of liquids, *Anal. Chim. Acta, 212*: 61.
36. Kanazawa, K. K., and Gordon, J. G. (1985). The oscillation frequency of a quartz resonator in contact with a liquid, *Anal. Chim. Acta, 175*: 99.
37. Nomura, T., and Mimatsu, T. (1982). Electrolyptic determination of traces of iodine in solution with a piezoelectric quartz crystal, *Anal. Chim. Acta, 143*: 237.
38. Nomura, T., and Iijima, M. (1981). Electrolytic determination of nanomolar concentrations of silver in solution with a piezoelectric quartz crystal, *Anal. Chim. Acta, 131*: 97.
39. Nomura, T., and Maruyama, M. (1983). Effect of metal ions on a piezoelectric quartz crystal in aqueous solution and the adsorptive determination of iron (III) as phosphate, *Anal. Chim. Acta, 147*: 365.
40. Nomura, T., and Nagamune, T. (1983). Internal electrolytic determination of silver in solution with a piezoelectric quartz crystal, *Anal. Chim. Acta, 155*: 231.
41. Yao, S.-Z, and Mo, Z.-H. (1987). Frequency properties of a piezoelectric quartz crystal in solutions and application to total salt determination, *Anal. Chim. Acta, 193*: 97.
42. Roederer, J. E., and Bastiaans, G. J. (1983). Microgravimetric immunoassay with piezoelectric crystals, *Anal. Chem., 55*: 2333.
43. Thompson, M., Dhaliwal, G. K., Arthur, C. L., and Calabrese, G. S. (1987). The

potential of the bulk acoustic wave device as liquid-phase immunosensor, *IEEE Trans. Ultrasonics, Ferroelectronics Frequency Control*, UFFC-34, *2*: 127.
44. Muramatsu, H., Dicks, J. M., Tamiya, E., and Karube, I. (1987). Piezoelectric crystal biosensor modified with protein A for determination of immunoglobulins, *Anal. Chem., 59*: 2760.
45. Muramatsu, H., Tamiya, E., Suzuki, M., and Karube, I. (1988). Viscosity monitoring with a piezoelectric quartz crystal and its application to the determination of endotoxin by gelation of Limulus amebocyte lysate, *Anal. Chim. Acta, 215*: 91.
46. Muramatsu, H., Tamiya, E., Suzuki, M., and Karube, I. (1989). Quartz crystal gelation detector for the determination of fibrinogen concentration, *Anal. Chim. Acta, 217*: 321.
47. Ishimori, Y., Karube, I., and Suzuki, S. (1981). Determination of microbial populations with piezoelectric membranes, *Appl. Environ. Microbiol., 42*: 632.
48. Grabbe, E. S., Buck, R. P., and Melroy, O. R. (1987). Cyclic voltametry and quartz microbalance electrogravimetry of IgG and anti-IgG reactions on silver, *J. Electroanal. Chem., 223*(1-2): 67.
49. Guilbault, G. G. (1984). Determination of formaldehyde with an enzyme-coated piezoelectric crystal detector, *Anal. Chem., 55*: 1682.
50. Hahn, E. E. (1988). Piezoelectric crystal detectors and their applications, Ph.D. Dissertation, University of New Orleans, New Orleans, Louisiana.
51. Ngeh-Ngwainbi, J., Foley, P. H., Kuan, S. S., and Guilbault, G. G. (1986). Parathion antibodies on piezoelectric crystals, *J. Am. Chem. Soc., 108*: 5444.
52. Rajakovic, L., Ghaemmaghami, V., and Thompson, M. (1989). Adsorption on film-free and antibody-coated piezoelectric sensors, *Anal. Chim. Acta, 217*: 111.
53. Okahata, Y., and Shimizu, O. (1987). Olfactory reception on a multibilayer-coated piezoelectric crystal in a gas phase, *Langmuir, 3*: 1171.
54. Lu, C., and Lewis, O. (1972). Investigation of film-thickness determination by oscillating quartz resonators with large mass load, *J. Appl. Phys., 43*: 4385.
55. Deakin, M. R., and Buttry, D. A. (1989). Electrochemical applications of the quartz crystal microbalance, *Anal. Chem., 61*: 1147.
56. Pincus, M. R., and Rendell, M. (1981). General quantitative treatment for the binding of divalent antibodies to antigens immobilized on a solid phase, *Proc. Natl. Acad. Sci. USA, 78*: 5924.

7

FET-Based Biosensors

Toshihide Kuriyama and Jun Kimura

NEC Corporation
Kanagawa, Japan

PRINCIPLE OF FET-BASED BIOSENSORS

A biosensor is composed of a receptor, which uses such biological substances as enzymes, antibodies, and microbes, and a transducer, which transforms physical or chemical changes accompanying the biological reaction into an electrical signal. As a transducer, a field effect transistor (FET) whose conductance is modulated by the gate voltage through the field effect of a semiconductor can be used. In the FET conductance is also affected by surface voltage, surface charge, and work function other than the gate voltage (electrical signal). A biosensor can be obtained by a combination of FET and biological material, with conductance changed by the biological reaction. Table 1 shows the FET factors affected and the type of FET used for biosensor design.

ISFET

An ISFET (ion-sensitive field effect transistor) selectively measures ion activity in an electrolyte (1,2). It works as an ISE (ion-selective electrode) with low-output impedance. By combining the ISFET with a membrane that contains a biological material, such as an enzyme or a microbe, a biosensor able to measure a specific organic substance can be obtained. Among the ISFET-based biosensors, research on ISFET with an immobilized enzyme membrane has been the most developed, with the ion concentration change accompanying the enzymatic reaction detected by the ISFET. For example, a urea ISFET sensor uses a

Table 1 FET Factors Affected by Biological Reactions

Factor	FET type	Chemicals	Reaction
Surface voltage	ISFET	pH, ion	Enzyme reaction Microbe metabolism
Work function	Pd-MOSFET	H_2, NH_3 gas	Enzyme reaction
Surface charge	Immuno-FET	Excess charge	Immunoreaction

reaction in which urea is hydrolyzed in the urease-immobilized membrane. When urea enters the membrane, the following reaction occurs:

$$H_2N-CO-NH_2 + 2H_2O + H^+ \longrightarrow 2NH_4^+ + HCO_3^-$$

In this reaction the following detection methods by ISFET can be used:

1. According to the reaction, the pH value in the membrane increases. This pH change at the ISFET surface is detected by a pH-sensitive ISFET.
2. The ammonium ion concentration also changes and can be detected by an ammonium-sensitive ISFET.

Pd-MOSFET

The Pt/Pd MOSFET and Ir/Pd MOSFET have a catalytic gate electrode and work as ammonia gas sensors (3). Ammonia gas is decomposed catalytically on the Pt or Ir and creates changed reaction intermediates, which give rise to a surface potential change. By combining the device with biological inducing ammonia concentration changes, a biosensor is obtained.

Immuno-FET

An immuno-FET (immunochemical FET) detects a charge excess at the FET surface due to an immunochemical reaction (4): antibodies are immobilized at the FET gate region surface, and antigens present in the sample solution form an antigen-antibody complex inducing the excess of charge. In the following sections, ISFET-based biosensors are described in detail.

ISFET-BASED BIOSENSOR

ISFET Principle

Figure 1 shows the structure of an *n*-channel MOSFET (metal oxide semiconductor field effect transistor) commonly used in an IC (integrated circuit). In the *n*-channel MOSFET electrons move from the n^+ source to n^+ drain doped regions, which are produced by doping impurities into a *p*-type silicon. There is a gate electrode over a channel region between the source and the drain. A thin

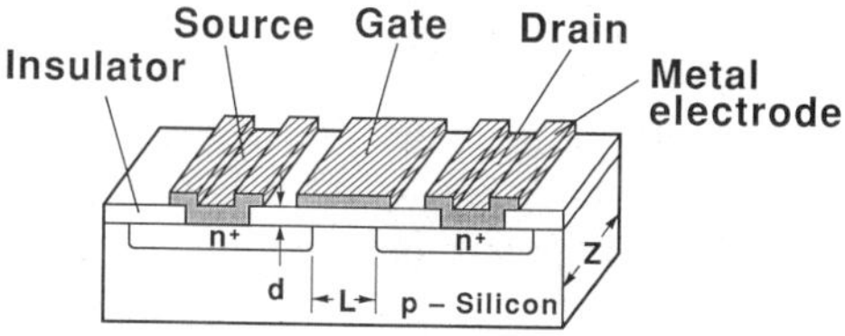

Figure 1 *n*-Channel MOSFET structure.

insulator layer, such as SiO_2, exists between the gate electrode and the silicon. Channel length values L range from 0.5 to several hundred μm; channel width values Z range from a few μm to several cm. The insulating layer thickness values range from 10 nm to 1 μm. By applying a positive voltage on the gate electrode against the *p*-type silicon, electrons in excess are generated at the SiO_2/Si interface and change the conductance between the source and the drain.

The ISFET principle was first reported in 1970 by Bergveld (5,6) and improved by Matsuo and Wise (7). The fundamental ISFET structure is shown in Figure 2. In the ISFET, instead of the gate electrode used in the MOSFET, an ion-sensitive membrane is formed on the channel region. Gate voltage is applied by a reference electrode, such as an Ag/AgCl electrode or a saturated calomel electrode, which is used to fix the solution potential.

A current, which flows between the source and the drain of the ISFET, is given by an equation similar to the equation employed for the MOSFET (8):

$$I = \frac{Z\mu}{L} C_i \frac{(V_g - V_t)V_d^2}{2} \qquad \text{for } V_d < V_g - V_t \tag{1}$$

$$I = \frac{Z\mu}{2L} C_i(V_g - V_t)^2 \qquad \text{for } V_d \geq V_g - V_t \tag{2}$$

$$V_t = V_{t_0} - V_s \tag{3}$$

where Z is the gate width, L the gate length, μ the carrier mobility, C_i the insulating layer capacitance, V_g the liquid potential, V_d the drain voltage, and V_t

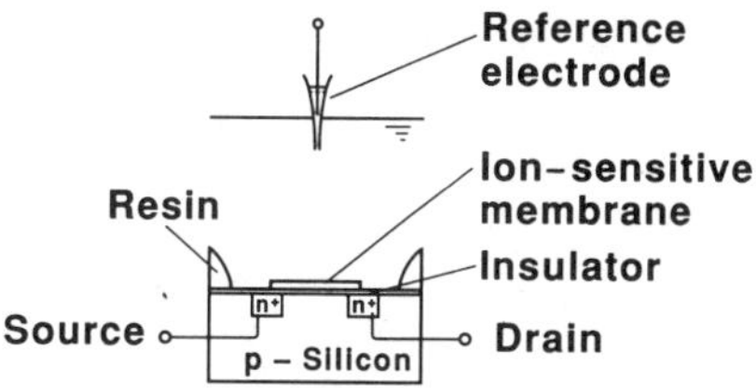

Figure 2 Fundamental ISFET structure.

the threshold voltage. V_{t_0} is the threshold voltage when V_s, the surface voltage generated at the surface of the ion-sensitive membrane, is zero.

Ion-sensitive membranes, which detect pH or hydrogen ion activity, are silicon nitride (Si_3N_4), alumina (Al_2O_3), zirconium oxide (ZrO_2), borosilicate glass, and tantalum oxide (Ta_2O_5) (2,9,10,14). These membranes also work as a protection layer against ion penetration. Surface voltage V_s at the ion-sensitive membrane is changed by pH. The value was calculated by a site binding model, and the response is almost Nernstian (59.2 mV/pH unit at 25°C) (10).

Other types of ion sensors are also obtained by combining an ion-sensitive membrane with the ISFET. ISFET that can detect Na^+, K^+, Ca^{2+}, NH_4^+, Ag^+, Cl^-, and Br^- were reported. A CO_2 gas sensor was made by combining a pH-sensitive ISFET with a CO_2 gas-permeable membrane.

The ISFET has several advantages over the conventional ISE:

1. The ISFET can be mass produced by Si IC technology, which makes it very small and cheap.
2. Multisensor capability is provided by a combination of various functional membranes on ISFET in a sensor chip.
3. Dielectric material can be used for an ion-sensitive membrane.
4. Output impedance is low, and pH response is fast.
5. Uniform sensor characteristics can be obtained by applying the Si IC process for fabricating sensing membranes, such as enzyme-immobilized membranes.

It is very important to insulate the ISFET from a solution because a very small current leak between Si and solution causes ISFET trouble. Several insulation technologies have been reported. These technologies are shown in Table 2.

Figure 3 shows an example of ISFET that uses an SOS (silicon on sapphire) wafer. An ISFET is made in a silicon island on the sapphire substrate. The ISFET surface consists of a thermally oxidized silicon (SiO_2) and a chemical vapor-deposited silicon nitride (Si_3N_4). The Si_3N_4 works as a pH-sensitive membrane and as a protection membrane against ion penetration. The SOS/ISFET is easily isolated from the other FET (ISFET and MOSFET) and from the solution being measured because sapphire is a good insulating material. When sensor chips are extracted from a wafer by cutting, there is no need to insulate the cut edges. Only bonding wires must be insulated. Thus its inherent electrical insulation property makes the SOS/ISFET suitable for the design of biosensors used for solutions.

ISFET Fabrication (16,17)

The ISFET is fabricated by a Si IC process. Fabrication processes for the SOS/ISFET are shown in Figure 4.

Table 2 ISFET Structure

	Insulating method	Characteristics	Reference
Resin coating	Cover bare silicon region with resin		20
Film resist	Cover with patterned film resist	Micropool	51
Needle type	Anisotropic etching of silicon wafer to obtain a Si needle, which is covered with insulation layers	Needle shape, very small sample can be measured	11 12 13
SOS[a]	ISFET is made in a silicon island on a sapphire wafer	Compatible with normal IC process, chemically stable, complete insulation	10 15 18
p-n Junction	Isolation by *p-n* junction	Compatible with normal Si process	14

[a]Silicon on sapphire.

Si Island Formation (Fig. 4A)

A SOS wafer is used that has a 0.6 μm thick silicon layer. The silicon layer is undoped and has a (100) surface orientation. The silicon layer is patterned onto an island by etching.

FET Fabrication (Fig. 4B)

Boron is doped into silicon by ion implantation. Phosphorus is implanted into the FET source and drain regions. A 100 nm thick SiO_2 layer is made by thermal oxidation of silicon. A small amount of phosphorus is also implanted on the gate region for threshold voltage control.

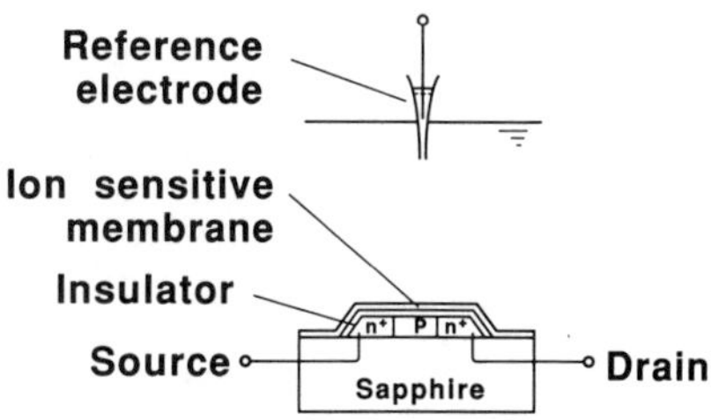

Figure 3 SOS/ISFET structure.

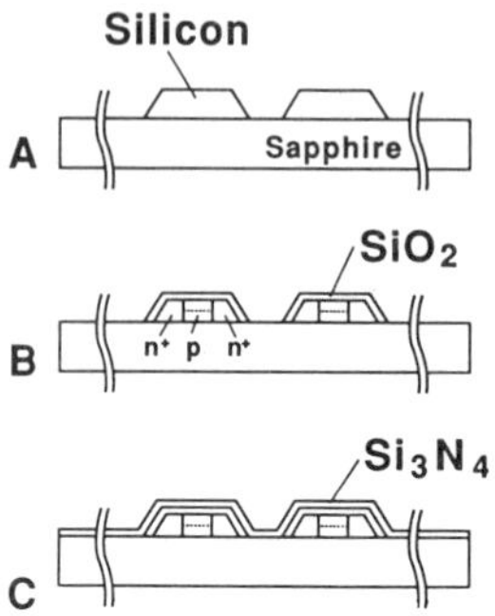

Figure 4 SOS/ISFET fabrication process.

Si_3N_4 Layer Deposition (Fig. 4C)

A 100 nm thick silicon nitride is made by low-pressure chemical vapor deposition. This layer works as a pH-sensitive membrane.

Immobilized Enzyme Membrane Fabrication

In the early stages immobilized enzyme membranes were fabricated manually. A tip of the ISFET was immersed into an enzyme solution, or an enzyme solution was dropped onto the ISFET sensing region. Later, by taking advantage of the ISFET planar structure, immobilized enzyme membranes were fabricated with good controllability by the methods used in IC fabrication. The membranes were deposited on an entire wafer by spin coating. By controlling the membrane material, viscosity, and spinning speed, an enzyme membrane with uniform thickness can be obtained. Enzyme membranes could be patterned to cover each ISFET sensing region on a wafer by using photolithographic techniques, such as photo–cross-linking or lift-off methods. An ink jet nozzle method was also applied to fabricate the enzyme membrane, making it possible to efficiently use expensive enzymes. Adhesion of immobilized enzyme membrane to the ISFET surface is an important factor to be controlled to obtain a stable biosensor. Coupling reagent, such as 3-aminopropyltriethoxysilane, was successfully used for surface treatment to improve membrane adhesion. Immobilized enzyme membrane fabrication methods are listed in Table 3. Immobilized enzyme membrane fabrication processes by IC technology are described later.

Ultraviolet (UV) Light Irradiation Method (17)

This method consists of two steps.

Immobilized Urease Membrane Fabrication (Fig. 5A). After treatment of the ISFET wafer in boiling toluene containing 5% 3-aminopropyltriethoxysilane for 24 hr an acetylcellulose layer containing glutaraldehyde and 1,8-diamino-4-

Table 3 Enzyme Membrane Fabrication Methods

Deposition	Characteristics	Patterning	References
Drop			20–22
Drop into micropool		Film photoresist	19, 24
Dip coat			23
Spin coat	Good control and uniform thickness in mass production	UV irradiation Photo–cross-link Life-off	15–17 25–27 28, 29
Ink jet nozzle	Small amount of enzyme multibiosensor for custom use		28, 30

aminomethyloctane is spin coated on the wafer. The wafer is washed with deionized water and immersed into a 1% glutaraldehyde solution and afterward in a urease solution. The immobilized urease membrane is about 10 μm thick and is slightly red because of Schiff coupling between the aldehyde radical of glutaraldehyde and 1,8-diamino-4-aminomethyloctane.

Immobilized Urease Membrane with Partial Inactivation (Fig. 5B). Immobilized urease is inactivated by ultraviolet light irradiation (wavelength 200–280 nm), except at the gate region for the urea-sensitive (ISFET). A quartz photomask is used for the patterning. Using this process an ISFET with an active urease membrane and an ISFET with an inactivated urease membrane were fabricated on the same chip.

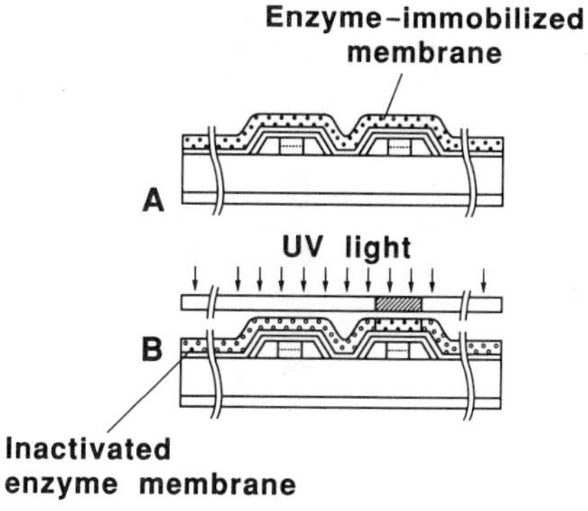

Figure 5 Membrane deposition method: UV light inactivation.

Photo–Cross-Linking Method (27)

This method consists of three steps.

Drop and Spin Coating of Enzyme Membrane. Glucose oxidase, bovine serum albumin, and photopolymer solutions are placed dropwise and spun at 2000 rpm for 2 min.

Ultraviolet Light Irradiation. The enzyme membrane around the gate surface is exposed to ultraviolet light irradiation through a UV filter for 5 sec.

Development and Fixation. The ISFET is immersed in aqueous glutaraldehyde solution for 5 min at room temperature to form a glucose oxidase membrane. After washing in water a cross-linking reaction by glutaraldehyde is followed by immersion into a glycine solution for 15 min for active site deactivation.

Lift-off Method (28)

Spin Coat and Patterning Positive Photoresist (Fig. 6A). A positive photoresist (Microposit 1300-27, Shipley Co., Ltd.) is spin coated on the SOS/ISFET wafer and removed from the enzyme membrane deposition region.

Surface Treatment. After spin coating with a 1% 3-aminopropyltriethoxysilane solution the wafer is left at 110°C for 5 min. It is immersed in a 5% glutaraldehyde solution for 15 min and then washed in water and dried.

Enzyme Solution Spin Coating (Fig. 6B). Bovine serum albumin solution (0.5 ml 300 mg/ml in 0.1 M PIPES-NaOH–buffered solution, pH 6.8), 0.25 ml urease solution (50 mg/ml), and 0.25 ml glutaraldehyde solution (2%) are mixed and spin coated on the wafer at 1500 rpm.

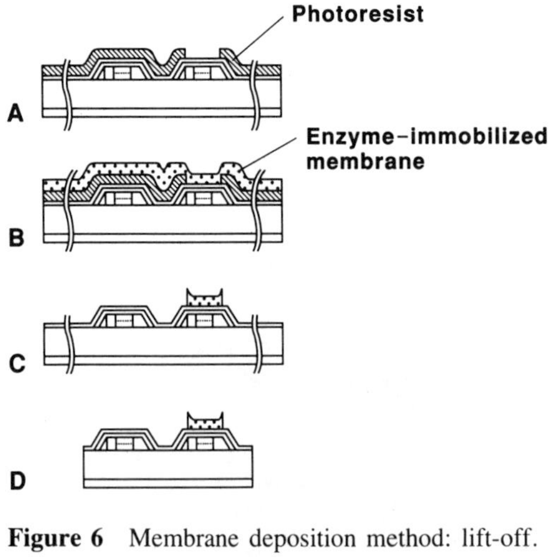

Figure 6 Membrane deposition method: lift-off.

Lift-off (Fig. 6C). After 30 min of immobilizing reaction the enzyme membrane, except on the enzyme FET (ENFET), is lifted off in acetone by ultrasonic vibration.

Lift-off Method for a Multibiosensor

Figure 7 shows the fabrication process by the lift-off method for a multibiosensor with immobilized glucose oxidase and urease membranes (29).

Surface Treatment [Fig. 7(1)]. A 1% 3-aminopropyltriethoxysilane aqueous solution is spin coated onto the ISFET wafer, followed by heat treatment at 110°C for 5 min to modify the wafer surface with amino groups.

Spin Coat and Patterning Positive Photoresist [Fig. 7(2)]. A positive photoresist (Microposit 1300-31, Shipley Co., Ltd.) is spin coated onto the wafer and then patterned to expose the gate region on which an immobilized glucose oxidase membrane is to be deposited.

Enzyme Solution Spin Coating [Fig. 7(3)]. A mixture of 0.25 ml 28% bovine serum albumin solution in 50 mM PIPES-NaOH (pH 6.8), 0.15 ml 4% glucose oxidase, and 0.1 ml 5% glutaraldehyde aqueous solution is spin coated on the wafer at 1500 rpm. The wafer is left in air for 2 hr to complete the cross-linking reaction.

Lift-off [Fig. 7(4)]. Ultrasonic vibration is applied to the wafer for 2 min in acetone. The photoresist layer between the membrane and the wafer surface dissolves into acetone. The loosened membrane is lifted off by ultrasonic vibra-

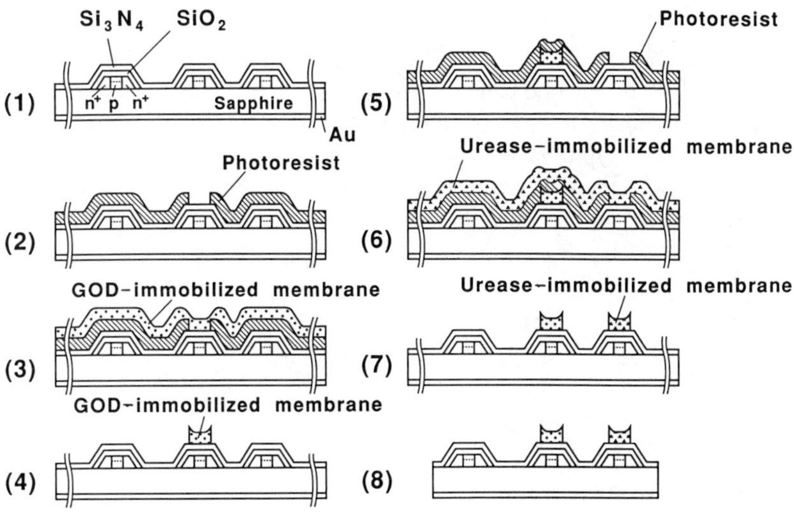

Figure 7 Membrane deposition method: light-off for multibiosensor.

tion, and only the part located on the gate region remains. The resulting wafer is washed in water and dried in a nitrogen gas flow.

Spin Coat and Patterning Positive Photoresist [Fig. 7(5)]. The positive photoresist is spin coated onto the wafer again. To prevent enzyme inactivation by heat treatment the photoresist is dried in a vacuum desiccator for 30 min, instead of being heat treated, and is patterned by the standard method to expose the gate region for the urea sensor.

Second Enzyme Membrane Deposition [Fig. 7(6), (7)]. The urease-immobilized membrane is deposited onto the gate region in a similar way. The enzyme mixture, spin coated on the wafer, comprises 0.25 ml 28% bovine serum albumin, 0.15 ml 5% urease, and 0.1 ml 5% glutaraldehyde.

Ink Jet Nozzle Method (30)

A setup for immobilized enzyme membrane fabrication is shown in Figure 8.

1. Ten enzyme drops are emitted onto different positions on one ISFET sensing region.
2. Preliminary immobilization is carried out by placing an ISFET wafer in a vapor chamber containing a 25% glutaraldehyde aqueous solution for approximately 5 min.
3. By repeating five times steps 1 and 2, 50 enzyme drops are immobilized on one ISFET sensing region.
4. Immobilization is completed by immersing the sensor device in 1% glutaraldehyde solution for about 5 min.

Measurement Method

In ISFET-based enzyme sensors, a differential measurement method has mostly been used because the method can compensate very effectively for temperature dependence, pH change in analyte solution, light sensitivity, and drift of the ISFET itself. When a reference electrode, such as an Ag/AgCl electrode or a saturated calomel electrode was used to fix the potential, the measurement device then designed cannot be miniaturized and maintenance of the reference electrode

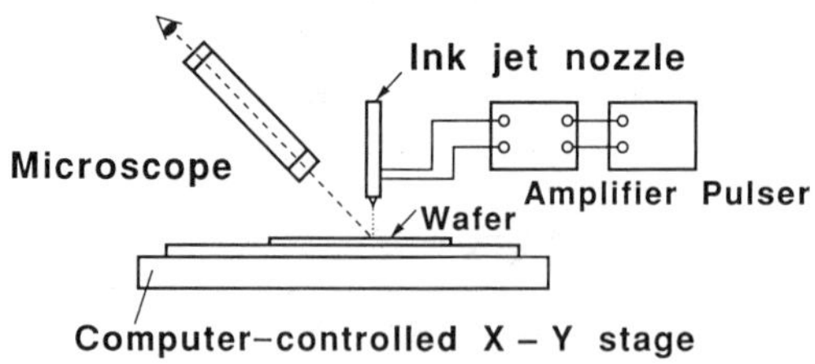

Figure 8 Membrane deposition method: ink jet nozzle.

was troublesome. To obtain a miniaturized biosensor the reference electrode should be of a solid-state type. When a metal reference electrode is used the liquid potential, however, cannot be fixed. The differential measurement method, which measures a differential output between ENFET (enzyme ISFET) and REFET (reference ISFET), can cancel the liquid potential fluctuation due to a metal pseudoreference electrode, and a stable measurement can be made. The ENFET, REFET, and metal reference electrode (pseudoreference electrode) can be fabricated on a sensor chip, which makes it possible to obtain a miniaturized biosensor. Table 4 lists the differential measurement methods.

Figure 9 shows the structure of a single-chip ISFET biosensor in which ENFET, REFET, and a pseudoreference electrode are integrated in a sensor chip. The biosensor is fabricated using the SOS/ISFET technique. It is 0.6 mm wide, 4 mm long and 0.35 mm thick. In the case of the urea sensor, an immobilized urease membrane is made on the surface of one ISFET, and an inactivated immobilized membrane is deposited on the surface of the other ISFET. A gold layer is deposited at the sapphire backsurface to work as a pseudoreference electrode.

Table 4 Differential Measurement for ISFET Biosensor

Year	Author (Reference)	Method
1980	Caras and Janta (20)	Differential current mode ENFET; cross-linked albumin-penicillinase membrane REFECT; cross-linked albumin membrane Reference electrode; Ag/AgCl (saturated KCl) electrode
1983	Miyahara et al. (23)	Differential output from source follower circuits ENFET; urease or acetylcholinesterase membrane REFET; glycine membrane Reference electrode; Ag/AgCl electrode
1983	Hanazato and Shiono (22)	Differential output from source follower circuits ENFET; glucose oxidase membrane REFET; bare Reference electrode; platinum pseudoreference electrode
1984	Kuriyama et al. (15)	Differential output from source follower circuits ENFET; urease membrane REFET; inactivated urease membrane Reference electrode; gold pseudoreference electrode integrated on a sensor chip
1985	Kimura et al. (18)	Integrated source follower circuits ENFET; urease membrane, glucose oxidase membrane REFET; albumin membrane Reference electrode; gold pseudoreference electrode integrated on a sensor chip

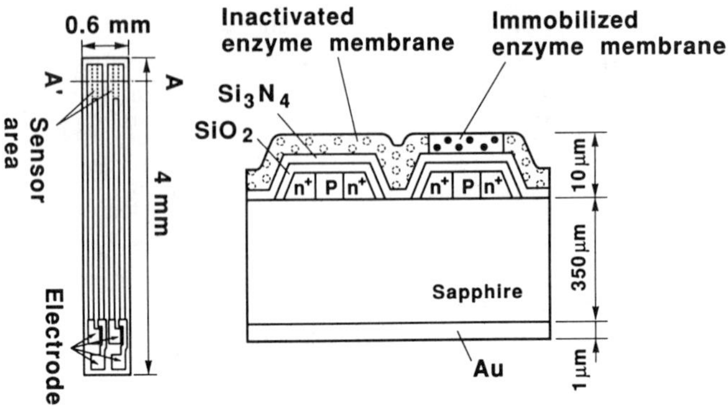

Figure 9 Single-chip biosensor structure.

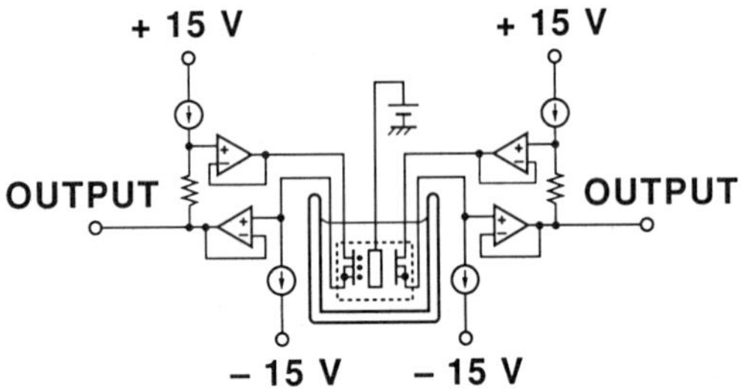

Figure 10 Measurement circuit: source follower circuit.

The measurement circuit is shown in Figure 10. The single-chip biosensor is shown within the dotted-line block. Both ENFET and REFET outputs are measured by using the source follower circuits, and a differential output is obtained.

A multibiosensor can be realized by fabricating plural ISFET with different functional membranes on a chip. By integrating MOSFET in the same sensor chip the source follower circuit can be incorporated. Figure 11 shows the structure for the multibiosensor with the source follower circuits. The sensor is capable of simultaneously measuring urea, glucose, and potassium ion.

Figure 12 shows the equivalent circuit for the multibiosensor. A pair of ISFET and MOSFET makes a source follower circuit, which enables direct measurement of ISFET gate voltage change by source voltage for the ISFET.

Characteristics and Performance

Table 5 gives a list of enzyme-catalyzed reactions used for ISFET biosensors. Some examples of microbial ISFET biosensors are also presented in Table 6.

The enzyme membrane material and thickness are the main factors that

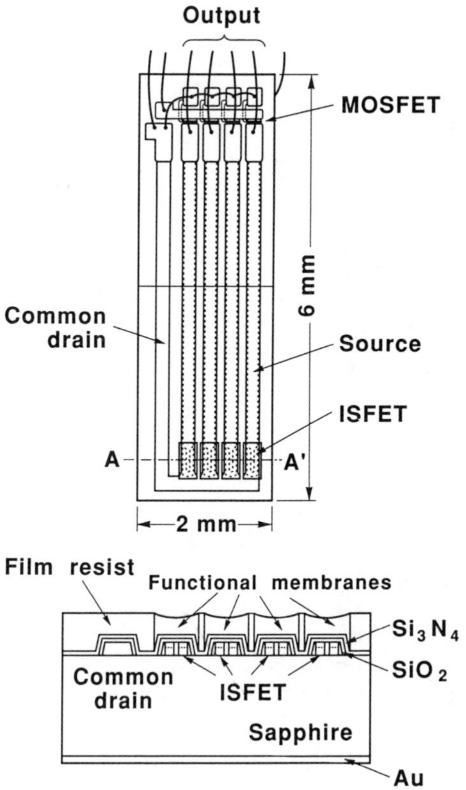

Figure 11 Single-chip multibiosensor structure.

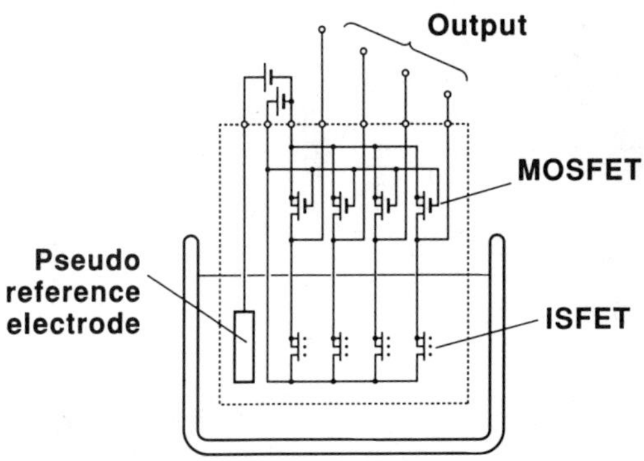

Figure 12 Equivalent circuit for a multibiosensor.

Table 5 Enzymatic Reactions Used for FET Biosensors

Substrate	Enzyme	Reaction	Reference
Penicillin	Penicillinase	Penicillin $\xrightarrow[\text{hydrolysis}]{}$ Penicilloic acid	20, 39
Urea	Urease	Urea + H^+ $\xrightarrow[\text{hydrolysis}]{}$ $2NH_4^+$ + HCO_3^-	15–19, 23–25, 28–32, 34
Glucose	Glucose oxidase	Glucose + O_2 $\xrightarrow[\text{oxidation}]{}$ gluconolactone + H_2O_2	17–19, 22, 24–30, 34–36
	Gluconolactonase	Gluconolactone $\xrightarrow[\text{hydrolysis}]{}$ gluconic acid	41
Triolein	Lipase	Triolein $\xrightarrow[\text{hydrolysis}]{}$ glycerol + oleic acid	26, 37
ATP	ATPase	ATP $\xrightarrow[\text{hydrolysis}]{}$ ADP + orthophosphate	33
ACh (acetylcholine)	ACh-esterase	ACh $\xrightarrow[\text{hydrolysis}]{}$ $(CH_3)_3N^+CH_2CH_2OH$ + $HOCOCH_3$	23

Table 6 Microbial ISFET Biosensors

Substrate	Microbe	Reaction	Reference
Glucose	Organic acid fermenting microorganism from methane fermentor	Glucose ⟶ gluconic acid	22
Ethanol	Acetic acid bacteria	Ethanol —(alcohol dehydrogenase)⟶ acetaldehyde Acetaldehyde —(aldehyde dehydrogenase)⟶ acetic acid	52

determine the response time, sensitivity, and lifetime of the biosensors. Stirring speed or flow speed of the measured solution also influences the response time. The output of ISFET biosensor depends on the buffer capacity of the measured solution as well as the pH value.

Figure 13 shows the response time of glucose sensors with various membrane thicknesses. As the membrane becomes thicker the response time decreases and the output increases (36). The relationship between the response time and the membrane thickness is shown in Figure 14 and the relationship between output and membrane thickness is shown in Figure 15.

Since the biosensor, which is composed of a pH-ISFET associated with an enzyme membrane, is a potentiometric sensor measuring pH changes at the surface, the pH and buffer capacity of the measured solution strongly influence the dynamic range sensitivity and linearity. The influence of buffer capacity has been examined both experimentally and theoretically (38–45). Figures 16 and 17 show the influence of buffer capacity in the case of a glucose sensor and of a urea

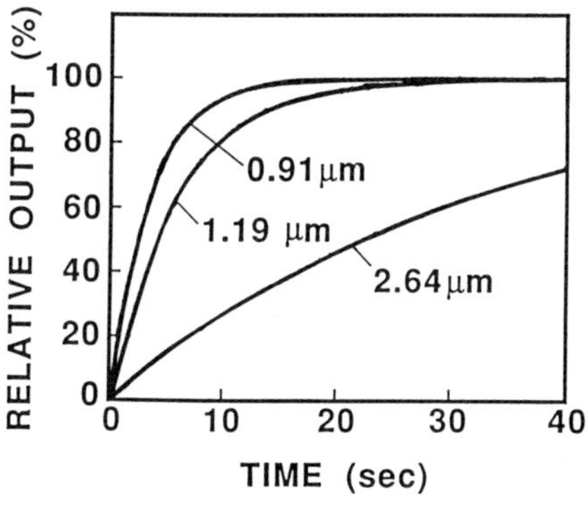

Figure 13 Glucose sensor response time as a function of the membrane thickness.

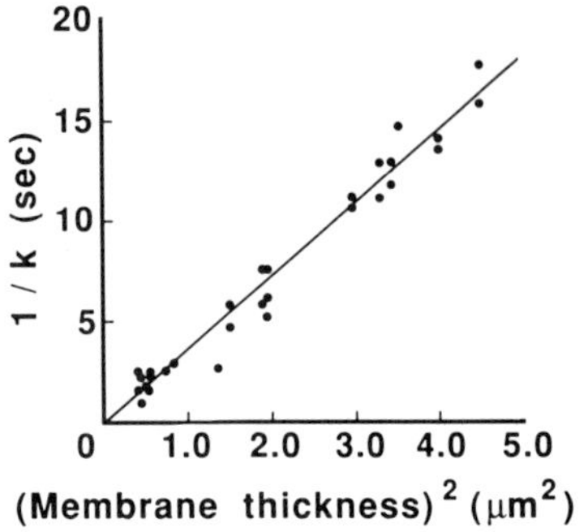

Figure 14 Relationship between response time and membrane thickness.

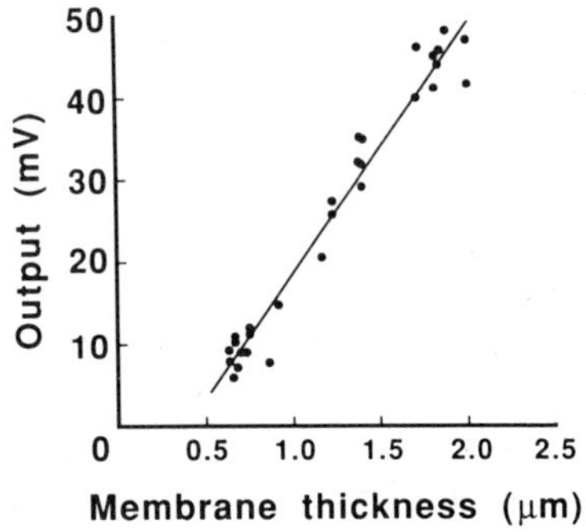

Figure 15 Relationship between output and membrane thickness.

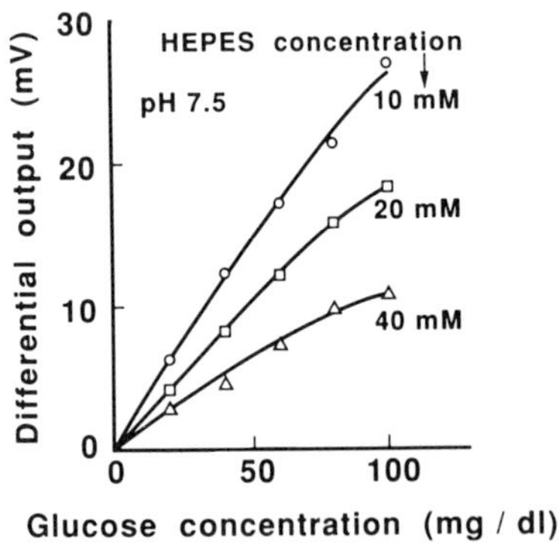

Figure 16 Influence of buffer capacity on the glucose sensor output.

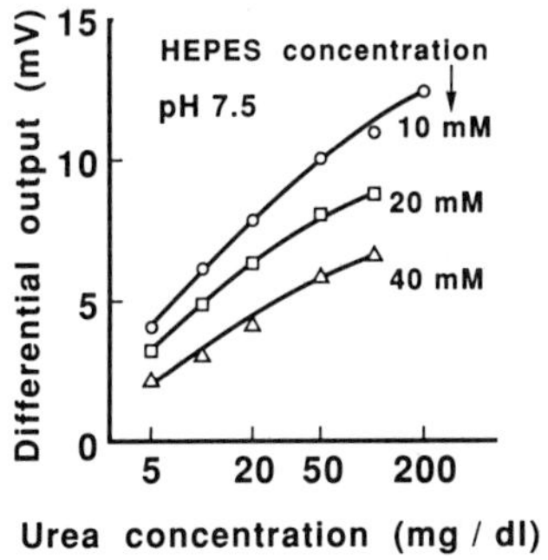

Figure 17 Influence of buffer capacity on the urea sensor output.

sensor, respectively. It is important to fix the pH and buffer capacity at a specific value by diluting the sample with a buffered solution to enable precise measurement and to increase the upper limit of concentration that can be determined.

The other measurement method for avoiding the influence of buffer capacity is a feedback method, which consists of using an ISFET and an actuator to maintain the pH in the enzyme-immobilized membrane at a constant value by electrohydrolysis (sensor-actuator system) (46,47). In Table 7 some typical sensors and their characteristics are shown.

Multibiosensor (29) characteristics were measured by adding concentrated glucose or urea solutions into 5 ml of 20 mM HEPES, 10 mM NaOH, and 0.4 M NaCl buffer (pH 7.5, stirred with a 4 mm diameter stirring bar rotating at 400 rpm. Figure 18 shows typical responses for the multibiosensor, which can detect glucose and urea. The arrows in the figure indicate additions of concentrated glucose or urea solution to the initial buffer medium. The final glucose or urea concentrations were 100 mg/dl. The response time was less than 30 sec. The glucose sensor did not respond to urea, and the urea sensor showed no response to glucose. Thus the multibiosensor can selectively detect glucose or urea.

Application

FET-based biosensors are expected to be applied to medicine, process control, and environmental protection. The miniaturization of ISFET biosensors would permit applications in the medical field, clinical diagnosis, artificial dialysis monitoring, and blood glucose control. Quasi-continuous blood glucose monitoring in transcutaneous effusion fluid has been attempted as one ISFET biosensor application. Recently, a very interesting result was reported in which the glucose concentration in the effusion fluid, collected through the skin by evacuation, was found to have a close correlation with that in the blood. The amount of effusion fluid, however, was too small for practical use.

Table 7 Typical Sensors and Their Characteristics

Type	Membrane	Immobilization	Response Time	Range (mg/dl)	Lifetime	Reference
Urea	Acetylcellulose	Glutaraldehyde	2 min	5–1000	27 days	23
	Acetylcellulose	Glutaraldehyde	30 sec	1–250	—	17
	PVA + stilbazolium	Photo–cross-link	90 sec	10–1000	Few days	24
	Albumin	Glutaraldehyde	2 min	1–100	30 days	19
	Albumin	Glutaraldehyde	5 sec	0.5–100	—	28
Glucose	PVP[a] + albumin	Photo–cross-link + glutaraldehyde	3 min	0–150	—	26
	Polyacrylamide		2 min	–250	2 weeks	41
	Albumin	Glutaraldehyde	—	5–50	30 days	19
	Albumin	Glutaraldehyde	30 sec	1–150	—	28

[a]Polyvinylpyrrolidone.

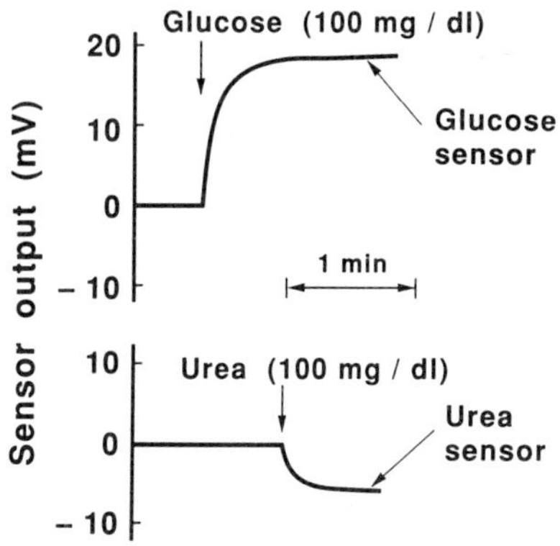

Figure 18 Multibiosensor responses to glucose and urea.

A combination of ISFET biosensor and effusion fluid is considered very attractive for continuous body fluid monitoring (48,49). The system was applied to human blood glucose monitoring, and Figure 19 shows an experimental setup for this purpose. Before the experiment a section of the human arm was treated by applying adhesive tape and then stripping it off about 40 times to eliminate the stratum corneum. Effusion fluid was gathered from the skin surface using a plastic cell with a 8 cm^2 stainless steel mesh area, as shown in Figure 20.

At 10 min intervals the effusion fluid was collected and measured by adding a 5 μl sample into a 20 μl buffered solution (20 mM HEPES, pH 7.5, 0.15 ionic strength). The ISFET biosensor was immersed in the buffered solution and vibrated to stir the sample solution. After measurement the sensor was washed in the buffered solution and wiped with tissue paper before other measurements were made. OGTT (oral glucose tolerance test) was carried out using the ISFET biosensor system. Figure 21 shows the correlation between the blood glucose concentration and the effusion fluid glucose concentration measured by the

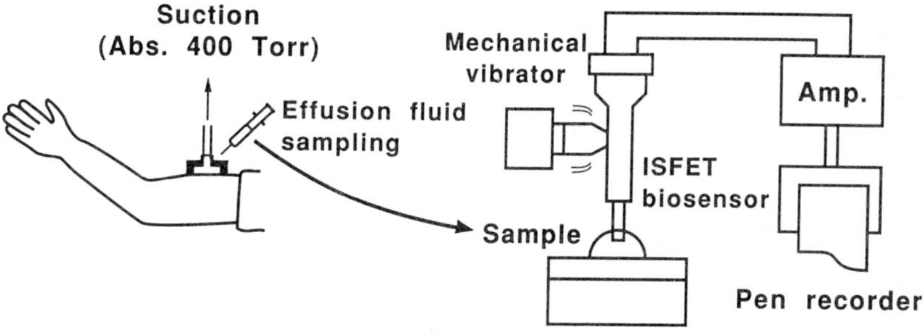

Figure 19 Experimental setup for glucose monitoring.

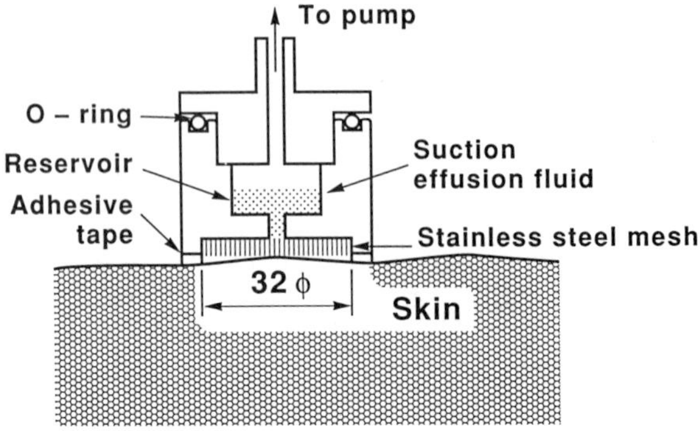

Figure 20 Suction cell structure.

ISFET biosensor during the test. The effusion fluid glucose concentration followed rather precisely the blood glucose concentration with a 10 min delay. The effect of sweat was evaluated by collecting it during the test; it was considered that about 6% of the total effusion fluid was due to sweat.

TRENDS AND PROSPECTS

In ISFET enzyme sensors the number of substrates that can be measured can be increased by integrating various types of functional membranes. One purpose of multifunction sensors is to replace the analyzers now used for clinical diagnosis. Not only the ISFET biosensor, but also other sensors, such as ISFET for ions, other amperometric biosensors, and immunosensors, are concerned, but it will

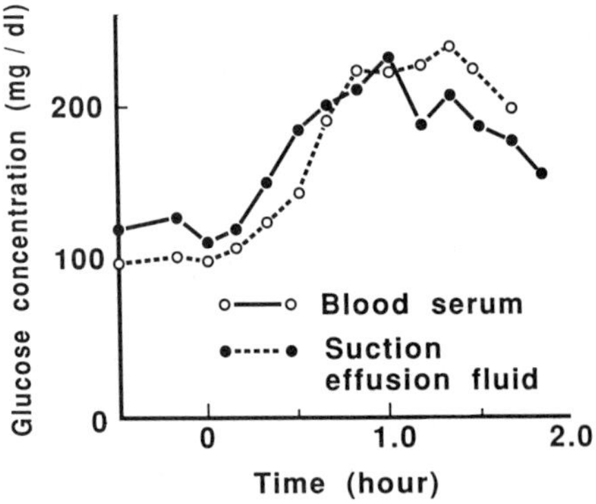

Figure 21 Correlation between the blood glucose concentration and effusion fluid glucose concentration during OGTT.

take much time to replace the analyzer. The other purpose is to measure important marker substrates of a particular disease. For example, a patient with kidney disease caused by diabetes needs to monitor both urea and glucose at the same time. For this purpose a limited number of functions are efficient for treatment.

A miniaturized system based on the FET biosensor, such as a portable monitoring system, will be developed. Since the FET biosensors can be miniaturized, the system using the sensor should be miniaturized to obtain all its advantages. Miniaturized liquid handling devices, such as a pump and a valve, are now under development using silicon micromachining technology. When these devices are in the practical stage, a portable closed-loop system, such as an artificial pancreas, will be realized.

REFERENCES

1. Bergveld, P. (1981). The operation of an ISFET as an electronic device, *Sensors Actuators, 1*: 17.
2. Matsuo, T., and Esashi, M. (1981). Methods of ISFET fabrication, *Sensors Actuators, 1*: 77.
3. Winquist, F., Lundström, I., and Danielsson, B. (1985). Biosensors based on ammonia sensitive metal-oxide-semiconductor structures. *Sensors Actuators, 8*: 91.
4. Schasfoort, R. B. M., Streekstra, G. J., Bergveld, P., Kooyan, R. P. H., and Greve, J. (1989). Influence on an immunological precipitate on D.C. and A.C. behaviour of an ISFET, *Sensors Actuators, 18*: 119.
5. Bergveld, P. (1970). Development of an ion-sensitive solid-state device for neurophysiological measurements, *IEEE Trans. Biomed. Eng., BME-17*: 70.
6. Bergveld, P. (1972). Development, operation, and application of the ion-sensitive field-effect transistor as a tool for electrophysiology, *IEEE Trans. Biomed. Eng., BME-19*: 342.
7. Matsuo, T., and Wise, K. D. (1974). An integrated field-effect electrode for biopotential recording, *IEEE Trans. Biomed. Eng., BME-21*: 485.
8. Sze, S. M. (1969). *Physics of Semiconductor Devices*, John Wiley & Sons, New York.
9. Abe, H., Esashi, M., and Matsuo, T. (1979). ISFET's using inorganic gate thin films, *IEEE Trans. Electron Devices, ED-26*: 1939.
10. Akiyama, T., Ujihara, Y., Okabe, Y., Sugano, T., and Niki, E. (1982). Ion-sensitive field-effect transistors with inorganic gate oxide for pH sensing. *IEEE Trans. Electron Devices, ED-29*: 1936.
11. Esashi, M., and Matsuo, T. (1978). Integrated micro multi ion sensor using field effect of semiconductor, *IEEE Trans. Biomed. Eng., BME-25*: 184.
12. Ohta, Y., Shoji, S., Esashi, M., and Matsuo, T. (1982). Prototype sodium and potassium sensitive micro ISFETs, *Sensors Actuators, 2*: 387.
13. Shoji, S., Esashi, M., and Matsuo, T. (1983). Micro ISFETs of 10 μm tip size. In: *Proceedings of International Meeting on Chemical Sensors*, Fukuoka, Japan, Sep-

tember 1983, Analytical Chemistry Symposia Series, Vol. 17, Elsevior, London, pp. 473–478.

14. Harame, D., Shott, J., Plummer, J., and Meindle, J. (1981). An implantable ion sensor transducer. In: *Proceedings of IEEE International Electron Devices Meeting*, Washington, D.C., December 1981, pp. 467–470.
15. Kuriyama, T., Kimura, J., and Kawana, Y. (1984). An integrated SOS/FET biosensor. In: *Proceedings of 16th International Conference on Solid State Devices and Materials*, Late News Abstracts, Kobe, Japan, August 31–September 2, 1984, pp. 66–67.
16. Kuriyama, T., Kimura, J., and Kawana, Y. (1985). A single chip biosensor, *NEC Res. Dev.*, No. 78, pp. 1–5.
17. Kuriyama, T., Kimura, J., and Kawana, Y. (1985). Development of biosensors with immobilized enzyme, *Chem. Economy Eng. Rev.*, *17*: 22.
18. Kimura, J., Kuriyama, T., and Kawana, Y. (1985). An integrated SOS/FET multibiosensor and its application to medical use. In: *Proceedings of 3rd International Conference on Solid-State Sensors and Actuators* (Transducers '85), Philadelphia, June 11–14, 1985, pp. 152–155.
19. Kimura, J., Kuriyama, T., and Kawana, Y. (1986). An integrated SOS/FET multibiosensor, *Sensors Actuators, 9*: 373.
20. Caras, S., and Janata, J. (1980). Field effect transistor sensitive to penicillin, *Anal. Chem.*, *52*: 1935.
21. Miyahara Y., Shokawa, S., Moriizumi, T., Matsuoka, H., Karube, I., and Suzuki, S. (1982). Biosensor using ISFET with immobilized enzyme membrane. In: *Proceedings of the 2nd Sensor Symposium*. Institute of Electrical Engineering of Japan, 1982, pp. 91–95.
22. Hanazato, Y., and Shiono, S. (1983). Bioelectrode using two hydrogen ion sensitive field effect transistors and a platinum wire pseudo reference electrode. In: *Proceedings of International Meeting on Chemical Sensors*, Fukuoka, Japan, September 19–22, 1983, Analytical Chemistry Symposia Series, Vol., 17, Elsevier, London, pp. 513–518.
23. Miyahara, Y., Matsu, F., Moriizumi, T., Matsuoka H., Karube, I., and Suzuki, S. (1983). Micro enzyme sensors using semiconductor and enzyme-immobilization techniques. In *Proceedings of International Meeting on Chemical Sensors*, Fukuoka, Japan, September 19–22, 1983, Analytical Chemistry Symposia Series, Vol. 17, Elsevior, London, pp. 501–506.
24. Miyahara, Y., Moriizumi, T., and Ichimura, K. (1985). Integrated enzyme FETs for simultaneous detections of urea and glucose, *Sensors Actuators*, 7: 1.
25. Moriizumi, T., and Miyahara, Y. (1985). Monolithic multi-function ENFET biosensors. In: *Proceedings of 3rd International Conference on Solid-State Sensors and Actuators* (Transducers '85), Philadelphia, June 11–14, 1985, pp. 148–151.
26. Hanazato, V., Nakako, G., Maeda, M., and Shiono, S. (1986). Application of water-soluble photocrosslinkable polymer to enzyme membrane for FET-biosensor. In: *Proceedings of 2nd International Meeting on Chemical Sensors*. Bordeaux, France, July 7–10, 1986, pp. 576–579.
27. Hanazato, Y., Nakako, G., Maeda, M., and Shiono, S. (1987). Glucose sensor based on a field-effect transistor with a photolithographically patterned glucose oxidase membrane, *Anal. Chim. Acta, 193*: 87.

28. Kuriyama, T., Nakamoto, S., Kawana, Y., and Kimura, J. (1986). New fabrication methods of enzyme immobilized membrane for ENFET. In: *Proceedings of 2nd International Meeting on Chemical Sensors*. Bordeaux, France, July 7–10, 1986, pp. 568–571.
29. Nakamoto, S., Ito, N., Kuriyama, T., and Kimura, J. (1987). A lift-off method for patterning enzyme-immobilized membranes in multi-biosensors, *Sensors Actuators, 13*: 165.
30. Kimura, J., Kawana, Y., and Kuriyama, T. (1988). An immobilized enzyme membrane fabrication method using an ink jet nozzle, *Biosensors, 4*: 41.
31. Anzai, J., Ohki, Y., Osa, T., Nakajima, H., and Matsuo, T. (1985). Urea sensor based on an ion sensitive field effect transistor. II. Effects of buffer concentration and pH on the potentiometric response, *Chem. Pharm. Bull., 33*: 2556.
32. Karube, I., Tamiya, E., Dicks, J. M., and Gotoh, M. (1986). A micro-sensor for urea based on an ion-selective field effect transistor, *Anal. Chim. Acta, 185*: 195.
33. Gotoh, M., Tamiya, E, Karube, I., and Kagawa, Y. (1986). A micro-sensor for adenosine-5′-triphosphate pH-sensitive field effect transistor, *Anal. Chim. Acta, 187*: 287.
34. Hanazato, Y., Nakako, G., and Shiono, S. (1986). Multi-enzyme electrode using hydrogen-ion-sensitive field-effect transistors, *IEEE Trans. Electron Devices, ED-33*: 47.
35. Hanazato, Y., Inatomi, K., Nakako, G., Maeda, M., and Shiono, S. (1988). Glucose-sensitive field-effect transistor with a membrane containing co-immobilized gluconolactonase and glucose oxidase. *Anal. Chim. Acta, 212*: 49.
36. Kimura, J., Saito, A., Ito, N., Nakamoto, S., and Kuriyama, T. (1989). Evaluation of an albumin-based, spin-coated, enzyme-immobilized membrane for an ISFET glucose sensor by computer simulation, *J. Membr. Sci., 43*: 291.
37. Nakako, M., Hanazato, Y., Maeda, M., and Shiono, S. (1986). Neutral lipid enzyme sensors based on ion-sensitive field effect transistor, *Anal. Chim. Acta. 185*: 179.
38. Caras, S. D., and Janata, J. (1985). pH-based enzyme field effect transistors. In: *Proceedings of 3rd International Conference on Solid-State Sensors and Actuators* (Transducers '85), Philadelphia, June 11–14, 1985, pp. 158–161.
39. Caras, S. D., and Janata, J. (1985). pH-based enzyme potentiometric sensors. Part 3. Penicillin-sensitive field effect transistor, *Anal. Chem., 57*: 1923.
40. Caras, S. D., Janata, J., Saupe, D., and Schmidt, K. (1985). pH-based enzyme potentiometric sensors. Part 1. Theory, *Anal. Chem., 57*: 1917.
41. Caras, S. D., Petelenz, D., and Janata, J. (1985). pH-based enzyme potentiometric sensors. Part 2. Glucose-sensitive field effect transistor, *Anal. Chem., 57*: 1920.
42. Eddowes, M. J. (1985). Response of an enzyme-modified pH-sensitive ion selective device; consideration of the influence of the buffering capacity of the analyte solution, *Sensors Actuators, 7*: 97.
43. Eddowes, M. J., Pedley, D. G., and Webb, B. C. (1985). Response of an enzyme-modified pH-sensitive ion-selective device; experimental study for a glucose oxidase-modified ion-sensitive field effect transistor in buffered and unbuffered aqueous solution, *Sensors Actuators, 7*: 233.
44. Eddowes, M. J., Pedley, D. G., and Webb, B. C. (1985). Corrigendum. *Sensors Actuators, 8*: 179.

45. Eddowes, M. J. (1987). Response of an enzyme-modified pH-sensitive ion selective device; analytical solution for the response in the presence of pH buffer, *Sensors Actuators, 11*: 265.
46. Chandler, G. K., and Eddowes, M. J. (1986). Enzyme-mediated pH sensitive FET devices; problems of the non linear response. In: *Proceedings of 2nd International Meeting on Chemical Sensors*, Bordeaux, France, July 7–10, 1986, pp. 531–533.
47. Van der Schoot, B. H., and Bergveld, P. (1987/88). ISFET based enzyme sensor, *Biosensors, 3*: 161.
48. Kimura, J., Narushi, I., Kuriyama, T., Kikuchi, M., Arai, T., Negishi, N., and Tomita, Y. (1989). A novel blood glucose monitoring method with an ISFET biosensor applied to transcutaneous effusion fluid, *J. Electrochem. Soc., 136*: 1744.
49. Ito, N., Saito, A., Miyamoto, S., Shinohara, S., Kuriyama, T., Kimura, J., Arai, T., Kikuchi, M., Kayashima, S., Nagata, N., and Takatani, O. (1989) A novel blood glucose monitoring system based on an ISFET biosensor and its application to a human 75 g oral glucose tolerance test. In: *Proceedings of 5th International Conference on Solid-State Sensors and Actuators* (Transducers '89), Montreaux, Switzerland, June 1989, pp. 302–303.
50. Murakami, T., Kimura, J., and Kuriyama, T. (1987). An integrated multi-biosensor for simultaneous amperometric and potentiometric measurements. In: *Proceedings of 4th International Conference on Solid-State Sensors and Actuators* (Transducers '87), Tokyo, Japan, June 2–5, 1987, pp. 804–807.
51. Ho, N. J., Kratochvil, J., Blackburn, G. F., and Janata, J. (1983). Encapsulation of polymeric membrane-based ion-selective field effect transistor, *Sensors Actuators, 4*: 413.
52. Kitagawa, Y., Tamiya, E., and Karube, I. (1987). Microbial-FET alcohol sensor, *Anal. Lett., 20*: 81.

8

Chemically Mediated Fiberoptic Biosensors

Bernhard P. H. Schaffar

AVL LIST GmbH
Graz, Austria

Otto S. Wolfbeis

Karl Franzens University
Graz, Austria

INTRODUCTION

A variety of fiberoptic biosensors (FOB) for determination of such analytes as glucose, lactate, cholesterol, penicillin, urea, and alcohol have been developed since the first fiberoptic biosensor was described in 1979 (1). This contribution gives an overview on the state of the art in the field of mediated FOB starting in 1979. Many of the sensors reported thus far are optical, but not fiberoptic. Some of them are included because they are based on sensing schemes related to fiberoptic biosensing. Because a generally accepted definition of a biosensor does not exist, the authors keep to the following: A biosensor is a device that allows the reversible detection of an analyte in a sample matrix using a selective biochemical action or reaction in which the analyte is preferably a small biomolecule.

The field of FOB can be divided into two main classes, the intrinsic or nonmediated FOB and the chemically mediated FOB. The principle of measurement with intrinsic FOB is to detect, in a biocatalyzed reaction, one partner whose optical properties change during a reaction. The reaction partner can be a substrate, a product, or a cosubstrate, such as NADH. FOB of this kind are reviewed in Chapter 9. If no species is involved in the biocatalyzed reaction whose optical properties can be monitored, a transducer is required to make an optical detection possible. Depending on the type of biocatalyzed reaction other transduction principles ("mediation schemes") must be applied. These may be subdivided into physical mediation, such as certain methods of competitive binding, and into chemical mediation, such as a change in the absorption of an indicator molecule because changes in the pH as a result of enzymatic action.

Intrinsic biosensors allow monitoring only analytes undergoing changes in their optical properties during bioreaction. Mediated FOB, in contrast, offer the advantage of using techniques and transduction schemes already well established in electrochemical biosensing. Thus existing fundamental techniques, including enzyme and dye immobilization, application of proper biocomponents, and use of suitable polymer support, can be exploited in designing biooptodes. A tremendous knowledge in this field does exist in the case of electrochemical sensing devices. This is supported by 528 papers written on amperometric and potentiometric (without ISE) sensors compared to 110 papers written on optical sensors between 1985 and 1987 (2). Also, several useful books have been published in the last decade covering mainly the field of electrochemical biosensors but usually including a few FOB as well (3–5).

Although optical and electrochemical transducer-based biosensors can be constructed in a similar way, they clearly display different working characteristics. At this point it is difficult, however, to decide whether optical or electrochemical biosensors are preferable. Each method has its respective merits and shortcomings, and the particular situation and problem usually determine which approach is more promising. To facilitate selection of the most proper method, it is useful to summarize the advantages and disadvantages of the various methods over others. The advantages of FOB over electrochemical biosensors include the following:

1. Optical sensors consume analyte only to reach an equilibrium state. This is in contrast to amperometric devices, in which the analyte is consumed. This allows measurements without suffering from diffusional processes, which strongly depend on temperature and membrane properties.
2. Optical sensors offer the ability to measure two or more parameters simultaneously using one sensor only, as has been shown by an optode for CO_2 and O_2 (6). A general advantage of fiberoptic sensors results from the tremendously high information density in optical fibers compared to an electrical wire).
3. There are no electrical interferences to the signal because the information carrier light is not influenced by magnetic, ionic, or electrical fields. Also, the use of FOB in measurements in vivo is supposedly safer because no electrical shock can be transmitted to the patient.
4. Using single fibers, FOB show the capability of miniaturization, which is important in constructing in vivo catheters or measurements in small sample volumes.
5. No reference cell, including the sensitive liquid-liquid junction, is required.

Among the disadvantages of FOB compared to electrodes, the following are most significant. First is the problem of internal referencing in methods based on light intensity measurements. Fluorimetric measurements, for example, give signals in arbitrary units because fluorescence F depends on the intensity of the exciting light I_0 according to *Parker's* law:

$$F = k\, I_0\, c\, \epsilon\, d\, \phi_f \tag{1}$$

where ϵ is the molar decadic extinction coefficient c the concentration of the analyte ϕ_f the quantum yield of the dye, d the penetration depth of light, and k a geometrical constant that takes into account the different geometries of optical arrangements. The intensity of the light source varies from one instrument to another, and a constant reference intensity is hardly available, except when expensive standard lamps are used for calibration. An optical sensor must therefore be calibrated at two points (because of varying slope and light source intensity). In some optical sensing applications an internal reference can be established to overcome these problems [two-wavelength detection (7), or decay time measurements (8)].

Second, ambient light can interfere. Therefore, all measurements must be performed in a dark environment, or optical isolations or pulse techniques must be applied to protect the sensor signal from ambient light.

Third, optical sensors in general show a limited dynamic range compared to electrochemical sensors. pH optodes typically have a dynamic range of 2 units when one indicator is used and up to 6 units when two indicators are used (7,9).

A more detailed discussion of the general properties of optical and fiberoptic sensors can be found in the respective books and reviews (10–14).

GENERAL REMARKS

Construction of an optical biosensor requires the cooperation of specialists from various fields, including optics, chemistry, electronics, and biochemistry. The design of the sensor and the choice of the proper sensing approach depend on both the analyte to be determined and the matrix. The choice of both the enzyme and the transducer depends on the sample material and other conditions, which may strongly vary from one situation to another. There are two principally different methods for fiberoptic biosensing. In the first the working chemistry is placed on the fiber itself. In the second the fiber is used as a light guide that views the optical properties of a sensing membrane that has been fabricated before it is attached to the fiber end.

Sensing Chemistry and Materials

An important issue is the proper choice of the materials used. The solid support, for instance, can function as a light guide, as a support on which to immobilize an enzyme, as a support for the indicator layer, and also as a mechanical support to make the sensor more rugged. The solid support can be both the fiber itself and a transparent polymer membrane. Using a transparent polymer as a support it is easier to replace the sensing layers.

A number of indicator have been known for years, but it is now more and more recognized that indicators for use in FOB must fulfill additional criteria, including photostability, simplicity of immobilization, high absorbances and fluorescence quantum yields, and an absorption in the visible part of the spectrum. In addition, the immobilization should result in an optimally immobilized indicator (i.e, whose properties, including its pK_a values or quenching constants, are not changed).

One of the ways to prevent ambient light from interfering is based on so-called optical isolation. Typical methods for optical isolation include covering the sensing layer with a red silicone rubber membrane (15) or a sheet of carbon black glued to a silicone rubber (16). Another approach to diminish the influence of ambient light is the use of pulsed light sources combined with pulsed detection triggered at the same frequency. The detector records the signal of ambient light between two pulses. This ambient light signal can be subtracted easily from the total signal as measured at the peak of the pulse. This approach, however, does not compensate for interference by the sample, for instance its intrinsic fluorescence or changes in its index of refraction.

The dynamic range of the FOB can be adjusted to some extent. Because FOB are usually constructed using an excess of enzyme (for better stability), the dynamic range is in the range of the K_m value of the enzyme used. If the working range of a FOB is to be larger than the K_m value for the substrate, a membrane acting as barrier for substrate molecules can be placed on top of the enzyme layer, thereby making the reaction diffusion controlled. A second function of this membrane is to protect the FOB from microorganisms, free proteases, or inhibitor molecules.

Sensing Techniques

Optical fibers can transport light over large distances, usually from a light source to the sensor tip. Reflected or fluorescent light is guided back to a photodetector. Figure 1 is a schematic cross section through an optical fiber. The basic require-

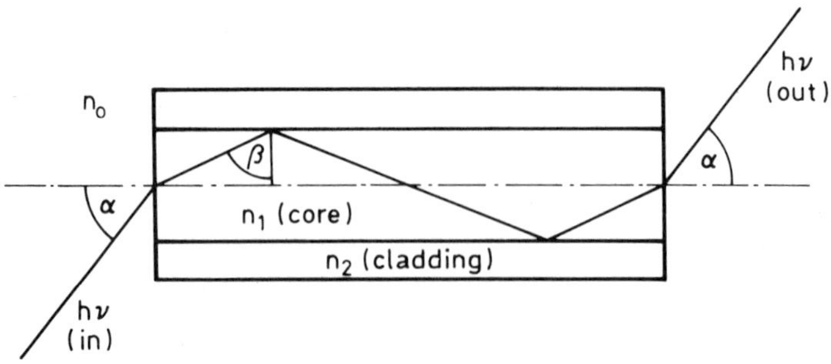

Figure 1 Path of light inside an optical waveguide consisting of core and cladding with refractive indices n_1 and n_2, respectively.

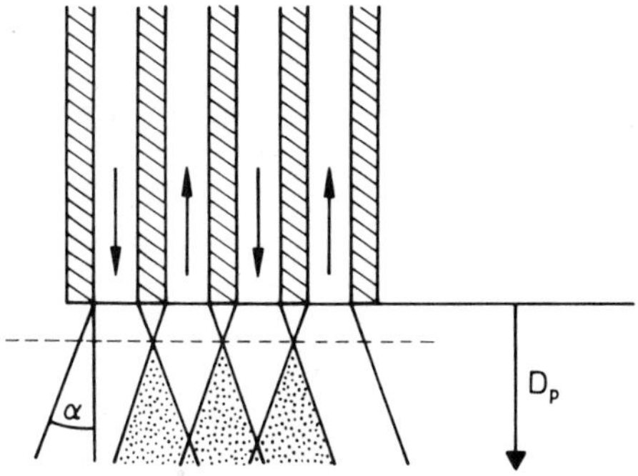

Figure 2 The end of a bifurcated fiber bundle demonstrating the lack of overlap of excitation and emission light within a critical distance. 2α is the numerical aperture, D_p the penetration depth of light.

ment to guide light in fibers is that the refractive index of the core is larger than the refractive index of the cladding. In some cases single fibers have been used in FOB (17), but usually fiber bundles are employed. Frequently bifurcated fibers are used, with one arm of the fiber bundle guiding light from the light source to the common end of the bundles and the other guiding light back to the detector. Figure 2 is a schematic of the common end of the bifurcated fiber bundle.

Single-fiber arrangements are more sophisticated (Fig. 3). A coupler/decoupler is required to separate ingoing from outcoming light. Lasers are the preferred light sources because of the small diameter of single fibers and, of course, the other advantages of lasers as light sources. Single-fiber–based FOB are imperative if measurements are to be performed in small sample volumes, as in many in vivo applications. Another field of application for single-fiber sensors is in long-distance measurements, in which the use of fiber bundles would be too expensive. In most other cases, however, the use of a fiber bundle is preferred because more light can be transported, resulting in better signal-to-noise ratios, and background from the intrinsic fluorescence of the fiber is strongly reduced because it does not enter the outgoing fiber.

In the most simple application plain fibers can be used to perform conventional spectrometry outside the photometer. In case of FOB, such methods as absorbance (18) and reflectance (19), as well as fluorescence (15,16,20,21), have been applied. The fluorescence approach is considered more sensitive, selective, and versatile but requires more expensive instrumentation. Aside from measurement of fluorescence intensity, fluorescence lifetime, energy transfer, or polarization may be utilized as well. Thus a fiberoptic oxygen sensor has been presented in which the fluorescence lifetime serves as the information carrier (8). The schematic of such a measurement is given in Figure 4. This approach shows some decisive advantages over measurement of light intensity (amplitude). It is intrinsically self-referenced and has negligible signal drift only because the decay

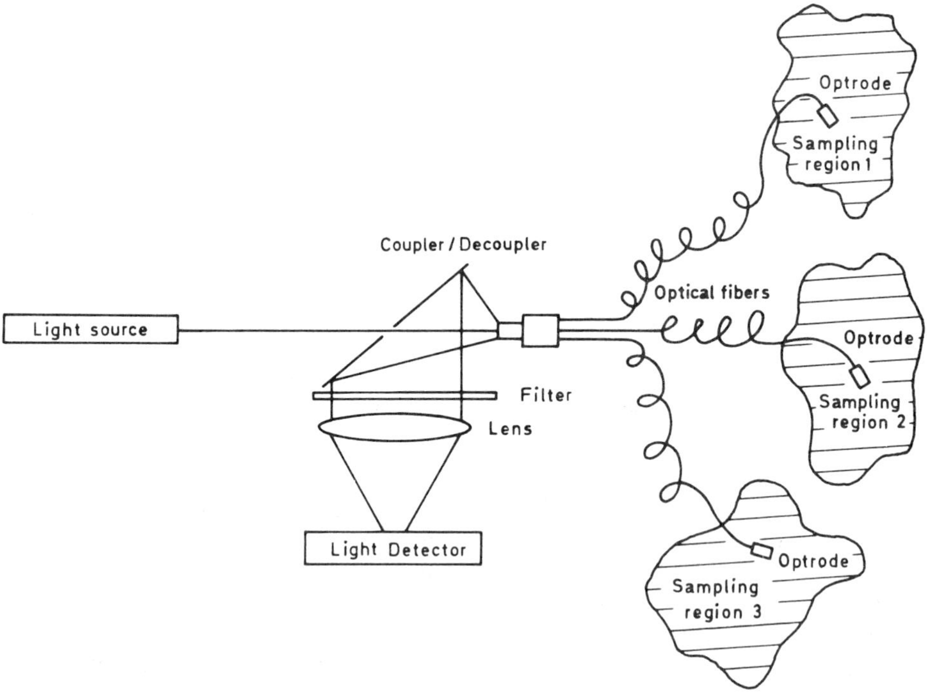

Figure 3 Optical arrangement to couple a laser beam with a single fiber and to collect the reflected or fluorescent light.

time is independent of the fluorophore concentration, and fluctuations in light source intensities and photodetector sensitivity play almost no role.

Transducer Types

Numerous types of fiberoptic chemical sensors are likely to be useful as transducers in FOB. Examples include the pH (18,30–34), ammonia (35), and oxygen optodes (16,21,22–29,36). The carbon dioxide optode, which has been described by many authors (6,20,37–39), has been used in FOB in one case only (39). The principles of these sensor types are explored in the following sections.

Oxygen Optodes

The feasibility of measuring oxygen partial pressure via the dynamic quenching of an indicator fluorescence by molecular oxygen was first described by Kautsky and Hirsch (40) in 1931. The process is described by the so-called Stern Volmer equation:

$$\frac{I_0}{I} = 1 + K_{sv}[Q] \qquad (2)$$

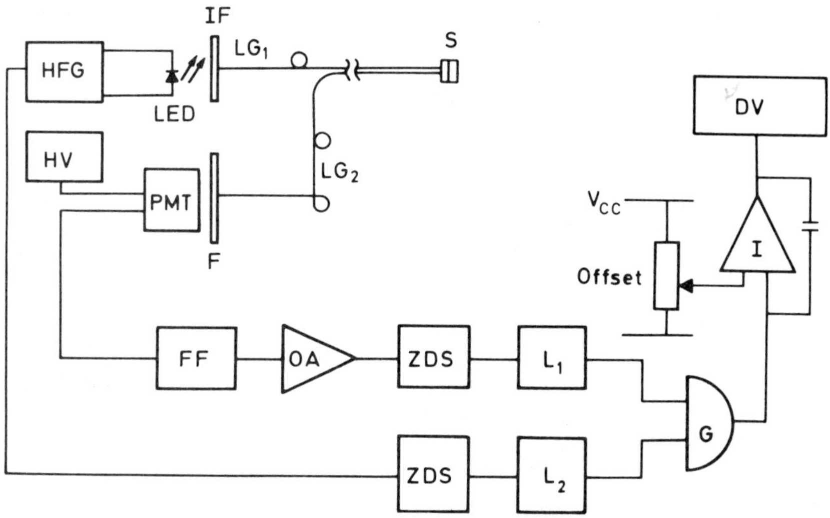

Figure 4 Experimental arrangement to perform decay-time measurements with a fiberoptic oxygen sensor. LED, blue light-emitting diode; HFG, high-frequency generator; IF, interference filter; LG_1 and LG_2, light guides; S, sensor with immobilized fluorescent dye; F, cutoff filter; PMT, photomultiplier tube; HV, high-voltage power supply; FF, frequency filter; OA, operational amplifier; ZDS, zero detection switch; L_1 and L_2, latches; G, *and* gate; I, integrator; DV, digital voltmeter.

where the ratio of the fluorescence intensities of a dye in the absence (I_0) and the presence (I) of oxygen is proportional to the oxygen concentration $[Q]$. K_{sv} is the so-called Stern-Volmer quenching constant, usually expressed as 1/torr O_2 or 1/%O_2 units. In practice this equation must be modified by subtracting false light I_s (21) from the measured intensities I' [Eq. (3)]. False light results from the combined effect of straylight, ambient light, and underground fluorescence.

$$\frac{I'_0 - I_s}{I' - I_s} = 1 + K_{sv}[Q] \qquad (3)$$

One of the first oxygen sensors based on this principle was described by Lübbers and Opitz in 1975 (20). Pyrenebutyric acid (PBA) dissolved in dimethylformamide (DMF) was kept in place between a quartz slide and a Teflon membrane. Oxygen from the analyte solution diffused through the Teflon membrane into the DMF to quench the fluorescence of PBA. The system was improved (41) by dissolving the oxygen-sensitive dye into silicone rubber membranes to achieve mechanically stable sensing layers that were placed at the end of a fiberoptic probe as can be seen in Figure 5. This optode type was steam sterilizable. In a

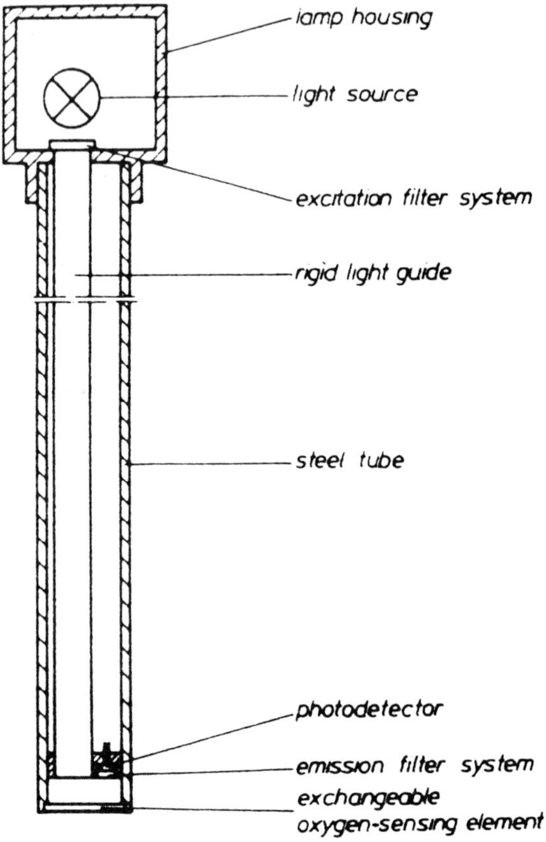

Figure 5 Cross section of an oxygen optode. (From Ref. 41.)

further refinement oxygen measurements could be performed in the temperature range from 300 to 500 K (42).

Limitations in the performance of these optode types include photobleaching of the dyes, the migration of the dye molecules from the silicone membrane into the analyte solution, and the required excitation wavelength of 350–400 nm, which makes them incompatible with glass or plastic fibers. Using a metalo-organic ruthenium complex as an oxygen indicator, some of these limitations were overcome (15). The resulting oxygen optode showed an excitation maximum at 460 nm, which makes possible the use of blue light-emitting diodes (LED) as light sources (8). A schematic of the experimental setup to detect oxygen concentrations based on fluorescence decay time measurements with this sensor is shown in Figure 4.

pH Optodes

pH optodes are based on the changes in the optical properties of an acid/base indicator. This principle has been widely used in colorimetric titrations and in developing pH paper strips. A miniaturized fiberoptic pH sensor was described in 1982 (19) in which phenol red was immobilized onto 10 μm polyacrylamide gel microspheres and packed into a cuprophane hollow fiber fixed on two plastic optical fibers. One fiber guides light to the immobilized indicator; the other guides the reflected and scattered light back to a detector. Hydrogen ions can pass the cuprophane membrane and cause a change in the reflectivity of phenol red. Another type of miniaturized reflection-based pH optode was presented in 1987 (43). It showed an enlarged dynamic range. Fluorescence-based pH optodes with a large dynamic range have also been reported (9,44).

Figure 6 shows the response of an HPTS (1-hydroxypyrene-3,6,8-trisulfonate)-based pH sensor toward changes in the pH. This sensor (7) was developed to measure pH in the physiological range. Such sensors are favorable for the construction of FOB because many enzymes have pH optima at around the physiological pH.

To overcome problems with inner filter effects, an interesting sensing scheme was described in 1987 (17). It consists of a fiberoptic pH sensor based on the

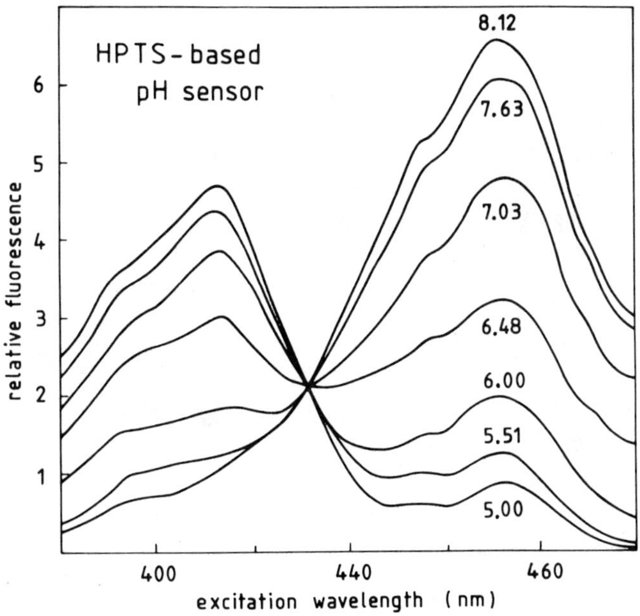

Figure 6 Excitation spectra of glass-immobilized HPTS at various pH values. Emission taken at 520 nm.

nonradiative energy transfer from photoexcited eosin (as a donor) to phenol red (acting as an acceptor of eosin fluorescence and as a pH indicator). Increasing the pH results in a decrease in the fluorescence intensity of eosin, because its fluorescence is more efficiently absorbed by phenol red. Thus changes in the absorption of phenol red are detected as changes in the fluorescence intensity.

Optodes for Carbon Dioxide and Ammonia

Almost all fiberoptic sensors for carbon dioxide are based on sensing pH changes in a buffer solution encapsulated in a gas-permeable membrane. The membrane also prevents ionic species, including protons from entering the internal buffer. Gaseous carbon dioxide or ammonia diffuses through the membrane material and changes the pH of the internal buffer. This in turn causes a change in the optical properties of the dissolved indicator according to the following fully reversible reactions:

$$CO_2 + H_2O \rightleftharpoons H^+ + HCO_3^- \quad (4)$$

$$NH_3 + H_2O \rightleftharpoons NH_4^+ + OH^- \quad (5)$$

To miniaturize the sensor a fiberoptic carbon dioxide sensor without a gas-permeable membrane has been described (37). It consists of a single fiber connected to a small capillary tube. The fluorophore (dichlorofluorescein) is fixed at the distal end of the fiber either by immobilization onto a glass bead, which is glued to the fiber, or with a liquid plug of indicator/buffer solution brought into the capillary tube. The surface of the capillary is hydrophobic, and a gas bubble at the tip of the capillary acts as a gas-permeable membrane to protect the pH indicator from ionic species.

To avoid both the internal buffer solution and the gas-permeable membrane, the pH-sensitive dye (fluorescein) was dispersed in polyethylene glycol and then incorporated in a membrane (38). This results in very short response times, for example 10 sec to reach 90% of the full response. Both these new sensor developments have limitations. The first (37) suffers from photobleaching; the second (38) is humidity sensitive because of the swelling of polyethylene glycol in water. Consequently this sensor can only be used to sense carbon dioxide at constant humidity.

Fiberoptic ammonia sensors are based on similar pH effects. In place of the internal bicarbonate buffer an ammonium chloride buffer is used (45). With carbon dioxide there is an increase in proton concentration, that is, a decrease in the pH of the internal buffer. With ammonia hydroxyl anion is formed, resulting in an increase in pH. Obviously the two gases would interfere if both were present in the same sample solution.

The Biologic Component

Many types of biological components can be used to design FOB. Among them there are antigens and antibodies, enzymes, whole cells, and microorganisms.

However, only enzymes have been used thus far to construct revesibly acting FOB. The antigen-antibody reaction is highly specific, but it is almost impossible to construct biosensors with a reversible response because of the high binding constants. Reaction of cells and microorganisms are considered rather unselective.

The variety of enzymes that conceivably can be combined with optical sensors is large. Table 1 lists some of the enzymes that have been used in FOB thus far. Many of the enzymes given in Table 1 have been used previously with electrochemical sensors (3–5).

Because biological components are unstable outside the cell, the long-term stability of biosensors is often limited by the biocomponent. An example of the long-term stability of a fiberoptic glucose biosensor is given in Figure 7. After an initial increase in the activity of the sensor due to washout effects of the enzyme gel matrix (resulting in a decreased diffusional barrier for glucose), the sensor activity decreases continuously with time. The immobilization of the enzyme on the transducer surface is known to be a critical step. As a result of enzyme immobilization the pH optimum of the enzyme can be broadened and its temperature-activity profile can be changed (21). A number of different methods to immobilize enzymes on fiberoptic sensors have been described in the literature; some are discussed briefly here with respect to their significance to FOB.

Table 1 Enzymes Applied in Fiberoptic Biosensors, Respective Optode Transducers, and References

	Transducer optode			
Enzyme	Oxygen	pH	CO_2	NH_3
Alcohol oxidase	27, 36, 53			
Ascorbic acid oxidase	50, 53			
Cholesterol oxidase	29			
Glucose oxidase	16, 21, 23–26, 36, 48, 53, 58	18, 31, 32		
Lactate oxidase	53			
Lactate monooxygenase	22, 28, 53			
Lipoxidase	53			
Phenolase	53			
Sulfite oxidase	53			
Tyrosinase	53			
Uricase	53			
Xanthine oxidase	24, 53			
Esterase		34		
Penicillinase		18, 30, 32–34		
Urease		18, 34, 47		35
Glutamate decarboxylase			39	

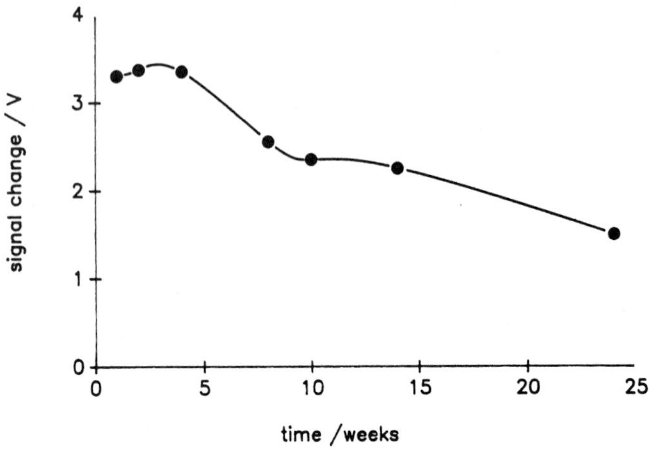

Figure 7 Long-term stability of the fiberoptic glucose biosensor based on an oxygen optode. For a schematic see Figure 8. (From Ref. 16.)

Physical entrapment. In a very useful method the enzyme was incorporated into a hydrogel, such as agar or polyacrylamide, both of which have excellent retaining capability and optical properties. The gel can be fixed to the transducer membrane by using a permeable dialysis membrane (30–32,36).

Cross-linking. The enzyme is cross-lined with a bifunctional species, like glutardialdehyde, to form a gel on either the fiber or the membrane surface (16,23,25,33,35).

Covalent binding. First, the enzyme is fixed covalently to a preactivated polymer matrix, such as a nylon net. The resulting enzyme-loaded membrane is then attached to the transducer to build up a biosensor (21,28,29). Another approach is to bind enzymes covalently to the transducer using classic protein chemistry, such as the carbodiimide method (18) or the avidin-biotin coupling method (34).

Adsorption. The enzyme is adsorbed onto supports like kieselgel or activated charcoal which can be attached to the transducer (16,27).

These procedures have both advantages and disadvantages. After cross-linking the enzyme the enzyme activity can drop to almost zero. Physical entrapment of enzymes, in contrast, suffers from leakage, and methods for covalent binding are usually complicated and time consuming. Therefore, no general recommendation can be given as to which method is the best. The method of choice depends on the kind of enzyme used, the kind of transducer, and the potential application of the biosensor. For all combinations of these variables a specific immobilization procedure must be applied.

Another critical step in constructing FOB is protection of the immobilized enzyme from species in the analyte solution (e.g., microorganisms, free proteases, or inhibitors). Except for inhibitors, a dialysis membrane can eliminate such limitations. To reduce the influence of inhibitors no solutions have been offered thus far, except when the substrate is the only species capable of penetrating the membrane such as in the case of ethanol through silicone rubber (27).

A number of biosensing schemes have been proposed thus far, but this needs comment. Most FOB have been tested with nonreal samples, such as stock solutions in distilled water and/or defined buffers in well-thermostated sensor cells. However, when measuring real samples the signal obtained is additionally affected by minor changes in the sample temperature, pH, ionic strength, and inhibitor or promotor concentrations. The temperature effect (26) and the pH effect (21) toward the response of a fiberoptic glucose biosensor have been described. To overcome problems with the analyte matrix the combination of FOB with FIA (flow injection analysis) as described later can be helpful (46–50).

TRANSDUCER-BASED FIBEROPTIC BIOSENSORS

In this section sensors are classified according to the type of transducer to point out that the choice of an appropriate transducer is of great importance to the performance characteristics of the FOB. Also, the transducer determines the class of biocomponent that can be used. The use of a fiberoptic oxygen sensor, for instance, requires a biocomponent from the group of oxido-reductases. Decarboxylases, in contrast, are of course combined with carbon dioxide sensors. Fiberoptic pH sensors are the most generally applicable transducers because most biochemical reactions cause a change in pH, but they suffer from a number of shortcomings.

Biosensors Based on Oxygen Optode Transducers

Oxygen sensors can be used as transducers for biosensors if the bioreaction follows the general scheme given in Equation (6). In this case the substrate is the analyte. It is oxidized under the consumption of oxygen and water to give hydrogen peroxide and the oxidized substrate (the product):

$$\text{Substrate} + O_2 + H_2O \xrightarrow{\text{oxidase}} \text{product} + H_2O_2 \qquad (6)$$

The consumption of oxygen is measured with the oxygen optode. It is necessary, however, that the enzyme layer on top of the oxygen sensor be permeable to the analyte. Second, it should be kept in mind that the sensor responds to the total oxygen concentration. Hence the response of the sensor to the substrate concentration is strongly affected by the oxygen supply of the analyte solution. Three strategies to overcome problems with the dissolved oxygen can be applied. One is to measure the oxygen concentration of the analyte solution by a second oxygen sensor (the so-called two-sensor technique); another is based on measure-e pO_2 gradient in the biosensor with the help of two different indicator species (22). Third, one can keep the oxygen concentration of the analyte solution constant, as in FIA instrumentation (48–50).

Glucose Biosensors

Glucose can be determined via the enzymatic oxidation of glucose using glucose oxidase (GOD). The reaction of glucose with oxygen can be described as

$$\text{Glucose} + O_2 + H_2O \xrightarrow{\text{GOD}} \text{gluconolactone} + H_2O_2 \qquad (7)$$

Hydrogen peroxide is usually removed through a second enzymatic reaction using catalase according to

$$2H_2O_2 \xrightarrow{\text{catalase}} O_2 + 2H_2O \qquad (8)$$

Hydrogen peroxide must be decomposed quickly because of its detrimental effect on the enzyme. Fortunately, the catalase required for this reaction is usually present as an impurity in the GOD. When ultrapure GOD fractions are used catalase must be added to the enzyme layer to achieve the desired long-term stability of the biosensor.

A cross section through the sensing layer of a fiberoptic glucose biosensor is shown in Figure 8. Here GOD was immobilized by adsorption onto a layer of carbon black followed by cross-linking with glutardialdehyde (16). The resulting sensor exhibited a good long-term stability, as can be seen in Figure 7. When working with biosensors based on oxygen optodes it should be taken into consideration that the response curve is linear only at the beginning because of the hyperbolic calibration function of oxygen given by the Stern-Volmer law [Eqs.(2) and (3)] and enzyme saturation at high substrate concentrations. This can be seen in Figure 9, were the intensity of the fluorescence signal is related to increasing glucose concentrations. The features of different types of fiberoptic glucose biosensors are summarized in Table 2.

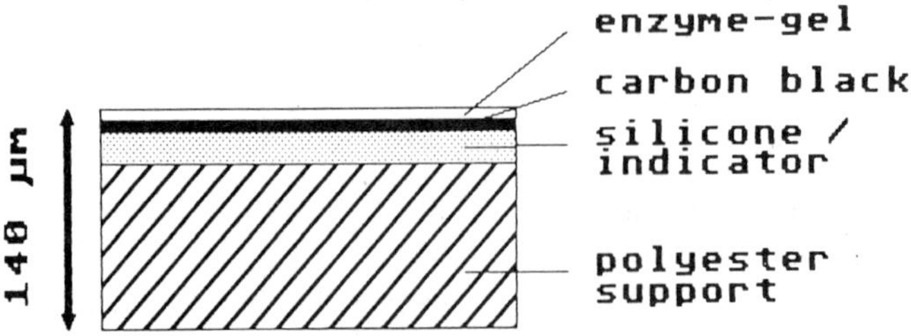

Figure 8 Cross section of the sensing layer of fiberoptic glucose biosensor, consisting of a polyester support, a silicone membrane with decacyclene as the fluorescent oxygen indicator, a layer of carbon black that serves as an optical isolation, and the GOD glutardialdehyde gel as the biocomponent.

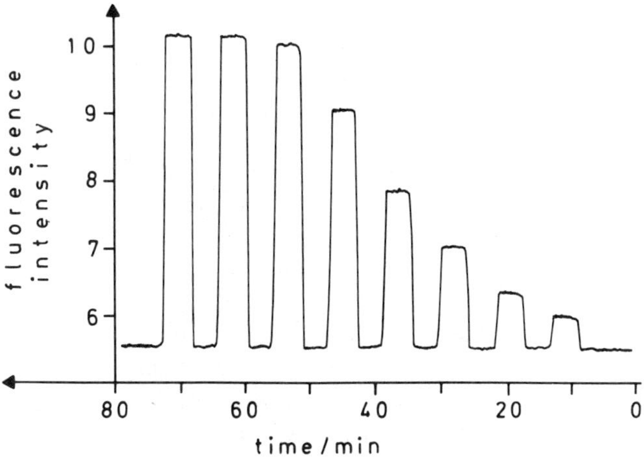

Figure 9 Response of the fiberoptic glucose biosensor (as described in Fig.8) to various glucose concentrations (0.1, 0.2, 0.4, 0.6, 0.8, 1.0, 1.2, and 1.4 mM glucose, respectively).

Lactate Biosensors

L-lactate has been determined with optical lactate biosensors using the enzyme lactate mono-oxygenase (LMO) (22,28). The reaction scheme of this oxidation is given in Equation (9):

$$\text{L-lactate} + O_2 \xrightarrow{\text{LMO}} \text{acetate} + CO_2 + H_2O \tag{9}$$

Table 2 Comparison of the Performance of Different Fiberoptic Glucose Biosensors Based on an Oxygen Optode

Enzyme source	Immobilization procedure	Analytical range (mM glucose)	Response time T_{90} (min)	Stability (months)	Reference
Aspergillus niger	Co–cross-linking	0–50	—	—	23
Aspergillus niger	Co–cross-linking	0–25	0.3	12	25
Aspergillus niger	Covalent	0–20	5	2	21
Aspergillus niger	Adsorption, cross-linking	0–2	0.2	6	16
Penicillium amagasakiense	Adsorption, cross-linking	0–4	0.2	4	16

The use of LMO appears to be advantageous over the use of lactate oxidase (LOD), which is due to the lack of the hydrogen peroxide production during the enzymatic reaction. Two types of optical lactate biosensors have been described in the past. The first is in combination with two different oxygen indicators (perylene and butyric acid) to measure the pO_2 gradient directly, which is established by the enzymatic reaction across the membrane (22). In the other (28), LMO was covalently immobilized onto a nylon net according to a procedure described in reference 51. Figures of merit for the latter type of biosensor are given in Table 3. A cross section through its sensing layer is shown in Figure 10.

In looking at respective data for the long-term stability of the corresponding glucose biosensor (Fig. 8), it can be seen that this biosensor exhibits a distinctly poorer stability (lifetime of 1–2 months only). This is due to the limited stability of the enzyme. A typical calibration plot for this lactate biosensor is given in Figure 11.

Alcohol Biosensors

The enzymatic determination of alcohols, such as ethanol or methanol, via oxygen optodes is usually based on the following enzymatic reaction [Eq. (8)], which is given for the specific case of alcohol oxidation using alcohol oxidase (AOD):

$$R-CH_2OH + O_2 + H_2O \xrightarrow{\text{AOD}} R-CHO + H_2O_2 \tag{10}$$

The hydrogen peroxide produced is decomposed by catalase, thus preventing the oxidation of the enzyme. Again, the consumption of oxygen is measured and related to the alcohol concentration. The immobilization of AOD has been performed by incorporation of the enzyme into an agar gel, resulting in a sensor with a dynamic range of 0–10 mM ethanol (36). Another approach (27) to design an alcohol biosensor utilizes the fact that lower alcohols are capable of permeating hydrophobic membranes. A schematic of a sensor covered with a hydro-

Table 3 Performance Characteristics of Two Types of Lactate Sensor

Lactate sensor	Type 1	Type 2
Nylon membrane	Immunodyne	Biodyne
Thickness of indicator layer, μm	10	25
Teflon layer, μm	—	12
Analytical range, mM	2–50	0.3–6.0
Response times t_{90}, min	2.25–3.0	2.5–5.0
Lifetime, months	1	2

Source: From Reference 28.

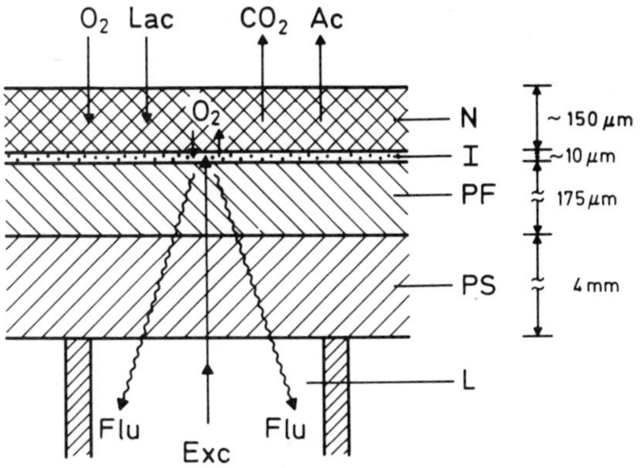

Figure 10 Cross section of the fiberoptic lactate sensor: L, light guide; PS poly(methyl methacrylate) support; PF, polyester film; I, indicator layer (decacyclene in silicone); N, nylon membrane with immobilized LMO

phobic membrane is given in Figure 12. The oxygen-sensitive indicator and AOD are immobilized by adsorbing them onto kieselgel beads suspended in a silicone membrane. Thus the enzyme is protected from inhibitors, microorganisms, or proteases present in the analyte solution.

The lifetime of this biosensor is rather poor in that the response is reduced to 10% of its initial activity after 2 weeks. The dynamic range of the sensor is from 0 to 500 mM ethanol. All FOB using AOD as the enzyme also lack selectivity,

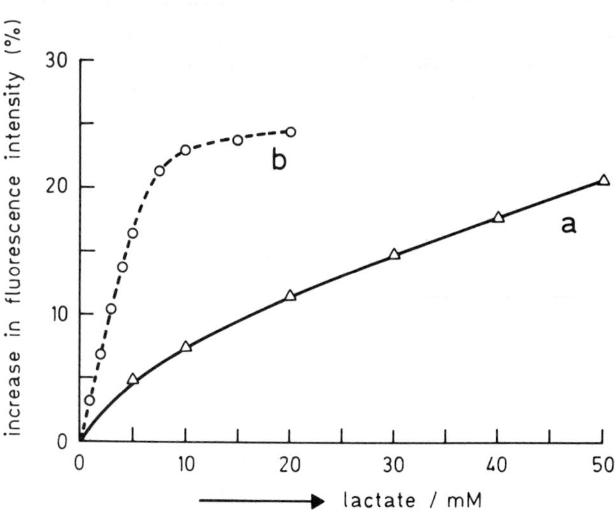

Figure 11 Calibration graphs for two types of lactate sensors: (a), type 1; and (b), type 2 according to Table 3.

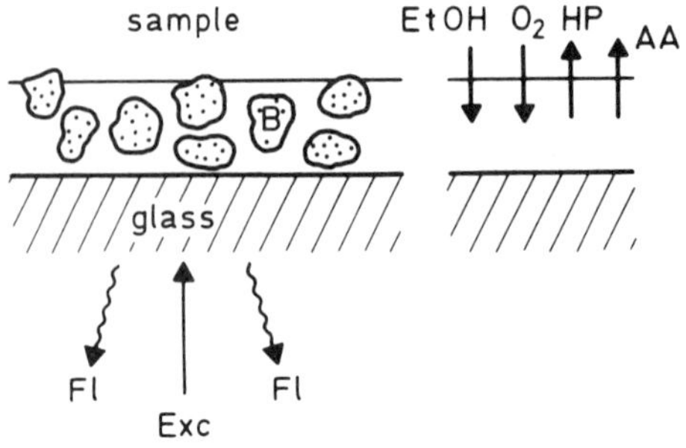

Figure 12 *(Left)* Cross section through an ethanol-sensitive membrane consisting of B, silica gel beads with adsorbed indicator, catalase, and AOD (alcohol oxidase) embedded in silicone rubber. *(Right)* Chemical species and the direction of their diffusion in the ethanol biosensor. EtOH, ethanol; HP, hydrogen peroxide; AA, acetaldehyde.

because AOD catalyzes the oxidation of methanol, ethanol, and even higher alcohols, as has been shown using electrochemical devices (52).

Other Types of Oxygen Sensor-Based Fiberoptic Biosensors

Aside from various kinds of oxidases and oxygenases that have been used in the FOB described thus far, several other less common oxidases and oxygenases have been applied in such sensors (Table 4) (53). All oxidases and oxygenases listed in Table 4 were immobilized by the same procedure, which is a cross-linking reaction with glutardialdehyde. Also, the same transducer was used, namely an fiberoptic oxygen sensor based on the quenching of the fluorescence intensity of decacyclene dissolved in silicone rubber. One can see that only FOB with GOD, ascorbic acid oxidase, and tyrosinase display a useful long-term stability, which is an important quality criterion for a biosensor. This does not mean, however, that this is the most appropriate immobilization protocol. More likely, each single biosensor requires an immobilization procedure of its own with respect to long-term stability of the resulting device.

A cholesterol biosensor has been described (29), similar in construction to the glucose (21) and lactate (28) biosensors but using the enzyme cholesterol oxidase (ChOD), which catalyzes the oxidation of cholesterol according to the equation

$$\text{Cholesterol} + O_2 \xrightarrow{\text{ChOD}} \text{4-cholestenone} + H_2O_2 \tag{11}$$

The response time of this sensor to cholesterol is rather slow, about 10 min for 90% of the steady-state signal. The sensor response is strongly affected by pH, especially small changes in the physiologic range. It may therefore become difficult to use this sensor with real samples, for instance in blood, because sample dilution with an appropriate buffer will decrease the cholesterol concentration such that it is out of the working range of this biosensor.

Table 4 Results Obtained with Various Oxidases and Oxygenases Immobilized by the Same Procedure[a]

Enzyme	Source	Specific activity (U/mg)	Substrate	Response time[b] (s)	Washing time[b] (s)	Analytic range (mM)	Long-term stability, T½ (days)
Glucose oxidase	*Aspergillus niger*	280	D-glucose	8	20	0.01–1	160
	Penicillium amagasakiense	120	D-glucose	20	40	0.02–5	60
Ascorbic acid oxidase	*Cucurbita* sp.	280	Ascorbic acid	30	30	0.01–1	>24[c]
Tyrosinase	Mushroom	4,990[d]	Tyrosine	10	20	0.01–0.5	>12[c]
Xanthine oxidase	Cow's milk	20 U/ml	Hypoxanthine	30	60	0.05–1	5
Lactate oxidase	*Pediococcus* sp.	20	L-lactate	8	30	0.01–0.6	3
Phenolase	Potato	570[d]	Phenol	4 min	10 min	0.5–100	>12[c]
Uricase	*Arthrobacter protophormiae*	20	Urea	1 min	2 min	0.1–10	>2[c]
Alcohol oxidase	*Pichia pastoris*	1,250 U/ml	Ethanol	1 min	2 min	1–10 vol%	3
	Candida boidinii	5	Ethanol	Response with stop-flow technique only			
Lactate monooxygenase	*Mycobacterium smegmatis*	15	L-lactate	Response with stop-flow technique only			
Lipoxidase	Soybean	134,000[d]	Linoleic acid	Response with stop-flow technique only			
Sulfite oxidase	Chicken liver	500 U/ml	Sulfite	No response			

[a]Adsorption to carbon black optical isolation on top of a fluorescence biooptode and subsequent cross-linking with glutaraldehyde using an oxygen optode as the transducer. *Source*: From Reference 53.
[b]Time required for a 90% signal change.
[c]Experiments were not carried further.
[d]Sigma units.

s Based on a pH Optode

s can be used as transducers for FOB when a change in the pH of the sample can be detected during the enzymatic reaction. This change can be observed via an immobilized indicator using absorbance, reflection, or fluorescence techniques. A variety of biocatalyzed reactions causing a pH change are known from the literature (18,30–34) and have been used for biosensing purposes. Typical enzyme reactions are given in the following equations:

$$\text{Glucose} + O_2 \xrightarrow{\text{GOD}} \text{gluconate}^- + H^+ + H_2O_2 \quad (12)$$

$$\text{Penicillin} \xrightarrow{\text{penicillinase}} \text{penicilloate}^- + H^+ \quad (13)$$

$$\text{Urea} + 3H_2O \xrightarrow{\text{urease}} 2NH_4^+ + HCO_3^- + OH^- \quad (14)$$

In the reactions in Equations (12) and (13) the pH decreases, but it increases during the enzymatic decomposition of urea [Eq. (14)]. A general drawback using pH optodes as transducers for FOB is the influence of both pH and buffer capacity of the analyte solution to the measured signal. Another problem is the effect of the oxygen content of the analyte solution on the response of the enzyme. Under usual experimental conditions the oxygen partial pressure in the sample can be controlled more easily than the pH and buffer capacity. Therefore FOB with oxygen transducers are preferred. Nevertheless, a variety FOB with pH transducers have been described and possibly will find application if the pH of the analyte is known and buffer capacities are small. If the buffer capacity of the analyte is strong, practically no response is detectable because each proton produced or consumed during enzymatic action is buffered by the solution. Some examples of FOB based on pH-sensing devices are described here.

Glucose Biosensors

An absorbance-based fiberoptic glucose biosensor was described in 1984 (18). Here, a pH-sensitive triphenylmethane dye, such as cresol green or bromothymol blue, was derivatized with glutathione and coimmobilized with glucose oxidase onto a transparent membrane. A red light-emitting diode was used as a light source and a photodiode as the detector. This sensor configuration was not coupled to fiberoptics, but easily could be. The dynamic range of the sensor is from 0 to 70 mM glucose, but the sensor must be regenerated after each measurement with 5 mM sodium phosphate buffer. Measurement and regeneration are completed within 4 min.

Another FOB using the pH-dependent fluorescence of HPTS (1-hydroxypyrene-3,6,8-trisulfonic acid) was constructed to determine glucose concentrations (31). This dye was immobilized on cellulose particles embedded (together with GOD) in an hydrogel matrix, as can be seen from the cross section through the sensing layer given in Figure 13. Again, the response of the sensor to glucose varies with the buffer capacity of the analyte solution. This is demonstrated by the calibration curves given in Figure 14 for 1.0 and 0.1 mM phosphate buffers. The response times of the biosensor are rather long (10 min).

A 1 mm diameter single fiber has been used to construct a fiberoptic glucose biosensor (32). On the distal end of that fiber a 200 μm membrane of fluoresceinamine (as the pH-sensitive dye) and GOD in polyacrylamide gel was fixed by means of a small piece of a hollow fiber. The response time was of the order of 10 min as a result of the thickness of the gel membrane. The linear dynamic range is approximately from 0 to 10 mM glucose. With this sensor it is possible to determine glucose concentrations over three orders of magnitude using the logarithmic calibration curve.

Penicillin Biosensors

A configuration similar to the single-fiber glucose biosensor has been used to design a penicillin biosensor (32). It is made from 10–20 μm thick membrane consisting of penicillinase and fluorescamine in polyacrylamide gel covalently bound to the distal end of the 240 μm single fiber. Compared to the glucose biosensor, the pH changes in the microenvironment of this sensor were more clearly expressed so that a thinner membrane could be used. This leads to response times of less than 1 min and an analytic range of 0.25–10 mM penicillin.

In another approach for a penicillin biosensor (30) the pH probe HPTS (hydroxypyrenetrisulfonate), penicillinase, and bovine serum albumin (BSA) were covalently cross-linked with glutardialdehyde to form a gel at the distal end

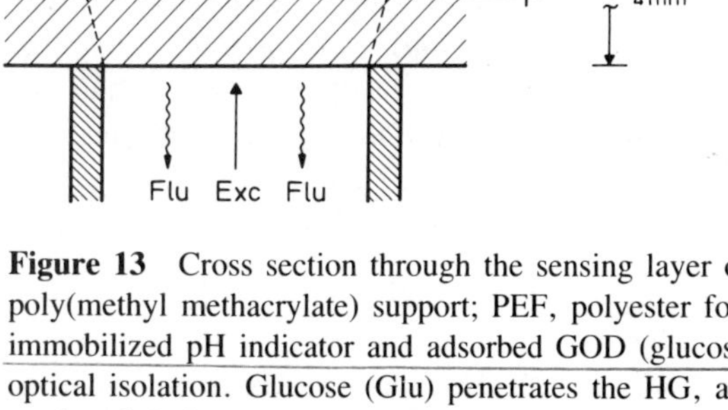

Figure 13 Cross section through the sensing layer of a fiberoptic glucose sensor: P, poly(methyl methacrylate) support; PEF, polyester foil; C, cellulose granules with immobilized pH indicator and adsorbed GOD (glucose oxidase); HG, hydrogel; OI, optical isolation. Glucose (Glu) penetrates the HG, and GA is the gluconic acid produced during the enzymatic reaction.

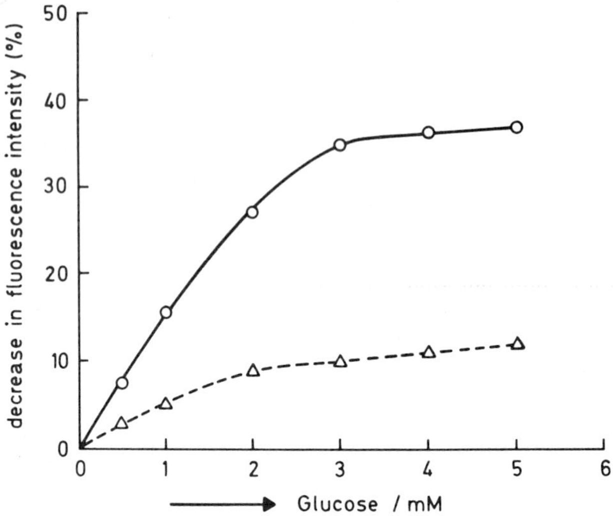

Figure 14 Calibration curve of the glucose sensor described in Figure 13. The phosphate buffers used were 0.0001 M (----) and 0.001 M containing 0.1 M sodium chloride (——).

of the single fiber. The resulting biosensor responded to penicillin over the 0.25–25 mM range. Of course the response of these sensor types is affected by the pH and buffer capacity of the analyte solution. Also, the flow and stirring rates have an effect on the sensor response.

Another single-fiber penicillin biosensor consists of an FITC (fluoresceineisothiocyanate)-labeled porous glass bead fixed at the distal end of the 125 μm fiber (33). The bead is covered by a membrane formed by cross-linking BSA and penicillinase with glutardialdehyde. The influence of various buffer systems on the response of the sensor has been investigated. As a result the working range of the sensor increases with increasing buffer capacities; for example with a 0.001 mM phosphate buffer the dynamic range is from 0.1 to 10 mM penicillin, but for 0.0001 mM phosphate buffers the range is from 0.1 to 1.0 mM penicillin.

A general method for preparing FOB was published in 1989 (34). Biotin-avidin noncovalent binding (vitamin and glycoprotein) can be used to immobilize FITC-labeled penicillinase to the surface of an optical fiber. This method has certain advantages over other immobilization procedures, such as physical entrapment, which suffers from loss of enzyme, or covalent immobilization, which is time consuming and requires some chemical skill because of the different nature of each enzyme.

Urea Biosensors

The biotin-avidin coupling method has also been used to construct an urea biosensor (34). Urea can be detected over the 0.1–10 mM range with response times of about 100 secs. pH, ionic strength, and the buffer capacity of the analyte solution must be kept constant. A 2 mM phosphate buffer of pH 6.5 containing 0.1 M magnesium sulfate was found to provide the best experimental conditions for measurements.

Biosensors Based on an Ammonia Optode

The enzyme reaction of Equation (14) has been used to construct a fiberoptic urea biosensor (35). A fluorescein derivative whose fluorescence is pH dependent was dissolved in a sodium chloride and ammonium chloride solution and was held in place at the end of an optical fiber system by means of a microporous Teflon membrane. Urease was coimmobilized with BSA via a cross-linking reaction with glutardialdehyde onto the Teflon membrane. Ammonia gas formed from urea permeates the Teflon membrane and changes the pH of the ammonium chloride buffer. This results in a change in the fluorescence intensity of the indicator. The dynamic range of this sensor type was 0.01–2.5 mM urea, with response times of about 3.3 min. One limitation is the short lifetime of the biosensor due to photodegradation of the indicators used. Therefore this sensor type requires the use of freshly prepared indicator and buffer solutions. Serum urea determinations have been performed. The data of the correlation plot of the results obtained by the fiberoptic urea biosensor compared very well with the results obtained by the standard enzymatic method.

OPTICAL BIOSENSORS BASED ON COMPETITIVE BINDING

An interesting sensing scheme not fitting the conventional definition of transducer-based FOB is based on competitive binding, resulting in a so-called affinity sensor first described in 1979 (1). Its sensing scheme (54–57) is similar to immunoassay principles but is based on a *reversible* competitive binding to a receptor between the analyte ligand and a fluorescently labeled ligand (Fig. 15). FOB of this type show an advantage over common transducer-based FOB in that the working mechanism is based on equilibrium binding rather than on reaction kinetics as in enzyme-based biosensors. This means that problems with the presence of reaction metabolites in the analyte solution, which interfere with the biosensor response, for example oxygen in oxygen optode-based biosensors, can be diminished.

The working principle is straightforward. In the absence of glucose the FITC-labeled dextran binds to the immobilized concanavalin A (ConA). ConA is a lectin with specific binding affinity for such carbohydrates as glucose. The excitation light practically excites no fluorescence because FITC-labeled dextran is bound to the ConA immobilized on a hollow dialysis fiber and is therefore out of the aperture of the fiber. When glucose permeates the hollow fiber it competitively binds to the immobilized ConA. Hence FITC-dextran diffuses into the cone ''seen'' by the fiber, resulting in an increase in fluorescence intensity. Two

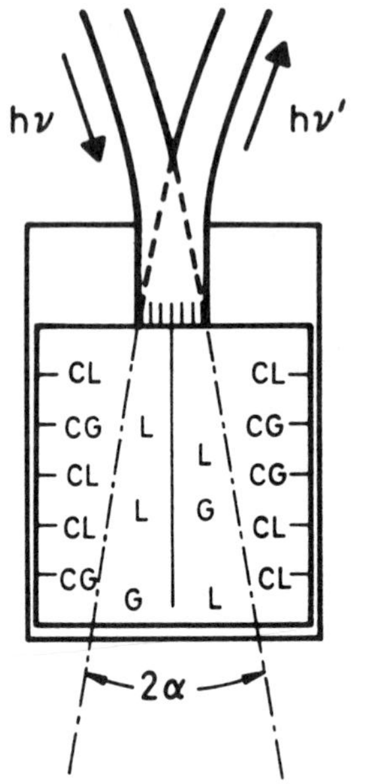

Figure 15 An affinity sensor for glucose determinations, with C concanavalin A; G, glucose; and L, FITC-labeled dextran. (Redrawn from Ref. 55.)

competitive binding equilibria are involved in this process, which are described by the following equations:

$$\text{Glucose} + \text{ConA} \longrightarrow \text{glucose-ConA} \qquad (15)$$

$$\text{FITC-dextran} + \text{ConA} \longrightarrow \text{FITC-dextran-ConA} \qquad (16)$$

Defined quantities of ConA and FITC-dextran are used in this sensing device, but the glucose concentration varies. This sensor works in the 0–50 mM glucose range, with a drift in the calibration curve of 15% in 15 days (54,55). The response time is about 7 min. The procedure for the immobilization of ConA on the hollow fiber has been published (56).

One drawback of this elegant method is the poor specificity of ConA. The sensor responds not only to D-glucose but also to D-fructose, sucrose, D-mannose, and other carbohydrates. The binding of ConA to fructose is about three times stronger than that to glucose. In other words, 10 mM fructose causes the same signal change as 30 mM glucose. In comparison to enzyme-based biosensors this sensor type exhibits a much poorer selectivity. However, this FOB has been used successfully in determinations in vivo of glucose in blood (54,55).

A modification of this sensor type was described in 1988 (57) in which energy transfer (ET) spectroscopy was applied, resulting in a homogeneous affinity sensor. The advantages of this technique include the possibility of an internal calibration via the second chromophore and the lack of the sophisticated chemical immobilization of the receptor on the hollow dialysis fiber. This homogeneous affinity sensor consists of a solution of both FITC-dextran and ConA labeled with rhodamine B (ConA-RhoB) in the hollow dialysis fiber. Here, ConA is not immobilized on the hollow fiber as in the heterogeneous affinity sensor described earlier but rather labeled with RhoB. In the absence of glucose the FITC-dextran is bound to the RhoB-ConA and the fluorescence of FITC is quenched by ET to RhoB. When glucose is present it competitively binds to ConA-RhoB and the FITC-dextran fluoresces. The working range of this sensor is from 0 to 12 mM glucose.

FIBEROPTIC BIOSENSORS IN FLOW INJECTION ANALYSIS

The basic idea that resulted in the combination of flow injection analysis and FOB is the need for reducing the interference of the analyte solution, which adversely affects the accuracy of the determination. As discussed earlier, ionic strength, the buffer capacity of the analyte solution, or varying oxygen supply is critical in biosensing. Also, interference caused by substances changing the performance of the biocomponent (e.g., enzyme inhibitors or promotors) can be minimized to some extent.

Using FIA it is possible to overcome many interferences by automatically diluting the analyte solution with a defined buffer solution. Thus the effects of varying pH, ionic strength, buffer concentration, and concentrations of substrates necessary for the enzymatic reaction or inhibitor concentrations can be easily controlled or widely reduced. Another important feature of the FIA approach is the possibility of overcoming problems with a limited dynamic range of FOB. By changing the dilution factor in the FIA device the dynamic range can be adjusted over a wide range.

The first attempts to realize the combination of optosensing and FIA were made in 1986 (46). A FIA was used to determine urea via pH changes caused by the decomposition of urea by urease. In this approach no FOB was used. Rather, a fiber was coupled to a pH-sensing device (a non-bleeding pH paper), and the biocomponent was dissolved in the carrier stream of the FIA apparatus. Although not a FIA/FOB combination as such, this sensor configuration demonstrated for the first time that optosensing can be used in combination with FIA.

In a more recent version of the urea FIA (47) a "true" FOB was used. Urease was coimmobilized with BSA onto the surface of a pH sensor. To achieve good sensitivity the stop-flow technique was applied, and a dynamic range up to 10 mM urea could be achieved. A FIA/FOB-based glucose assay has been described more recently (48). A fast-responding fiberoptic glucose biosensor was used (see Fig. 8) that exhibits a response time of about 10 sec over a small dynamic range, 0.01–2.0 mM glucose. To increase the analytical range a FIA system allowing the performance of the zone-sampling technique has been used. Variable dilution

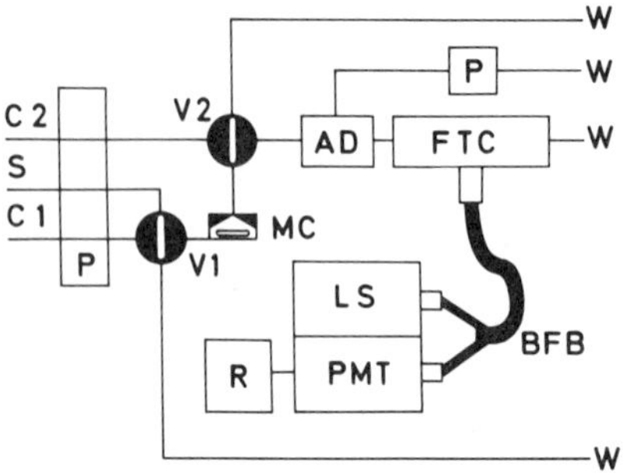

Figure 16 FIA arrangement for zone sampling. (a) *Optical part*: LS, light source; BFB, bifurcated fiber bundle; FTC, flow-through cell with the fiberoptic glucose biosensor; PMT, photodetector; and R, recorder. (b) *FIA part:* C1 and C2, carrier streams; S, sample; V1, and V2, valves; MC, mixing chamber; AD, air damper; P, pumps.

factors can be achieved simply by changing the time span between the first and the second injection. The schematic of the FIA arrangement is shown in Figure 16. The corresponding calibration curve is given in Figure 17. Glucose determinations in wine and fruit juices using this technique have been performed. The results obtained with the FIA compared to results obtained by an official food laboratory were excellent.

In similar approaches FOB with oxygen transducers were applied to the determination of lactate (49) and ascorbic acid (50). Lactate oxidase or ascorbic acid oxidase, respectively, was immobilized. The resulting calibration curves are given in Figures 18 and 19. In both cases analyses were performed with real samples, namely ascorbic acid in fruit juices (50) and lactate in various dairy products (49).

Although the FIA/FOB combination showed excellent analytical results, some drawbacks are obvious. First, good FIA instrumentation is expensive and rather sophisticated. Second, real-time measurements with FIA devices are almost impossible. Even with 60 samples assayed per hour, only a quasi–real-time analysis can be performed. Despite this, FOB combined with FIA certainly will find application in the routine analysis of complex media, and FOB (without FIA) will find application sensing in vivo or in real-time measurements in remote control.

TRENDS AND PROSPECTS

A tremendous amount of work on FOB has accumulated over the past 10 years. These efforts have not yet been translated into commercially available products, but fiberoptic sensors for blood gas analysis have been on the market for years. However, given the potential utility of biosensing devices in numerous fields, it

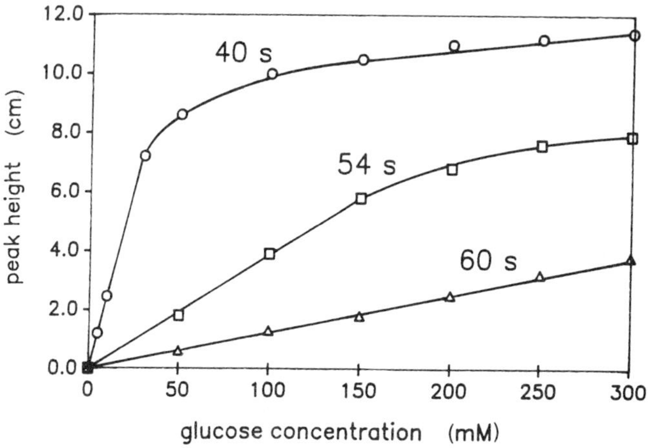

Figure 17 Calibration graphs for the FIA system described in Figure 16 obtained with three different zone-sampling modes. Zone-sampling times of 40, 54, and 60 sec; respectively, were adjusted.

is very likely that FOB will be available within the foreseeable future. Research activities are going strong and, most likely, will even be increased in the future.

Current trends include the miniaturization of FOB, the use of biocomponents other than enzymes, the use of carbon dioxide optodes as transducers, and adapting new optoelectronic devices, such as laser diodes. Miniaturization of FOB will lead to sensors for in vivo application, which is important in medical

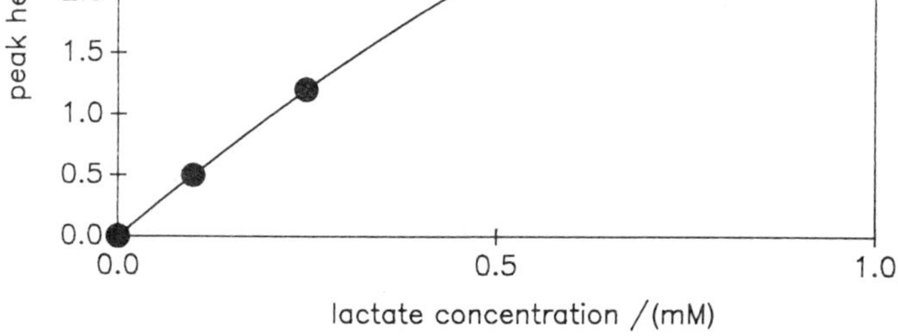

Figure 18 Calibration graph obtained with a fiberoptic lactate biosensor combined with the FIA described in Figure 16.

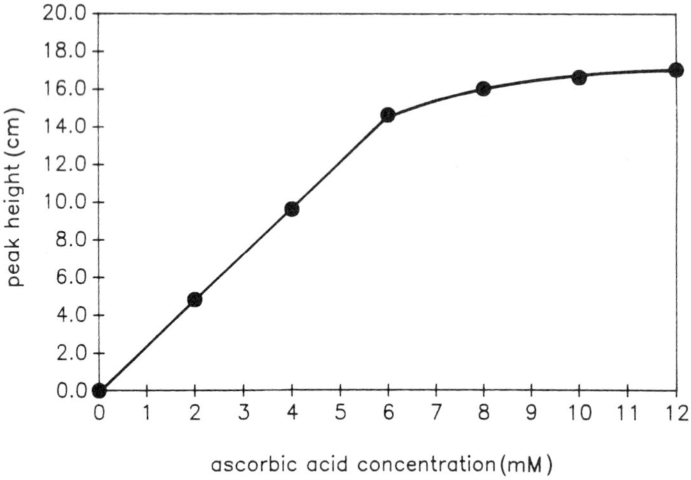

Figure 19 Calibration graph of a fiberoptic ascorbic acid biosensor combined with the FIA device described in Figure 16.

care, and to disposable sensing devices. Especially for biosensors that cannot be sterilized thermally, disposable sensors are greatly needed. Another trend for miniaturization of FOB is based on the use of single fibers (17,30,32–34). Another possibility to miniaturize FOB is the use of the Langmuir-Blodgett technique, which allows reproducible deposition of indicators (58) and enzymes (59) in a definite, molecular layer structure directly onto the waveguide surface. Finally, the application of the evanescent wave technique (60) will provide additional advantages over conventional sensing techniques.

The most obvious limitation in present-day enzyme-based biosensors is the unsatisfying stability of the biocomponent. Ways to achieve improved stability include (1) the development of better immobilization procedures for enzymes, (2) the use of thermophilic enzymes or bacteria possessing improved stability, and (3) the stabilization of enzymes by keeping them inside their cells (61,62) because the enzyme then is allowed to live in its "normal" environment. This approach has been shown to be feasible for electrochemical transducers, but selectivity is limited because of the presence of more than one active enzyme in the biocompartment.

General improvements in optosensing can be expected from the use of semiconductor lasers (63), as well as from the concept of distributed fiberoptic sensors (64) in which many transducer spots are placed on a single fiber performing a multisided measurement.

REFERENCES

1. Schultz, J. S., and Sims, G. (1979). Affinity sensors for individual metabolites, *Biotechnol. Bioeng. Symp.*, 9: 65.

2. Janata, J., and Bezegh, A. (1988). Chemical sensors, *Anal. Chem., 60*: 62R.
3. Turner, A. P. F., Karube, I., and Wilson, G. S. (Eds) (1987). *Biosensors: Fundamentals and Applications,* Oxford University Press, Oxford.
4. Schmid, R. D., Guilbault, G. G., Karube, I., Schmidt, H. L., and Wingard, L. B. (Eds) (1987). *Biosensors: International Workshop 1987, GBF monographs Vol. 10,* VCH Verlagsgesellschaft, Weinheim.
5. Schmid, R. D., and Scheller, F. (Eds) (1989). *Biosensors: Application in Medicine, Environmental Protection and Process Control,* GBF monographs Vol. 13, VCH Verlagsgesellschaft, Weinheim.
6. Wolfbeis, O. S., Weis, L. J., Leiner, M. J. P., and Ziegler, W. E. (1988). Fiber-optic fluorosensor for oxygen and carbon dioxide, *Anal. Chem., 60*: 2028.
7. Offenbacher, H., Wolfbeis, O. S., and Fürlinger, E. (1986). Fluorescence optical sensors for continuous determination of near neutral pH values, *Sensors Actuators, 9*: 73.
8. Lippitsch, M. E., Pusterhofer, J., Leiner, M. J. P., and Wolfbeis, O. S. (1988). Fibre-optic oxygen sensor with the fluorescence decay time as the information carrier, *Anal. Chim. Acta, 205*: 1.
9. Posch, H. E., Leiner, M. J. P., and Wolfbeis, O. S. (1989). Towards a gastric pH-sensor: An optrode for the pH 0-7 range, *Fresenius' Z. Anal. Chem., 334*: 162.
10. Wolfbeis, O. S. (1988). Fiber optical fluorosensors in analytical and clinical chemistry. In: *Molecular Luminescence Spectroscopy, Methods and Applications,* Part 2 (S. G. Schulman, ed.), John Wiley & Sons, New York, pp. 129–281.
11. Wolfbeis, O. S. (1989). Novel techniques and materials for fiber optical sensing. In: *Optical Fiber Sensors,* Proceedings of the 6th International Conference, OFS '89, Paris (H. J. Arditty, J. P. Dakin, and R. T. Kersten, Eds), Springer-Verlag, Berlin.
12. Wolfbeis, O. S. (1987). Fibre-optic sensors in biomedical sciences, *Pure Appl. Chem., 59*: 663.
13. Hirschfeld, T., Callis, J. B., and Kowalski, B. R. (1984). Chemical sensing in process analysis, *Science, 226*: 312.
14. Boisdè, G., and Perez, J. J. (1988). Une nouvelle génération de capteurs: les optodes, *C. R. Acad. Sci. [III], 5*: 303.
15. Wolfbeis, O. S., Leiner, M. J. P., and Posch, H. E. (1986). A new sensing material for oxygen measurement with the indicator embedded in an aqueous phase, *Microchim. Acta, III*: 359.
16. Schaffar, B. P. H., and Wolfbeis, O. S. (1990). A fast responding fibre optic glucose biosensor based on an oxygen optrode, *Biosensors, 5*: 137.
17. Jordan, D. M., Walt, D. R. and Milanovich, F. P. (1987). Physiological pH fiber-optic chemical sensor based on energy transfer, *Anal. Chem., 59*: 437.
18. Goldfinch, M. J., and Lowe, C. R. (1984). Solid-phase optoelectronic sensors for biomedical analysis, *Anal. Biochem., 138*: 430.
19. Peterson, J. I., and Goldstein, S. R. (1982). A miniature fiberoptic pH sensor potentially suitable for glucose measurements, *Diabetes Care, 5*: 272.
20. Lübbers, D. W., and Opitz, N. (1975). Die pCO_2/pO_2 Optode: Eine neue Meβsonde zur Messung des pCO_2 oder pO_2 von Gasen und Flüssigkeiten, *Z. Naturforsch., 30C*: 532.
21. Trettnak, W., Leiner, M. J. P., and Wolfbeis, O. S. (1988). Fibre optic glucose biosensor with an oxygen optrode as the transducer, *Analyst, 113*: 1519.
22. Lübbers, D. W., Völkl, K. P., Grossmann, U., and Opitz, N. (1981). Lactate

measurements with an enzyme optode that uses two oxygen fluorescence indicators to measure the pO_2 gradient directly. In: *Progress in Enzyme and Ion Selective Electrodes,* (D. W. Lubbers, H. Acker, R. P. Buck, G. Eisenmann, M. Kessler and W. Simon, Eds.), Springer-Verlag, Berlin, pp. 67–73.
23. Uwira, N., Opitz, N., and Lübbers, D. W. (1984). Influence on enzyme concentration and thickness of the enzyme layer on the calibration curve of the continuously measuring glucose optode, *Adv. Exp. Med. Biol., 169*: 913.
24. Völkl, K. P., Grossmann, U., Opitz, N., and Lübbers, D. W. (1984). The use of the O_2^- optode for measuring substances such as glucose by using oxidative enzymes for biological applications. *Advances in Physiological Sciences, Vol. 25, Oxygen Transport to Tissue* (A. G. B. Kovach, E. Dora, M. Kessler, and I. A. Silver, Eds), pp. 99–100.
25. Kroneis, H. W., and Marsoner, H. J. (1987). Enzyme sensors using fluorescence based oxygen detection, In: *Biosensors*, International Workshop 1987, GBF monographs Vol. 10 (R. D. Schmid, G. G. Guilbault, I. Karube, H. I. Schmidt, and L. B. Wingard, Eds), VCH Verlagsgesellschaft, Weinheim, pp. 303–304.
26. Völkl, K. P., Opitz, N., and Lübbers, D. W. (1988). Temperature dependence of enzyme optodes as exemplified by the glucose optode. In: *Oxygen Transport to Tissue X* (M. Mochizuki, C. R., Honig, T. Koyama, T. K. Goldstick, and D. F. Bruely, Eds), Plenum, New York, pp. 199–204.
27. Wolfbeis, O. S., and Posch, H. P. (1988). Optical sensors part 20. A fibre optic ethanol biosensor, *Fresenius' Z. Anal. Chem., 332*: 255.
28. Trettnak, W., and Wolfbeis, O. S. (1989). A fibre optic lactate biosensor with an oxygen optrode as the transducer, *Anal. Lett., 22*: 2191.
29. Trettnak, W., and Wolfbeis, O. S. (1990). A fiberoptic cholesterol biosensor with an oxygen optrode as the transducer, *Anal. Biochem., 184*: 124.
30. Kulp, T. J., I. Camins, S. M. Angel, C. Munkholm, and Walt, D. R. (1987). Polymer immobilized enzyme optrodes for the detection of penicillin, *Anal. Chem., 59*: 2849.
31. Trettnak, W., Leiner, M. J. P., and Wolfbeis, O. S. (1989). Fibre-optic glucose sensor with a pH optrode as the transducer, *Biosensors, 4*: 15.
32. Kulp, T. J., Camins, I., and Angel, S. M. (1988). Enzyme based fiber optic sensors, *Proc. SPIE, 906*: 134.
33. Fuh, M. R. S., Burgess, L. W., and Christian, G. D. (1988). Single fiber-optic fluorescence enzyme-based sensor, *Anal. Chem.,60*: 433.
34. Luo S., and Walt, D. R. (1989). Avidin-Biotin coupling as a general method for preparing enzyme-based fiberoptic sensors, *Anal. Chem.,61*: 1069.
35. Rhines, T. D., and Arnold, M. A. (1989). Fiber-optic biosensor for urea based on sensing of ammonia gas, *Anal. Chim. Acta, 227*: 387.
36. Völkl, K. P., Opitz, N., and Lübbers, D. W. (1980). Continuous measurement of concentrations of alcohol using a fluorescence photometric enzymatic method, *Fresenius' Z. Anal. Chem., 301*: 162.
37. Hirschfeld, T., Miller, F., Thomas, S., Miller, H., Milanovich, F., and Gaver, R. W. (1987). Laser-fiberoptic "optrode" for real time in vivo blood carbon dioxide level monitoring, *J. Lightwave Technol., LT-5*: 1027.
38. Kawabata, Y., Kamichika, T., Imasaka, T., and Ishibashi, N. (1989). Fiber-optic sensor for carbon dioxide with a pH indicator dispersed in a poly(ethylene glycol) membrane, *Anal. Chim., Acta, 219*: 223.

39. Dremel, B. A. A., Schmid, R. D., and Wolfbeis, O. S. (1991). Comparison of two fibre-optic L-glutamate biosensors based on the detection of oxygen or carbon dioxide, and application of such sensors for glutamate determination in combination with flow-injection analysis, *Anal. Chim. Acta.*, (in press).
40. Kautsky, H., and Hirsch, A. (1935). Nachweis geringster Sauerstoffmengen durch Phosphoreszenztilgung, *Z. Anorg. Allg. Chem.*, 222: 126.
41. Kroneis, H. W., and Marsoner, H. J. (1983). A fluorescence-based sterilizable oxygen probe for use in bioreactors, *Sensors Actuators, 4*: 587.
42. Opitz, N., Graf, H. J., and Lübbers, D. W. (1988). Oxygen sensor for the temperature range 300 to 500 K based on fluorescence quenching of indicator-treated silicone rubber membranes, *Sensors Actuators, 13*: 159.
43. Boisdè, G., and Perez, J. J. (1987). Miniature chemical optical fiber sensors for pH measurements, *Proc. SPIE, 798*: 238.
44. Fuh, M. R. S., Burgess, L. W., Hirschfeld, T., Christian, G. D., and Wang, F. (1987). Single fibre optic fluorescence pH probe, *Analyst, 112*: 1159.
45. Wolfbeis, O. S., and Posch, H. E. (1986). Fibre-optic fluorescing sensor for ammonia, *Anal. Chim. Acta, 185*: 321.
46. Yerian, T. D., Christian, G. D., and Ruzicka, J. (1986). Enzymatic determination of urea in water and serum by optosensing flow injection analysis, *Analyst, 111*: 865.
47. Yerian, T. D., Christian, G. D., and Ruzicka, J. (1988). Flow injection analysis as a diagnostic technique for development and testing of chemical sensors, *Anal. Chim. Acta., 204*: 7.
48. Dremel, B. A. A., Schaffar, B. P. H., and Schmid, R. D. (1989). Determination of glucose in wine and fruit juice based on a fibre-optic glucose biosensor and flow injection analysis, *Anal. Chim. Acta, 225*: 293.
49. Dremel, B. A. A., Trott-Kriegeskorte, G., Schaffar, B. P. H., and Schmid, R. D. (1989). L-lactic acid determination in milk products based on a fibre optic biosensor and flow injection analysis (FIA). In: *Biosensors: Application in Medicine, Environmental Protection and Process Control*, GBF monographs Vol. 13 (R. D. Schmid, and F. Scheller, Eds), VCH Verlagsgesellschaft, Weinheim, pp. 225–228.
50. Schaffar, B. P. H., Dremel, B. A. A., and Schmid, R. D. (1989). Ascorbic acid determination in fruit juices based on a fibre optic ascorbic acid biosensor and flow injection analysis. In: *Biosensors Application in Medicine, Environmental Protection and Process Control,* GBF monographs Vol. 13 (R. D. Schmid and F. Scheller, Eds), VCH Verlagsgesellschaft, Weinheim, pp. 229–232.
51. Bardeletti, G., Sechaud, F. and Coulet, P. R. (1986). A reliable L-lactate electrode with a new membrane for enzyme immobilization for an amperometric assay of lactate, *Anal. Chim. Acta, 187*: 47.
52. Scheller, F. W., Pfeiffer, D., Schubert, F., Renneberg, R., and Kirstein, D. (1987). Application of enzyme based amperometric biosensors to the analysis of "real" samples. In: *Biosensors: Fundamentals and Applications* (A. P. F. Turner, I. Karube, and G. S. Wilson, Eds), Oxford University Press, Oxford, p. 331.
53. Schaffar, B. P. H. (1988). Unpublished results.
54. Mansouri, S., and Schultz, J. S. (1984). Optical glucose sensor based on reversible competitive binding. In: *Proceedings of the IEEE/NSF Symposium on Biosensors,* Los Angeles (A. R. Potvin and M. R. Neuman, Eds), pp. 112–115.
55. Mansouri, S., and Schultz, J. S. (1984). A miniature optical sensor based on affinity binding, *Bio/Technology, 1984*: 885.

56. Srinivasan, K. R., Mansouri, S., and Schultz, J. S. (1986). Coupling of concanavalin A to cellulose hollow fibers for use in glucose affinity sensors, *Biotechnol. Bioeng., 28*: 233.
57. Meadows, D., and Shultz, J. S. (1988). Fiber-optic biosensors based on fluorescence energy transfer, *Talanta, 35*: 145.
58. Schaffar, B. P. H., and Wolfbeis, O. S. (1989). New optical chemical sensors based on the Langmuir-Blodgett technique, *Proc. SPIE, 990*: 122.
59. Okahata, Y., Tsuruta, T., Ijiro, K., and Ariga, K. (1988). Langmuir-Blodgett films of an enzyme-lipid complex for sensor membranes, *Langmuir, 1988*: 1373.
60. Wolfbeis, O. S. (1988). Fiber optical fluorosensors in analytical and clinical chemistry. In: *Molecular Luminescence Spectroscopy, Methods and Applications,* Part 2 (S. G. Schulman, Ed), John Wiley & Sons, New York, Chap. 3, pp. 129–282.
61. Arnold, M. A., and Rechnitz, G. A. (1987). Biosensors based on plant and animal tissue. In: *Biosensors: Fundamentals and Applications* (A. P. F. Turner, I. Karube, and G. S. Wilson, Eds) Oxford University Press, Oxford, pp. 30–59.
62. Karube, I. (1987). Micro-organism based sensors. In: *Biosensors: Fundamentals and Applications* (A. P. F. Turner, I. Karube, and G. S. Wilson, Eds), Oxford University Press, Oxford, pp. 13–29.
63. Okazaki, T., Imasaka, T., and Ishibashi, N. (1988). Optical-fiber sensor based on the second harmonic emission of a near-infrared semiconductor laser as a light source, *Anal. Chim. Acta, 209*: 327.
64. Lieberman, R. A., Blyler, L. L., and Cohen, L. G. (1990). A distributed fiber optic sensor based on cladding fluorescence, *IEEE J. Lightwave Technol., 8*: 212.

9

Fluorophore- and Chromophore-Based Fiberoptic Biosensors

Mark A. Arnold

University of Iowa
Iowa City, Iowa

INTRODUCTION

This chapter focuses on fiberoptic biosensors (FOB) in which a chemical mediation step is not required to link the biochemical reaction with the optical measurement. For these nonmediated FOB the optically measured species actively participates in the biochemical reaction. These FOB are constructed by immobilizing either an enzyme or an antibody at the distal end of a fiberoptic probe. For enzyme-based sensors the immobilized enzyme catalyzes either the production or consumption of a fluorophore or chromophore and the resulting fluorescence or absorbance is measured through the fiberoptic probe. Dehydrogenase reactions coupled with the fluorometric detection of the reduced form of nicotinamide adenine dinucleotide (NADH) have been used extensively for this purpose. For antibody-based FOB complexation between the immobilized antibody and a fluorescent or chromophoric antigen is monitored. Typically the rate of complexation is related to the concentration of the antigen.

ENZYME-BASED NONMEDIATED FIBEROPTIC BIOSENSORS

The general design of an enzyme-based nonmediated FOB includes an enzyme layer immobilized at the distal end of a fiberoptic sensing probe. The enzyme catalyzes a reaction that either produces or consumes an optically measurable species. The fiberoptic probe is used to measure this species through either an absorbance or fluorescence measurement. When the measured species is produced by the enzyme reaction, a steady-state condition is established when the rate of product generation is balanced by the rate at which this product diffuses away from the sensing region. To a large extent heterogeneous reaction kinetics control the response. The rate of product generation can be controlled either by the rate of the reaction or by the rate at which the substrate diffuses to the enzyme active region. In either case a steady-state concentration of the measured species results in a corresponding steady-state signal related to the concentration of the analyte through a calibration curve.

The fiberoptic probe can be either a single fiber or a bundle of fibers. A bifurcated fiberoptic bundle is the most convenient type of probe to use because it provides high optical throughput and it can be handled with ease. The enzyme is immobilized at the common end of the bifurcated probe. One arm is connected to the source optics, and the other is connected to the detection optics. The specifics of the source and detection optics depend on the application and the species to be measured. In any case, the intensity from the sensor tip is monitored as a function of time to identify the corresponding steady-state signal.

Chromophore Detection

The principle of a chromophore-based nonmediated FOB was first demonstrated with a model system that used alkaline phosphatase as the immobilized enzyme, *p*-nitrophenylphosphate as the substrate, and *p*-nitrophenoxide as the detected product (1). Alkaline phosphatase catalyzes the hydrolysis of *p*-nitrophenylphosphate to generate *p*-nitrophenoxide and phosphate according to the reaction

This FOB is constructed by covalently attaching the enzyme to a mesh of nylon and then positioning this membrane at the common end of a bifurcated fiberoptic probe. A schematic view of the sensing region is given in Figure 1. A miniature transmission cell is created at the sensor tip by holding the nylon membrane a short distance from the end of the probe. Incident radiation exits the fiberoptic probe and travels the short distance through the solution to the membrane. A portion of this incident light is backscattered by the membrane, collected by the probe, and detected by a photomultiplier tube (PMT) detector.

p-Nitrophenoxide has an absorbance maximum at 404 nm with a corresponding molar absorptivity of 19050/M/cm. The light intensity at this wavelength is monitored simply by placing an interference filter with a peak transmission at 404.7 nm just before the PMT detector. As *p*-nitrophenoxide is formed in the

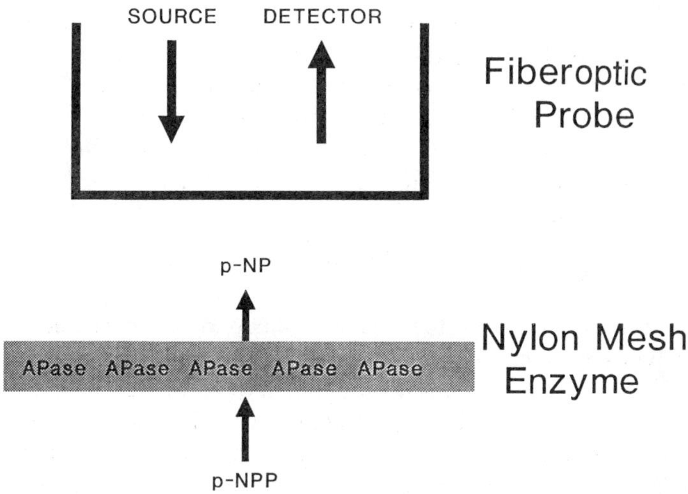

Figure 1 The *p*-nitrophenylphosphate (p-NPP) biosensor based on the production of the chromophore *p*-nitrophenoxide (p-NP) by immobilized alkaline phosphatase (APase).

sensing region a fraction of the incident radiation is absorbed, thereby decreasing the intensity of the measured light. The amount of light transmitted is measured relative to the intensity measured in the absence of the chromophore ($T = I/I_0$), and the effective absorbance is calculated from the transmittance $A = -\log T$.

The measured absorbance is related in a linear manner to the concentration of *p*-nitrophenylphosphate in the sample solution. The calibration curve is linear over the concentration range from 10 to 380 μM. No response is obtained from *p*-nitrophenylphosphate without alkaline phosphatase.

Although the model system based on alkaline phosphatase effectively demonstrates the feasibility of nonmediated FOB based on chromophoric species, no other such FOB have been reported. This lack of activity is a reflection of the fact that few biologically significant analytes exist for which a suitable enzyme reaction is available to produce a chromophore. On the other hand, FOB based on the detection of a fluorophore offer a wide range of analytically relevant possibilities.

Fluorophore Detection

NADH-Based Biosensors

General Concept. The principle of a fluorescence-based, nonmediated FOB has been demonstrated by using immobilized lactate dehydrogenase (LDH) in conjunction with the fluorometric detection of NADH (2,3). Relative to many fluorescent compounds used for analytical purposes, NADH is considered a weak fluorophore. The molar absorptivity and quantum efficiency of NADH are approximately 6.22×10^{-3}/M/cm and 0.02, respectively (4). Nevertheless, submicromolar concentrations of NADH can be measured by exciting with 350 nm radiation while monitoring the resulting emission at 450 nm.

LDH catalyzes the reaction

$$\text{Lactate} + \text{NAD}^+ \rightleftharpoons \text{pyruvate} + \text{NADH}$$

This reaction has been used to demonstrate the feasibility of biosensors based on the production and consumption of NADH. In the NADH production mode NAD^+ must be present in the sensing region and the addition of lactate results in the conversion of this NAD^+ to NADH. Higher fluorescence intensities are expected for higher concentrations of lactate in the sample. In the NADH consumption mode NADH must be present in the sensing region and the addition of pyruvate causes a decrease in the NADH level, which is detected as a decrease in the fluorescence signal.

The LDH-based FOB is constructed in a similar fashion as the alkaline phosphatase-based sensor. The key differences are the immobilization of LDH

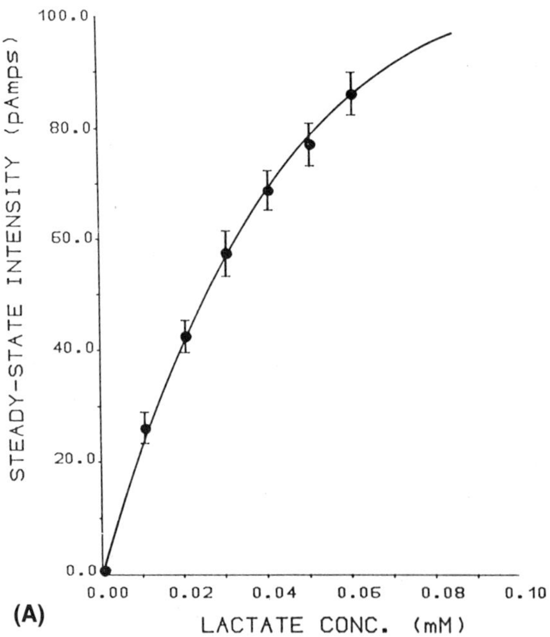

Figure 2 Responses curves by the lactate dehydrogenase biosensor for lactate (A) and pyruvate (B).

on the nylon membrane and the measurement of a fluorescence signal. An interference filter isolates the excitation radiation from the light source, and this radiation is focused into a quartz optical fiber. This fiber guides the excitation radiation to the sensing region. The enzyme (LDH) is covalently attached to a mesh of nylon, and this active membrane is positioned a short distance from the probe tip. The excitation radiation exits the quartz fiber and illuminates the LDH-containing membrane. NADH in this illuminated region is excited, and a fraction of the emitted radiation is collected by a ring of plastic or glass optical fibers. The collected radiation is guided by the fibers to a detection unit, where 450 nm radiation is isolated by a second interference filter. The intensity of the resulting fluorescence is measured with a common PMT detector.

Figure 2 shows calibration curves for lactate and pyruvate when the LDH-based FOB is operated in the NADH production and consumption modes, respectively. As expected, the measured fluorescence intensity increases as the concentration of lactate increases because of the generation of NADH. Likewise, the expected decrease in fluorescence is observed for higher concentrations of pyruvate owing to the consumption of NADH. The principal analytical response properties of this sensor are summarized in Table 1.

The analytical performance of the LDH-based FOB depends on the pH of the

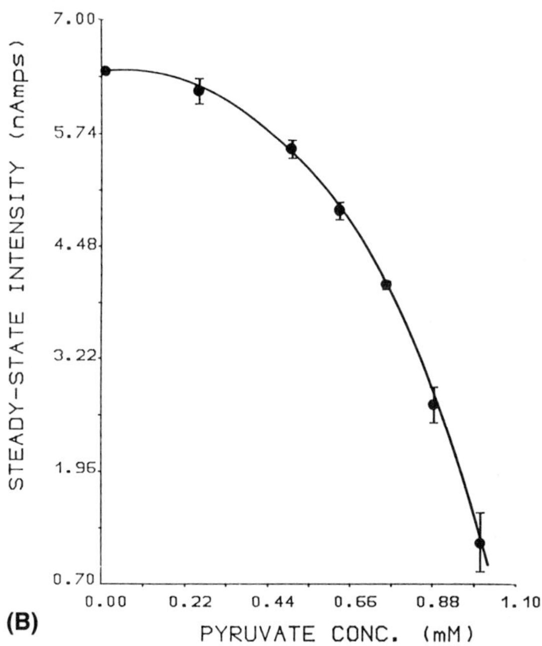

(B)

Table 1 Response Properties of LDH-Based FOB

Property	Lactate response	Pyruvate response
Cofactor concentration, mM	3	1
Detection limit, μM	2	25
Dynamic range, μM	2–50	0.3–0.9
Response times, min	5–12	3–10
Recovery times, min	0–3	0–2
Lifetime, days	4	7

sample solution, the concentration of the cosubstrate, and amount of active enzyme immobilized at the sensor tip (enzyme loading). The best response is obtained when the pH of the sample matches the optimum pH of the enzyme-catalyzed reaction. For LDH the optimal pH for the oxidation of lactate is 8.6; that for the reduction of pyruvate is 7.4. Another critical parameter is the concentration of the cosubstrate. In the NADH production mode the best analytical performance is achieved with high levels of NAD^+ at which the sensor response is independent of the cosubstrate concentration. The situation is more complicated in the NADH consumption mode. In this case the response from the sensor corresponds to a decrease in fluorescence intensity superimposed on a relatively high background signal. The limit of detection is strongly related to the concentration of NADH, with poor detection limits resulting from high levels of NADH. Finally, enzyme loading is an important consideration. Higher amounts of active enzyme provide greater stability, longer sensor lifetimes, better selectivity, and a wider dynamic range (4).

The potential of biosensors based on the direct detection of NADH is the large number of dehydrogenase enzymes that are available for sensor development. For example, Narayanaswamy and Sevilla reported a FOB for glucose based on the immobilization of glucose dehydrogenase at the end of a fiberoptic probe (5). The production of NADH according to the following reaction is monitor by a fluorescence measurement:

$$\text{Glucose} + NAD^+ \longrightarrow \text{gluconic acid} + \text{NADH}$$

The analytical signal used in this example is the change in fluorescence intensity during the first 5 min after exposing the sensor to the sample. This intensity change is related to the concentration of glucose in a linear manner. The detection limit for glucose is 0.6 mM. Many other dehydrogenase enzymes are available for sensor development.

Nucleotide Immobilization. The major disadvantage of the NADH-based biosensor scheme is the need to add the cosubstrate to the sample solution. One way to overcome this limitation is to immobilize the cosubstrate along with the enzyme. The concept of biosensors based on immobilized cosubstrates has been around for many years (6), and this concept has been successfully demonstrated for biosensors based on electrochemical detection (7).

Scheper and Buckmann have reported examples of fiberoptic sensing schemes in which a macromolecular form of NAD^+ is retained in an enzyme chamber (8,9). Sensing schemes for lactate, pyruvate, and ethanol have been reported. Figure 3 is a schematic representation of their sensing region. The analytical reactions occur in a small chamber at the distal end of a fiberoptic probe. The probe supplies the excitation radiation and collects the emitted radiation from the fluorescence of NADH in this chamber. Both the enzymes and the cosubstrates are retained in the reaction chamber by an ultrafiltration membrane. As the sample passes through this chamber LDH catalyzes the oxidation of lactate to form pyruvate with the concomitant reduction of NAD^+ to form NADH. Under the operating conditions used in these experiments pyruvate must be removed to drive the LDH reaction toward the formation of NADH. Removal of pyruvate is accomplished by a second reaction that is catalyzed by the enzyme glutamate-pyruvate transaminase (GPT).

The system is operated in an automated flow arrangement in which a discrete volume of each sample flows through the chamber. NADH accumulates accord-

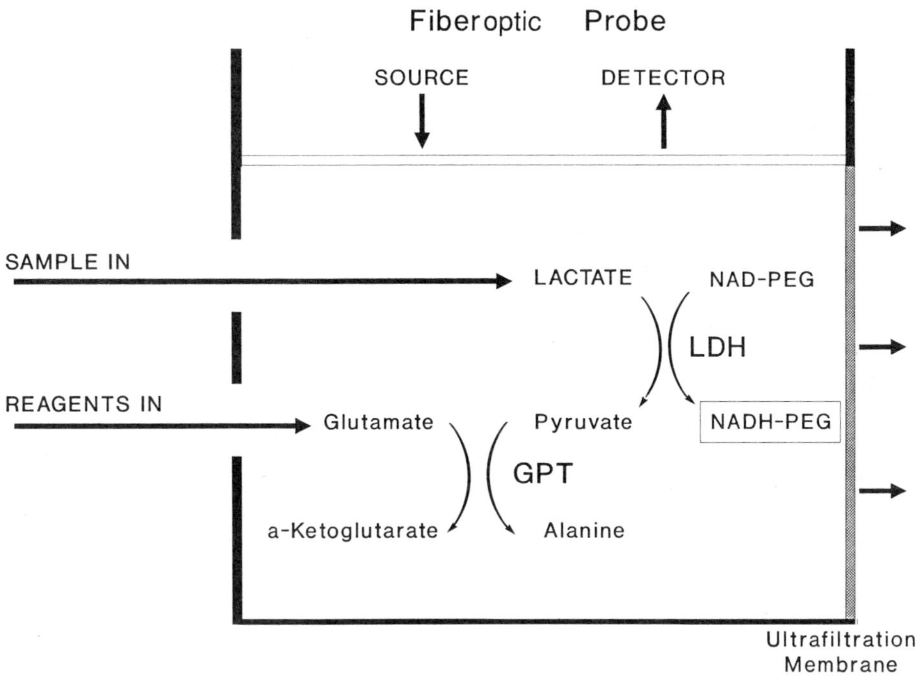

Figure 3 The flow-through reaction chamber used for the fiberoptic determination of lactate with immobilized NAD^+/NADH.

ing to the amount of lactate present and the corresponding fluorescence is measured. A calibration curve relates this fluorescence value to the concentration of lactate in the sample. NADH produced by the lactate in the sample must be oxidized back to NAD^+ before the next sample passes through the chamber. NAD^+ is regenerated by passing a high concentration of pyruvate through the chamber.

NAD^+ and NADH are chemically modified to allow physical retention by the ultrafiltration membrane. Polyethylene glycol conjugates of NAD^+ (PEG-NAD^+) are prepared with molecular weights of approximately 20,000 daltons. These conjugates are easily retained by the membrane, and they still react efficiently with the dehydrogenase enzyme. As a result there is no need to add the cosubstrate to the sample solution. Glutamate, on the other hand, is required to drive the analytical reaction and must be added to the sample.

A noteworthy feature is the accumulation of NADH as lactate passes through the chamber. The accumulation of NADH offers lower detection limits and higher precision compared to conventional biosensors. For samples with low concentrations of lactate accurate measurements can be obtained simply by passing a large volume of the solution through the chamber. NADH continues to accumulate as lactate enters the chamber, thereby building up to a measurable level. In contrast, a biosensor depends on the establishment of a steady-state condition and the concentration of NADH is limited by the concentration of lactate in the sample. A significant amplification can be achieved with an accumulation sensor by increasing the volume of the sample solution. In addition, the accumulation of NADH is inherently more precise than the steady-state condition required by a conventional biosensor. Hence greater reproducibility is expected for the accumulation design.

Internal Enzyme Biosensors. The need to add NAD^+ to the sample solution can also be eliminated by using an internal enzyme biosensor configuration (10). In this configuration a perm-selective membrane is used to separate the sample solution from an internal enzyme solution, which contains all the reagents needed for the analytical reaction. The rate of this analytical reaction is monitored and related to the analyte concentration.

The concept of the internal enzyme biosensor has been demonstrated with a FOB for ethanol. The sensor design and the processes involved in the sensor response are shown schematically in Figure 4. A thin microporous Teflon membrane separates the sample from the internal enzyme solution. The principal components of this solution are the enzyme alcohol dehydrogenase (ADH) and NAD^+. Ethanol vapor diffuses across the gas-permeable Teflon membrane and enters the internal solution, where it is oxidized to acetaldehyde with the formation of NADH. Semicarbazide is used to drive this reaction toward the formation of NADH by removing the acetaldehyde as it is formed. A fiberoptic probe is positioned in the internal solution to follow the production of NADH by a fluorescence measurement. Conditions are set so that the rate at which ethanol

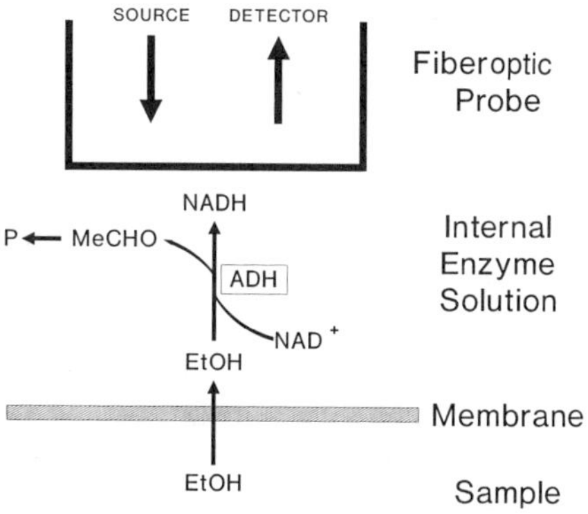

Figure 4 Configuration of the internal enzyme biosensor for ethanol.

diffuses into the internal solution is rate limiting, thereby maintaining a first-order relationship between the rate of NADH production and the concentration of ethanol in the sample.

The principal advantage of the internal enzyme concept is the possibility of making in situ measurements without perturbing the system. This advantage is possible because the solution in which the analytical reaction occurs is physically separated from the sample solution. The analyte must be able to cross the boundary before it can be detected. This type of sensor is restricted to volatile analytes when a gas-permeable membrane is used to separate the two solutions. With the development of other perm-selective membranes the concept of internal enzyme sensors could be extended to other classes of analytes.

Whole-Cell Biosensors. The intracellular levels of NADH (NADPH) in microbial cells can be measured by fluorescence spectroscopy. Such measurements have been suggested as a way to establish the status of bioreactors (11,12). This concept centers around the idea that a stress on the microbial cells will result in a decrease in the ability of these cells to oxidize NADH back to NAD^+. Scheper et al. (11) found that a decrease in the oxygen content in the growth medium causes an increase in the fluorescence of intracellular NADH. Presumably, restricted levels of oxygen limit the capacity of these cells to oxidize NADH by the oxidative phosphorylation process. In addition, compounds that exert their toxic effects by blocking the electron transport chain, such as cyanide, can likewise be detected as an increase in fluorescence (12). A miniature fiberoptic fluorescence probe has been used in these experiments. The fluorescence from intracellular NADH has been measured as a sample of the cells in the reactor flow through the sensing region of the probe.

A similar system has been reported in which the fluorescence from endogenous chlorophyll of algae is used to detect chemicals that are toxic to these cells (13,14). Herbicides and pesticides typically inhibit the electron transport of the photosynthetic process. By blocking this electron transport path, relaxation of excited chlorophyll is more likely through a fluorescence pathway. The result is an enhancement in the fluorescence of cellular chlorophyll. Experimentally, fluorescence measurements are made on a sample of cells taken from a reactor. These cells are mixed with an aliquot of a water sample, and the fluorescence is monitored continuously over a period of 2 min. The measured fluorescence is compared to a blank sample, and the magnitude of the excess fluorescence is related to the strength of the herbicide.

Although these devices are not true biosensors because the biological element is not directly coupled to the sensing device, the concept of toxin sensing by immobilizing cells at the surface of a fiberoptic probe is clearly possible. Indeed, biosensors for the detection of carcinogens and mutagens have been developed based on electrochemical detection (see discussion in Ref. 15). Similar devices using optical detection can be easily envisioned.

Intrinsic Fluorescence of Enzymes

Trettnak and Wolfbeis reported a novel biosensing scheme based on changes in the intrinsic fluorescence of enzymes that contain flavin-based cofactors (16–18). These researchers have discovered significant increases in the intrinsic fluorescence of glucose oxidase (GOD) upon binding glucose (16–18). Similarly, the fluorescence of lactate monooxygenase (LMO) increases upon binding lactate (17,18).

The proposed response mechanism is based on differences in the fluorescence properties of the reduced and oxidized forms of the cofactors involved. These enzymes are flavoproteins, with flavin-adenine dinucleotide (FAD) used in the GOD reaction and flavin mononucleotide (FMN) used in the LMO reaction. The following simplified reaction sequences describe the two enzyme systems:

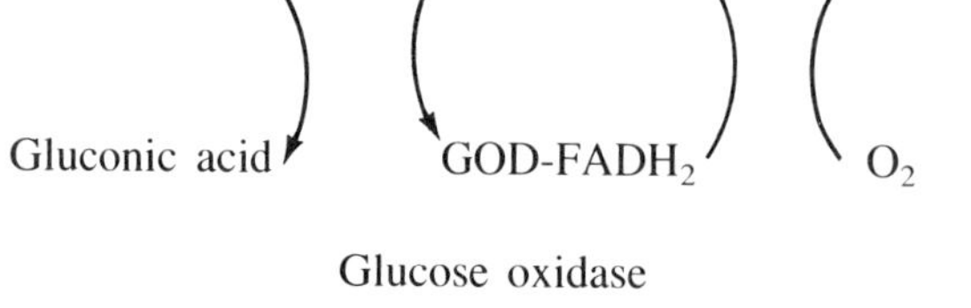

Glucose oxidase

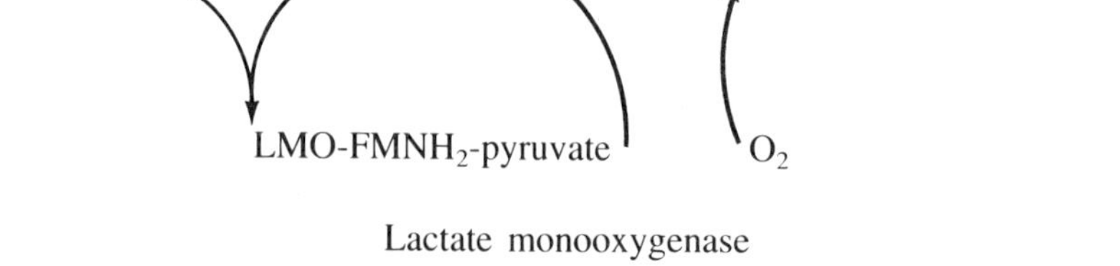

Lactate monooxygenase

Glucose is initially oxidized by the enzyme to produce gluconic acid and the reduced form of the flavin, $FADH_2$. Subsequently, $FADH_2$ is oxidized back to FAD by reacting with oxygen. Likewise, pyruvate and $FMNH_2$ are formed temporarily during the oxidation of lactate. In both cases the fluorescence of the reduced form of the nucleotide is monitored. Presumably the observed fluorescence signal is observed when the rate of flavin reduction is faster than the subsequent oxidation.

Experimentally a sharp change in fluorescence is observed over a narrow concentration range of the substrate. Specifically, the change in fluorescence is observed over the concentration ranges from 1.5 to 2.0 mM and from 0.5 to 1.0 mM for glucose and lactate, respectively. By decreasing the amount of oxygen in the solution, lower concentrations of substrate are needed to induce the change in fluorescence. In both systems several minutes are required to establish a steady-state condition, thereby providing a stable response. Response times increase as the substrate concentration decreases. Both sensors operate over a wide pH range.

The suggested analytical configuration for the measurement of glucose and lactate is a flowing system with the enzyme physically retained at the tip of a fiberoptic probe. The sample flows past the enzyme layer for a specified period of time (i.e., 65 sec), and the fluorescence of the enzyme is measured as a function of time. The area under the fluorescence-versus-time peak is calculated and used as the analytical signal. Kinetic measurements of this type extend the dynamic range to nearly one order of magnitude.

A key feature of biosensors based on the intrinsic fluorescence of the enzyme is the reversibility of the system. Unlike NADH-based FOB, there is no need to add reagents to the sample as long as endogenous oxygen levels are suitable and constant. Disadvantages include the narrow dynamic range and the rather high limits of detection.

Sensors to Measure Enzyme Activity

Most biosensors are designed to measure the concentration of the enzymatic substrate by immobilizing the enzyme at the tip of the transducer element. Wolfbeis turned the situation around to develop a FOB designed to measure the enzyme activity by immobilizing the substrate at the tip of the fiberoptic probe (19). The concept has been demonstrated for a group of esterase enzymes based on the immobilization of a fatty acid ester of 1-hydroxypyrene-3,6,8-trisulfonate (HPTS). A schematic diagram of this sensing scheme is shown in Figure 5. The HPTS ester is immobilized on an anion-exchange membrane at the distal end of a single optical fiber. This fiber supplies the excitation radiation to the sensor tip and also collects a fraction of the resulting emitted radiation. Although the HPTS ester is nonfluorescent, the enzymatic hydrolysis of this ester produces a highly fluorescent species that is retained on the membrane. The rate of hydrolysis is measured by following the fluorescence intensity increase as a function of time.

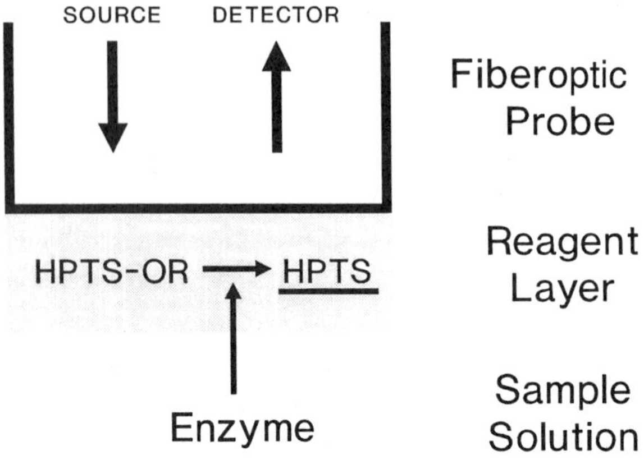

Figure 5 The fiberoptic biosensor for measuring enzyme activity.

Faster rates of hydrolysis are measured for larger amounts of enzyme activity in the sample. Rates are linearly related to enzyme activity.

The small size and probelike configuration of this enzyme activity sensor suggest that a probe of this type may be useful for direct in vivo measurements of enzyme activities. Potential toxic effects caused by the products of the enzymatic reaction are minimized in this example because the principal product is retained at the sensor tip and is not released into the sample solution. The nonreversible nature of this type of sensor, however, restricts its application to special situations in which periodic "spot tests" are acceptable.

ANTIBODY-BASED NONMEDIATED FIBEROPTIC BIOSENSORS

In addition to enzyme-based systems, considerable effort has been given to the development of FOB based on biological complexation reactions (15). The central idea is to immobilize a specific binding agent in the sensing region of a fiberoptic probe and to monitor the binding reaction through an optical measurement. Typically, a chemical mediator is required to follow this binding reaction. Fluorescent conjugates are by far the most popular type of mediator for this type of sensor. Examples include the Schultz type of glucose sensor, in which glucose and fluorescently labeled dextran compete for a fixed number of concanavalin A binding sites (20), and the reversible immunosensors reported recently by Miller and Anderson, in which a fluorescence energy-transfer detection strategy is used in conjunction with competitive binding between the antigen and fluorescein-labeled antigen for a limited number of binding sites associated with a Texas red-labeled antibody (21). In comparison, relatively few nonmediated FOB based on complexation reactions have been reported. Nevertheless, nonmediated antibody-based FOB have been developed based on light scattering, absorbance, and fluorescence measurements.

The most general approach to nonmediated biosensing is the light-scattering scheme introduced by Sutherland and coworkers (22). In this system a selected antibody is adsorbed on the surface of a silica slide. The resulting slide serves as an optical waveguide through which 440 nm radiation is transmitted. These researchers found that the intensity of this transmitted light decreases as the complexation reaction between the antibody and antigen proceeds. They hypothesize that the formation of the antigen-antibody complex scatters the radiation in the evanescent field at the waveguide/solution interface, thereby reducing the amount of light transmitted through the waveguide. Preliminary results suggest detection limits of 7 nM are possible for proteinaceous antigens.

The most attractive feature of the evanescent-field light-scattering arrangement is the generic nature of this approach. Other key features include the elimination of the requirements to add labeled conjugates and to perform a physical separation. Indeed, that the evanescent field penetrates only a short distance into the aqueous medium (i.e., less than a wavelength or approximately 250 nm in this example) greatly simplifies the experimental protocol.

Immunosensing based on the absorbance of light in the evanescent field during the binding of a chromophoric antigen has also been demonstrated by the Sutherland research team (23,24). The feasibility of this approach has been established by detecting the antibody binding of methotrexate on the surface of a quartz optical waveguide. In this case the antibody is covalently attached to the quartz surface, and 310 nm radiation is transmitted through the waveguide (evanescent field penetration depth $\approx$ 90 nm). Methotrexate contains several chromophoric groups, including a benzoyl group that absorbs in the region of 300 nm. The amount of light transmitted through the waveguide decreases when methotrexate is added to a solution in contact with the treated quartz waveguide. The rate of absorbance change is related to the concentration of methotrexate in solution. A detection limit of 0.26 μM is estimated.

Overall, nonmediated sensors based on evanescent field measurements are less sensitive than their mediated counterparts. Nonspecific adsorption at the waveguide/solution interface is a potential problem that may limit the general success of this approach for routine analyses. Even so, internal reflection spectroscopy is clearly a potentially useful tool to study the fundamental chemistry of biological complexation reactions.

Nonmediated antibody-based FOB have also been developed in which the natural fluorescence of the antigen is measured (25). This approach has been demonstrated by Sepaniak and coworkers (26) for the benzo[a]pyrene metabolite benzopyrenetetraol (BPT). In this work monoclonal antibodies for BPT are retained in a small reservoir at the distal end of an optical fiber probe. BPT in the solution enters this reservoir and complexes with the antibody. The fluorescence of bound BPT is measured through the optical fiber. The antibody concentrates the BPT at the probe tip, thereby providing a strong signal for low concentrations. This arrangement also offers selectivity enhancement over spectrally similar compounds because only BPT accumulates in the sensing region. Unfortunately, operation of this sensor under equilibrium conditions is not practical because the response times are on the order of hours. The device can be used in a fixed-time assay, however, in which the extent of fluorescence increase is measured over a period of 15 min. The relationship between intensity change and BPT concentration is linear with a detection limit of 0.5 nM.

Recently this research team reported the construction of antibody-based FOB in which a capillary reagent delivery system is used to regenerate the chemistry in the sensing region of the probe (26). This technology allows the in situ replacement of old antibody with an aliquot of fresh antibody. Semicontinuous monitoring of BPT is possible with this approach.

TRENDS AND PROSPECTS

Clearly, a vast amount of research must be done before the broad merits of nonmediated FOB will be firmly established. The field is in its infancy and many exciting results are anticipated. Several conclusions can be made, however, based on the work completed to date.

In the area of enzyme-based biosensors, the major advance has been sensors based on the detection of NADH. At present these sensors could find numerous applications in biomedical research. It is not likely, however, that this approach will offer solutions to general sensing problems until the limitation of reagent addition is solved. That either NAD^+ or NADH must be added to the sample solution is a major impediment to the implementation of this technology. Cofactor immobilization is only a slight improvement because the sensor is still limited by the need to regenerate the active form of the nucleotide. Perhaps the use of continuous reagent release polymers recently described by Walt and coworkers (27) can be adapted for NADH-based biosensors. In this way the nucleotide can be slowly and continuously supplied in the sensing region with minimal alterations to the sample. Alternatively, Cammann and coworkers (28) recently described a dye that can be coimmobilized with the dehydrogenase enzyme and serve as a reversible redox couple in place of NAD^+/NADH. Unfortunately, no details have been released concerning the chemistry of this system.

Biosensors for the measurement of enzymes is an interesting approach that has seen little activity since it was described several years ago. Such sensors can provide spatially resolved information about the distribution of enzyme activity within a sample. Clearly, the full merits and potential of this approach must be established.

Finally, it must be concluded that nonmediated antibody-based biosensors are inferior to their mediated counterparts. The use of fluorescence labels allows both competitive binding and sequential addition immunoassays. Furthermore, detection limits and sensitivity are superior when strong fluorophores are used as labels.

REFERENCES

1. Arnold, M. A. (1985). Enzyme-based fiber-optic sensor, *Anal. Chem., 57*: 565.
2. Wangsa, J., and Arnold, M. A. (1988). Fiber-optic biosensors based on the fluorometric detection of reduced nicotinamide adenine dinucleotide, *Anal. Chem., 60*: 1080.
3. Wangsa, J. (1989). Chemical sensors: Accuracy of sodium glass electrodes and the development of NADH-based fiber-optic biosensors, Ph.D. Dissertation, University of Iowa.
4. Carr, P. W., and Bowers, L. D. (1980). *Immobilized Enzymes in Analytical and Clinical Chemistry,* Wiley-Interscience, New York.
5. Narayanaswamy, R., and Sevilla, F. (1988). An optical fiber probe for the determination of glucose based on immobilized glucose dehydrogenase, *Anal. Lett., 21*: 1165.
6. Mosbach, K. (1974). Immobilized cofactors and multienzyme systems, *Adv. Exp. Med. Biol., 42*: 187.
7. Pau, C. P., and Rechnitz, G. A. (1984). Bound cofactor/dual enzyme electrode system for L-alanine, *Anal. Chim. Acta, 160*: 141.
8. Scheper, T., and Bückmann, A. F. (1990). A fiber-optic biosensor based on fluorometric detection using confined macromolecular nicotinamide adenine dinucleotide derivatives, *Biosensors Bioelectronics, 5*: 125.
9. Schelp, C., Scheper, T., and Bückmann, A. F. (1989). The use of fluorescence sensors as optical biosensors, In: *Biosensors: Applications in Medicine, Environmental Protection and Process Control* (R. D. Schmid and F. Scheller, Eds), GBF Monographs Vol. 13, VCH Publishers, New York, p. 263.
10. Walters, B. S., Nielsen, T. J., and Arnold, M. A. (1988). Fiberoptic biosensor based on an internal enzyme concept, *Talanta, 35*: 151.
11. Reardon, K. F., Scheper, T., Anders, K. -D., Müller, W., and Bückmann, A. F. (1990). Novel application of fluorescence sensors, *Appl. Biochem. Biotechnol., 24/25*: 363.
12. Scheper, T., Anders, K. D., Freitag, R., Hundeck, H. G., Müller, W., Schelp, C., Bückmann, A. F., and Reardon, K. F. (1989). Biosensor systems for process control in biotechnology. In: *Biosensors: Applications in Medicine, Environmental Protection and Process Control* (R. D. Schmid and F. Scheller, Eds), GBF Monographs Vol. 13, VCH Publishers, New York, p. 253.

13. Noack, U., Boenisch, J., and Schneider, T. (1989). Algae toximeter: A biosensor for water monitoring. In: *Biosensors: Applications in Medicine, Environmental Protection and Process Control* (R. D. Schmid and F. Scheller, Eds), GBF Monographs Vol. 13, VCH Publishers, New York, p. 243.
14. Gohl, W., Krause, H., and Schmidt, C. (1989). Algal fluorescence bioassay with correlation of spontaneous and delayed fluorescence. In: *Biosensors: Applications in Medicine, Environmental Protection and Process Control* (R. D. Schmid and F. Scheller, Eds), GBF Monographs Vol.13, VCH Publishers, New York, p. 247.
15. Arnold, M. A., and Meyerhoff, M. E. (1988). Recent advances in the development and analytical applications of biosensing probes, *CRC Crit. Rev. Anal. Chem., 20*: 149.
16. Trettnak, W., and Wolfbeis, O. S. (1989). Fully reversible fiber-optic glucose biosensor based on the intrinsic fluorescence of glucose oxidase, *Anal. Chim. Acta, 221*: 195.
17. Trettnak, W., and Wolfbeis, O. S. (1989). A fully reversible fiber optic lactate biosensor based on the intrinsic fluorescence of lactate monooxygenase, *Fresenius' Z. Anal. Chem., 344*: 427.
18. Wolfbeis, O. S., and Trettnak, W. (1989). A new method for determination of enzyme substrates by monitoring the intrinsic fluorescence of immobilized enzymes. In: *Biosensors: Applications in Medicine, Environmental Protection and Process Control* (R. D. Schmid and F. Scheller, Eds), GBF Monographs, Vol. 13, VCH Publishers, New York, p. 213.
19. Wolfbeis, O. S. (1986). Fiber-optic probe for kinetic determination of enzyme activities, *Anal. Chem., 58*: 2874.
20. Schultz, J. S., Mansouri, S., and Goldstein, I. J. (1982). Affinity sensor: A new technique for developing implantable sensors for glucose and other metabolites, *Diabetes Care, 5*: 245.
21. Miller, W. G., and Anderson, F. P. (1989). Fiber optic immunochemical sensors for continuous monitoring of hapten concentrations, *Proc. SPIE, 4*: 30.
22. Sutherland, R., Dahne, C., and Place, J. (1984). Preliminary results obtained with a no-label, homogeneous immunoassay for human immunoglobulin G, *Anal. Lett., 17*: 43.
23. Sutherland, R. M., Dahne, C., Place, J. F., and Ringrose, A. S. (1984). Optical detection of antibody-antigen reactions at a glass-liquid interface, *Clin. Chem., 30*: 1533.
24. Sutherland, R. M., and Dahne, C. (1987). IRS devices for optical immunoassays. In: *Biosensors: Fundamentals and Applications* (A. P. F. Turner, I. Karube, and G. S. Wilson, Eds), Oxford University Press, New York, p. 655.
25. Tromberg, B. J., Sepaniak, M. J., Alarie, J. P., Vo-Dinh, T., and Santella, R. M. (1988). Development of antibody-based fiber-optic sensors for detection of a Benzo[a]pyrene metabolite, *Anal. Chem., 60*: 1901.
26. Sepaniak, M. J., Tromberg, B. J., Alarie, J. P., Bowyer, J. R., Hoyt, A. M., and Vo-Dinh, T. (1989). Design considerations for antibody-based fiber-optic chemical sensors: In: *Chemical Sensors and Microinstrumentation* (R. W. Murray, R. E. Dessy, W. R. Heineman, J. Janata, and W. R. Seitz, Eds), ACS Symposium Series 403, Washington, D.C., p. 318.

27. Walt, D. R., Munkholm, C., Yuan, P., Luo, S., and Barnard, S. (1989). Design, preparation, and application of fiber-optic chemical sensors for continuous monitoring. In: *Chemical Sensors and Microinstrumentation* (R. W. Murray, R. E. Dessy, W. R. Heineman, J. Janata, and W. R. Seitz, Eds), ACS Symposium Series 403, Washington, D.C., p. 252.
28. Cammann, K., Jahn, K., Rohen, A., Weiss, T., Wilken, H., and Winter, B. (1989). Comparison of electrochemical with optical biosensors. In: *Biosensors: Applications in Medicine, Environmental Protection and Process Control* (R. D. Schmid and F. Scheller, Eds), GBF Monographs, Vol. 13, VCH Publishers, New York, p. 201.

10

Bioluminescence- and Chemiluminescence-Based Fiberoptic Sensors

Loïc J. Blum and Sabine M. Gautier

CNRS-Université Claude Bernard Lyon 1
Villeurbanne, France

INTRODUCTION

Bio- and chemiluminescence reactions are now widely used for analytical purposes because of the great sensitivity of these methods. Modern instrumentation allows light to be measured at a very low level, and by choosing proper reaction conditions the light intensity can be linearly related to the concentration of one of the reactants over a range of several orders of magnitude and with such a sensitivity that a few picomoles are easily detectable.

The bioluminescence reaction of the firefly allows ATP measurement, whereas bioluminescence bacterial systems are used for NADH or NADPH determination. The chemiluminescent reactions of interest are mainly those involving hydrogen peroxide, since in this case the luminescent process can be coupled with H_2O_2-generating enzyme-catalyzed reactions.

Although the potentialities of immobilized luminescence reagents have been widely explored during the past decade, little attention has been paid to their use in convenient devices and the development of biosensors, in the strict meaning of the definition, that is, small probe types of devices incorporating biologically active materials based on both bioluminescence and chemiluminescence reactions, is just breaking now.

Since in these luminescent reactions the end product of interest is light, fiberoptics associated with a light detector appears to be one of the most convenient transducers for designing a bioluminescence- or chemiluminescence-based true biosensor. Thus in this chapter such fiberoptic based sensors are reviewed in detail. However, considering a larger definition of a biosensor, some analytical devices, which include immobilized luminescent reagent phases, may be mentioned and are also briefly reviewed.

BIOLUMINESCENCE REACTIONS

Bioluminescence, or ''living light,'' a very diverse and unique natural phenomenon, is the emission of light by living organisms. Bioluminescence is a special case of chemiluminescence in which a protein, generally an enzyme (a luciferase), is involved. Among the numerous individual species of luminous organisms, several thousands are known, but the chemistry of only a few of them has been studied in detail and only the firefly and the marine bacteria bioluminescent systems have really been evaluated and exploited for analytical purposes.

The Firefly System

The bioluminescence reaction of the American firefly *Photinus pyralis* has been extensively studied and well reviewed by DeLuca and McElroy (1–3). *In vitro,* firefly luciferase (EC 1.13.12.7) catalyzes the production of light in the presence of ATP, Mg^{2+}, molecular oxygen, and a specific substrate, luciferin. The complex mechanism of the reaction is not yet completely understood, but the overall stoichiometry has been established:

$$\text{ATP} + \text{luciferin} + O_2 \xrightarrow[Mg^{2+}]{\text{luciferase}} \text{AMP} + \text{oxyluciferin} + \text{PPi} + CO_2 + h\nu \qquad (1)$$

When ATP is the limiting component of the reaction the light intensity is proportional to its concentration. The color of the light emission is yellow-green with a peak at 560 nm. The luciferase can be obtained only from the firefly. Luciferin, D-(—)-2-(6'-hydroxy-2'-benzothiazolyl)-Δ^2-thiazoline-4-carboxylic acid, was initially obtained also from the natural source, but since the elucidation of its structure and the achievement of its synthesis (4) synthetic preparations of this substrate have been widely used.

Marine Bacterial System

Bioluminescence is also encountered in some marine bacteria, which are easily cultured (5,6), such as *Vibrio fischeri, Vibrio harveyi, Photobacterium phosphoreum,* or *Photobacterium leiognathi.* The luminescence process is catalyzed by a luciferase (EC 1.14.14.3). In the different luminescent bacteria the same light emission mechanism is involved and the luciferases are similar (6). The light is produced [reaction (2)] in the presence of reduced flavine mononucleotide (FMN), molecular oxygen, and a long-chain aldehyde (*R*-CHO) of eight or more carbon atoms. Decanal is generally used as the aliphatic aldehyde substrate.

$$FMNH_2 + O_2 + R\text{-}CHO \xrightarrow{\text{luciferase}} FMN + R\text{-}COOH + H_2O + h\nu \quad \lambda_{max} = 490\text{nm} \tag{2}$$

Luminescent bacteria contain also a NAD(P)H-FMN oxidoreductase (EC 1.6.8.1), which catalyzes the reaction.

$$NAD(P)H + H^+ + FMN \xrightarrow{\text{oxidoreductase}} NAD(P)^+ + FMNH_2 \tag{3}$$

This reductase has also been isolated from various species of nonluminous aerobic and anaerobic bacteria (7).

Then, using oxidoreductase coupled to the bacterial luciferase, NAD(P)H can be assayed according to the reaction scheme presented in Figure 1. Provided that NAD(P)H is the limiting substrate of the bienzymatic system, the variation in light intensity is proportional to its concentration.

CHEMILUMINESCENT REACTIONS

Several synthetic compounds, such as luminol, lophine, lucigenin, or oxalate derivatives, can undergo oxidation in the presence of hydrogen peroxide to produce light. Among these chemiluminescent reagents, only luminol and oxalate derivatives have really been used for the detection of enzymatically generated hydrogen peroxide.

Luminol Chemiluminescence

The chemiluminescence reaction of luminol (5-amino-2,3-dihydrophthalazine-1,4-dione) with hydrogen peroxide requires the use of a catalyst and/or a cooxidant (8). The oxidation reaction that occurs under alkaline conditions is

$$\text{Luminol} + 2H_2O_2 + OH^- \xrightarrow[\text{cooxidant}]{\text{catalyst or}} \text{3-aminophthalate} + N_2 + 3H_2O + h\nu \tag{4}$$

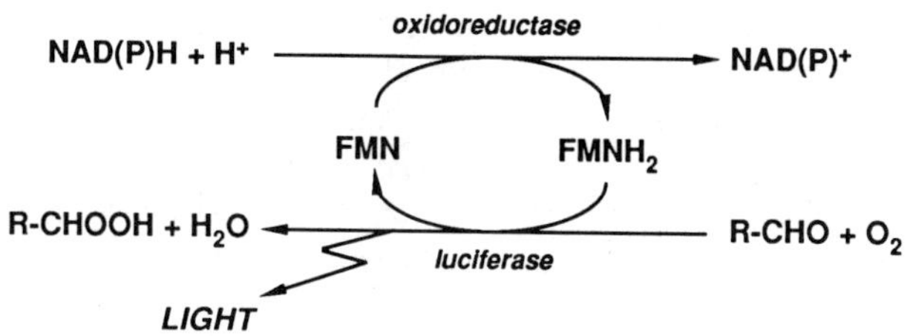

Figure 1 Coupled bacterial bioluminescence reaction.

Ferricyanide, which is both a catalyst and a cooxidant, oxidizes luminol. It is then reduced to ferrocyanide and reoxidized to ferricyanide. Hemin (or hematin) can also be used as a catalyst or cooxidant. Peroxidase (EC 1.11.1.7), which remains unchanged after the reaction, is an example of a catalyst. The use of this enzyme has the advantage over ferricyanide in that the chemiluminescent reaction can proceed at near neutral pH values (9). Another way to produce light from luminol and hydrogen peroxide is electrogenerated chemiluminescence (10,11). Using a positively biased electrode luminol is oxidized, and in the presence of hydrogen peroxide the light emission occurs.

Peroxyoxalate Chemiluminescence

Esters of oxalic acid, such as bis-(2,4,6-trichlorophenyl)oxalate (TCPO) or bis-(2,4,5-trichloro-6-pentoxycarbonyl)oxalate (CPPO), react with hydrogen peroxide to form a high-energy intermediate, 1,2-dioxetanedione (12). This intermediate is capable of transferring its energy to a fluorescent acceptor (fluorescer), which then emits light by returning to the ground state (Fig. 2). Diphenylanthracene and perylene are the most commonly used fluorescers.

The light emitted, proportional to the concentration of hydrogen peroxide, has a λ_{em} dependent on the type of the fluorescer used. Peroxyoxalate chemiluminescence has the advantage of a greater pH range than luminol, but the most efficient peroxyoxalate derivatives are soluble only in organic solvents and dissolution problems of these solvents in aqueous solvents are encountered. This can limit the use of this chemiluminescence reaction for a direct coupling to an $_2O_2$-generating enzymatic reaction.

ANALYTICAL POTENTIAL OF THE LUMINESCENT REACTIONS

Sensitivity and Specificity

The sensitivity of luminescent assays is generally higher than that of spectrophotometric or fluorometric assays and depends on three main factors: the quantum yield of the reaction, the rate of the reaction, and the efficiency of the light detector.

The quantum yield is defined as the ratio between the total number of photons emitted and the number of luminescent reacting molecules. Its value is close to one for the firefly luciferase reaction, whereas it is about 0.2 for bacterial bioluminescence. The luminol-H_2O_2 chemiluminescence reaction is less sensitive since its quantum yield is equal to 0.01. Peroxyoxalate chemiluminescence is known to be the most efficient chemiluminescent reaction, with a quantum yield of about 0.25.

The reaction rate is an important factor because the light detector measures a light flux (light intensity), that is, the number of photons emitted per unit time.

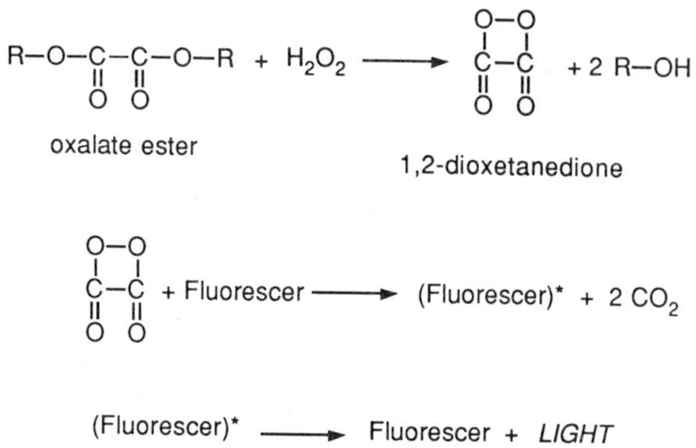

Figure 2 Chemiluminescence reaction of peroxyoxalate.

Thus the light intensity is a measurement of the instantaneous reaction rate, and the faster the reaction proceeds, the higher is the light intensity.

The efficiency of the light detector mainly depends on its absolute sensitivity, that is, its quantum yield, and on its spectral sensitivity. A photomultiplier tube has a lower detection limit than other types of detectors and is still the detector of choice for bio- and chemiluminescence analysis. However, silicon photodiodes can be used if sensitivity is not critical for a particular application. It is important to consider, too, that the detection limit is imposed not only by the sensitivity of the instrument but also by the background light emission inherent in each luminescent reaction.

Bioluminescence Analysis

Using soluble firefly luciferase under optimal reaction conditions, the detection limit is approximately 1×10^{-12} M ATP and the upper limit of the linear range is equal to 1×10^{-6} M.

The firefly luciferase has a remarkable specificity for ATP since the light emission does not occur with other nucleotide triphosphates (13).

Concerning the bacterial luciferase reaction, some reduced flavine analogs can be utilized by luciferase, but with such lower efficiency compared to $FMNH_2$ that the bioluminescence reaction can be considered highly specific for reduced FMN (14).

The two most commonly used light-emitting enzyme systems for the bioluminescent determination of NAD(P)H were isolated from *V. harveyi* and *V. fischeri*. Whereas the oxidoreductase from *V. fischeri* can react with either NADH or NADPH (15), two specific FMN reductases, one for NADH and the other for NADPH, have been isolated from *V. harveyi* (16). Thus in a mixture of NADH and NADPH there are two ways to selectively determine one of these compounds: either by using the specific FMN reductase or by destroying selectively one of the reduced pyridine nucleotides. With soluble reagents the bioluminescent assay of NAD(P)H is linear from 1×10^{-12} to 1×10^{-6} M.

Chemiluminescence Analysis

The specificity of the chemiluminescent reactions is less than that of the bioluminescent assays since the light emission reaction can proceed with oxidants other than hydrogen peroxide. In biochemical or clinical analysis however, the chemiluminescent reactions are coupled with an enzyme-catalyzed process leading to H_2O_2 and specific for a definite analyte, thus making selective the chemiluminescent coupled assay.

The chemiluminescent reactions also constitute a less sensitive analytical tool than the bioluminescent assays since the detection limit is generally at the nanomolar level for both hydrogen peroxide and analytes measured by coupled chemiluminescent assays.

Applications

Numerous papers have been published on the applications of firefly bioluminescence, including, in addition to the direct measurement of ATP, determinations of enzymes and metabolites *via* coupled ATP-converting reactions, and, after intracellular ATP extraction, quantitation of antibiotics, determination of biomass, and detection of bacterial infection (17). More recently, cloning of the firefly luciferase gene and its expression in cells from different organisms have generated applications in the fields of molecular and cell biology (18).

A range of sensitive assays for enzymes, substrates, and cofactors has been developed using the bacterial bioluminescence reaction. These are generally based on the use of a specific dehydrogenase coupled to the oxidoreductase-luciferase light-emitting reaction. Thus many compounds or enzymes coupled to the production or disappearance of NAD(P)H can be assayed with the bacterial luminescent system (19). Rapid microbiological assays are also possible using bacteriophages containing the *lux* genes from *V. fischeri,* that is, the genes coding for the proteins necessary for luminescence (20). This assay technique has the advantage over microbiological tests based on bioluminescent ATP assays because it is specific for a given strain of a particular bacterial species.

As stated earlier, the main interest of the chemiluminescent reactions in biochemical or clinical analysis is the possibility of coupling them with H_2O_2-generating enzyme-catalyzed reactions. In this way the substrates involved in some enzymatic reactions catalyzed by specific oxidases (OD) can be assayed by chemiluminescent detection. Some of such oxidase reactions are as follows:

$$\beta\text{-D-glucose} + O_2 + H_2O \xrightarrow{\text{glucose OD}} \text{D-glucono-}\delta\text{-lactone} + H_2O_2 \quad (5)$$

$$\text{Cholesterol} + O_2 \xrightarrow{\text{cholesterol OD}} \Delta\text{-4-cholestenone} + H_2O_2 \quad (6)$$

$$\text{L-lactate} + O_2 \xrightarrow{\text{lactate OD}} \text{pyruvate} + H_2O_2 \quad (7)$$

$$\text{Pyruvate} + HPO_4^{2-} \xrightarrow{\text{pyruvate OD}} CO_2 + \text{acetylphosphate} + H_2O_2 \quad (8)$$

In addition to these simple enzymatic reactions generating hydrogen peroxide, multienzymatic systems leading to the production of H_2O_2 can be used for the analysis of a particular metabolite for which no specific oxidase is available. For example, three consecutive enzymatic reactions are involved in producing peroxide and allowing the chemiluminescent determination of creatinine:

$$\text{Creatinine} + H_2O \xrightarrow{\text{creatininase}} \text{creatine} \quad (9)$$

$$\text{Creatine} \xrightarrow{\text{creatinase}} \text{sarcosine} + \text{urea} \quad (10)$$

$$\text{Sarcosine} \xrightarrow{\text{sarcosine OD}} \text{glycine} + \text{formaldehyde} + H_2O_2 \quad (11)$$

Availability of Reagents

As efficient and elegant in its conception as a biosensor may be wide utilization can be impaired by the lack of reagents at one's disposal. It is therefore desirable to obtain the different reagents easily, including substrates and enzymes, necessary for operating the biosensor. For bioluminescent ATP measurement highly purified and standardized reagents, that is luciferin and the enzyme luciferase, are now commercially available from different suppliers, making the firefly assay technique very attractive when high sensitivity is desired.

Substrates for the bacterial bioluminescence system are also widely available. Some suppliers propose tetradecanal (myristic aldehyde) instead of decanal as the aldehyde substrate. The enzymatic preparations commercially sold contain both luciferase and oxidoreductase, but purified preparations of FMN reductase can be obtained.

Peroxidase, used for the luminol chemiluminescent reaction, is widely available from different suppliers. The other chemiluminescent reagents can be obtained without difficulty from chemical suppliers.

SOLID-PHASE LUMINESCENCE REAGENTS

The first successful preparation of active immobilized luciferase was reported by Erlanger *et al.* (21). Insoluble bacterial luciferase from *V. fischeri* and *P. leiognathi* was prepared by coupling the enzyme to the azide of polyacrylic acid.

These insoluble derivatives were then used in studies of the mechanism of bacterial bioluminescence.

Later, several groups explored the analytical potentialities of immobilized bioluminescence enzymes from both the firefly and the bacterial system, as well as the potential applications of immobilized reagents for chemiluminescence analysis. For this purpose supports of different forms, including membranes, beads, gels, and coils, have been used with various immobilization procedures (adsorption, entrapment, cross-linking, and covalent binding to an activated matrix).

Two main approaches have been used to develop analytical systems associated with an immobilized reagent phase for the assay of different compounds and enzymes. In the first approach, light measurements were performed in assay cuvettes containing the immobilized compounds and suitable reagents located in a batch luminometer. The other approach is the use of flow analysis methods involving immobilized reagents in association with modified analytical devices. The main characteristics and performance of these systems and their applications are summarized in Tables 1 through 3 for the batchwise systems and in Tables 4 through 6 for the flow systems. In the bioluminescence-based systems firefly luciferase (Tables 1 and 4) or bacterial luciferase and oxidoreductase (Tables 2 and 5) are immobilized alone or coimmobilized with suitable enzymes, whereas in the different chemiluminescent-based systems described (Tables 3 and 6) the immobilized reagent phases consist of either immobilized luminescent compounds or immobilized catalysts that are directly involved in the light-emitting reaction or in a coupled enzymatic reaction.

FIBEROPTIC BIOSENSORS

The basic principle of these biosensors is the use of a single fiber or a fiber bundle to bring the light emitted during a bio- or chemiluminescence process from an immobilized reagent phase to a light detector. Since the analytical reaction generates its own light, no light source or monochromator is required. The instrumentation needed is then simpler than that used for fiberoptic sensors based on spectrophotometry and fluorimetry. Details of these sensors can be found in Chapters 8 and 9.

Chemiluminescence-Based Fiberoptic Sensors

In the first report of a fiberoptic biosensor, Freeman and Seitz described the use of a thin layer of peroxidase immobilized in a polyacrylamide gel at the tip of a fiberoptic sensor that was 2 feet long with a 1/8 inch diameter (90). The probe was immersed in a 10 ml stirred solution of buffer and luminol. A light-tight shield was placed around the enzymatic tip of the sensor to exclude ambient light. The other end of the fiberoptic sensor was placed in front of a photo-

Table 1 Characteristics and Performance of Batchwise Systems with Immobilized Firefly Luciferase

Analyte	Support and activation technique	Linearity or detection limit	Precision	Stability	References
ATP	Alkylamine glass beads	$1 \times 10^{-10} - 1 \times 10^{-6}$ M	—	3 weeks	22
Creatine kinase	+ glutaraldehyde	1.27–9.5 U	—	(4°C)	
ATP	CNBr Sepharose-4B	$5 \times 10^{-8} - 1 \times 10^{-3}$ M	—	—	23
ATPase		0.5 μg/ml	—		
Puruvate kinase		3 μg/ml	—		
ATP	CNBr Sepharose-4B	0.1 pM	6.4% at 1×10^{-3} M	60% after 1 month	24
ATPase	and	0.5–13μg/ml	—		
Pyruvate kinase	CNBr cellophane films	3–10 μg/ml	—	50% after 5 days	
ATP	CNBr Sepharose-4B and CL-6B	$0.4 \times 10^{-9} - 4 \times 10^{-6}$ M	3.2% at 1×10^{-5} M	Stable at −196°C	25
ATP	Cyanuric chloride and $NaIO_4$− activated cellulose films	$1 \times 10^{-10} - 1 \times 10^{-6}$ M	—	—	26
ATP	Acylazide-activated collagen films	$1 \times 10^{-11} - 1 \times 10^{-6}$ M	3% at 1×10^{-10} M	20% after 8 months (dehydrated form)	27
Creatine kinase	CNBr Sepharose	0.5–1000 U/L	15–20% at 2 U/L	Half-life 40 days	28, 29

Table 2 Characteristics and Performance of Batchwise Systems with Immobilized Bacterial Bioluminescence Enzymes

Analyte	Support and activation technique	Linearity or detection limit	Precision	Stability	Reference
NADH	Diazo coupling on arylamine glass beads	$1 \times 10^{-8} - 5 \times 10^{-4}$ M	3%	2 weeks (4°C)	30
NADPH		$1 \times 10^{-7} - 2 \times 10^{-3}$ M	3%		
D-glucose	Diazo coupling on arylamine glass beads	0.15–15 nmol	—	—	31
Ethanol		0.004–0.015‰	—		
Alcohol dehydrogenase		0.015–3 pmol	—		
Glucose-6-phosphate dehydrogenase		0.0015–0.1 pmol	—		
Hexokinase		0.1–2 pmol	—		
Malate dehydrogenase		0.007–0.7 pmol	—		
Lactate dehydrogenase		0.003–0.7 pmol	—		
D-glucose	Diazo coupling on arylamine glass beads	20 pmol to 10 nmol	—	—	32
Glucose-6-phosphate		1 pmol to 2 nmol	—		
Androsterone	CNBr Sepharose-4B	0.3 pmol to 2 nmol	1–3%	—	33
Glucose-6-phosphate		10 pmol to 20 nmol	—		
NADH		1 pmol to 2 nmol	—		
Testosterone		0.8 pmol to 1 nmol	1–3%		
H_2O_2	Calcium alginate gel	$4 \times 10^{-7} - 4 \times 10^{-6}$ M	—	Half-life 20 hr	34
Creatine kinase	BSA gel	1–10 U/ml	10%	3 months (4°C)	35
7α–Hydroxy bile acids	CNBr Sepharose-4B	$1.5 \times 10^{-6} - 5 \times 10^{-5}$ M	6-8% intraassay 8–10% interassay	Stable at 4°C	36

NADH	CNBr Sepharose-4B	0.2 pmol to 1 nmol	5.8% at 2×10^{-6} M	2 months (−196°C)	25
L-alanine	CNBr Sepharose4B	50 pmol to 10 nmol	2.5–4%	Several months (4°C)	37
6-Phospho-gluconate		10 pmol to 100 nmol	2.5–4%		
D-glucose		10 pmol to 1.5 nmol	2.5–4%		
L-glutamate		0.1–5 nmol	2.5–4%		
L-lactate		0.1–100 nmol	2.5–4%		
L-malate	(With malate DH)	0.1–10 nmol	5%		
L-malate	(With malic enzyme)	10 pmol to 10 nmol	2.5–4%		
NAD^+		1 pmol to 1 nmol	2.5–4%		
$NADP^+$		0.2 pmol to 0.2 nmol	2.5–4%		
12α-Hydroxy bile acids	CNBr Sepharose-4B	4 pmol to 2 nmol	8%	2 months (4°C)	38
3α-Hydoxy bile acids	CNBr Sepharose-4B	1 pmol to 20 nmol	6.2–8.2%	4 months (4°C)	39
NADH	Acylazide-activated collagen films	$1 \times 10^{-9} - 2 \times 10^{-5}$ M	5% at 1×10^{-8} M	3 weeks (4°C)	40
Chenodiol	CNBr Sepharose-4B	$1 \times 10^{-6} - 32 \times 10^{-6}$ M	5%	—	41
Formate	CNBr-activated agarose	$1 \times 10^{-12} - 1 \times 10^{-7}$ M	—	30 days (4°C)	42
NAD^+		$1 \times 10^{-12} - 1 \times 10^{-7}$ M	—		
NADH		$1 \times 10^{-12} - 1 \times 10^{-7}$ M	—		
FMN	CNBr-activated agarose	$1 \times 10^{-10} - 1 \times 10^{-7}$ M	—	—	43
Glucose-1-phosphate		$1 \times 10^{-9} - 1 \times 10^{-5}$ M	—		
Glucose-6-phosphate		$1 \times 10^{-9} - 1 \times 10^{-5}$ M	—		
NAD^+		$1 \times 10^{-12} - 1 \times 10^{-7}$ M	—		
NADH		$1 \times 10^{-12} - 1 \times 10^{-7}$ M	—		

Table 3 Characteristics and Performance of Chemiluminescence Batchwise Systems with Immobilized Reagents

Analyte	Support and activation technique	Immobilized compound	Chemiluminescent reagent	Linearity or detection limit	Reference
α-Chymotrypsin	Activated cross-linked agarose beads (Affigel 10)	Isoluminol coupled to specific peptides	Isoluminol/hematin	0.2 μg/L	44
Human thrombin				0.5 μg/L	
Trypsin				20 ng/L	
Thiol compounds	Thiopropyl–Sepharose-6B	Mercaptoacetylluminol	Luminol/ferricyanide	5 pmol	45
H_2O_2	Photo–cross-linkable resin	Peroxidase	*Cypridina* luciferin analog	0.2 μM	46
Hypoxanthine		Xanthine oxidase		0.02 μM	
Xanthine	Prepolymer	Xanthine oxidase		0.02μM	
Cholesterol	Ion-exchangeable cellulose beads + glutaraldehyde	Cholesterol oxidase	*Cypridina* luciferin analog	2 μM	46
β-D-glucose		Glucose oxidase		0.4 μM	
Uric acid		Urate oxidase		2 μM	
β-D-glucose	Dialysis membrane	Luminol	Luminol/peroxidase	28 nmol	47
	Filter membrane	Peroxidase/glucose oxidase			
β-D-glucose	Polyacrylamide gel	Peroxidase/glucose oxidase	Luminol/peroxidase	0.1–1.5 M	48
H_2O_2		Peroxidase		1–10 mM	
Cholesterol	Preactivated polyamide membrane	Peroxidase	Luminol/peroxidase	$1 \times 10^{-6} - 2.5 \times 10^{-4}$ M	49
H_2O_2				$1 \times 10^{-8} - 1 \times 10^{-4}$ M	
H_2O_2	Entrapment in a silk fibroin membrane	Peroxidase	Luminol/peroxidase	2 mM	50

Table 4 Characteristics and Performance of Flow Systems with Immobilized Firefly Luciferase

Analyte	Support and activation technique	Linearity or detection limit	Precision	Half-life or number of assays	Reference
ATP	CNBr Sepharose-CL-6B	3–3000 pmol	3.6%	≈140 assays	51
NAD^+	CNBr Sepharose-4B or CL-6B	10 nM to 10 μM	—	—	52
ATP	CNBr Sepharose-4B and silanized glass beads + glutaraldehyde	10 pM to 0.1 μM	<2% (whole range)	2 months	53
Creatine kinase		10–400 U/L	≈4% (whole range)		
Creatine phosphate		10 μM to 0.1 M	1–7%		
ATP	O-alkylated nylon coils	0.3–100 pmol	5–8% intra- and interassay	15 days (25°C)	54
ATP	Epoxy methacrylate beads	1–500 pmol	4.9–9.6% intra- and interassay	3 days (25°C)	55
Creatine kinase	O-alkylated nylon coils + glutaraldehyde	0.1–100 U/L	4.1–9.7% intra- and interassay	—	56

Table 5 Characteristics and Performance of Flow Systems with Immobilized Bacterial Bioluminescence Enzymes

Analyte	Support and activation technique	Linearity or detection limit	Precision	Half-life or number of assays	Reference
NADH	CNBr Sepharose-4B	1 pmol to 10 nmol	/	400 assays	57
Chenodeoxycholylglycine	CNBr Sepharose-4B	8–325 pmol	3.8–6.9%	—	51
Cholylglycine		8–325 pmol	2.6–2.9%	677 assays	
Cholyltaurine		4–750 pmol	4.0–5.5%	—	
Glucose-6-phosphate		1.5–1500 pmol	3.6–7.6%	721 assays	
NADH		6–600 pmol	2.3%	220 assays	
NADH	O-alkylated nylon coils + glutaraldehyde	1–2500 pmol	5.0–10% intra- and interassay	500 assays	58
7α-hydroxy bile acids	O-alkylated nylon coils + glutaraldehyde	10–2500 pmol	5.0–10% intra- and interassay	500 assays	59
Ethanol	Diazo coupling on arylamine glass beads	76–3800 pmol	8.1%	—	60
L-lactate		76–3800 pmol	7.5%	—	
NADH		0.13–1300 pmol	3%	700 assays	
6-Phospho-gluconate	CNBr Sepharose-4B and CL-6B	1 nM to 10 μM	—	—	52
NADH		0.1 nM to 1 μM	2–10%	—	
Acetaldehyde	O-alkylated nylon coils + glutaraldehyde	50–1000 pmol	10%	30 days	61
Ethanol		50–2500 pmol	10%	60 days	
Glycerol		50–500 pmol	10%	60 days	
3α-, 7α-, or 12α-hydroxy bile acids		1 pmol	8–10%	30 days	
NADH		1–2000 pmol	5–9%	20 days	

FMN	CNBr Sepharose-4B	1 nM to 1 μM	1.6%	—	62
$FMNH_2$		1 nM to 1 μM	1.5–6.6%		
NADH		1 nM to 1 μM	1.2–4.1%		
NADPH		10 nM to 1 μM	2.3–3.4%		
L-alanine	O-alkylated nylon coils	5–1500 pmol	—	2 months	63
L-alanine (serum)	+ glutaraldehyde	50–1500 pmol	1.9–7.8%		
L-alanine (urine)		20–500 pmol	6.0–9.5%		
Alcohol dehydrogenase	CNBr Sepharose-4B	0.03–30 pmol	—	—	64
Ethanol		10 μM to 1 mM	1.8%	200 assays	
Isoleucine	O-alkylated nylon coils	10–2000 pmol	2.6–8.1%	2 months	65
Leucine	+ glutaraldehyde	10–2000 pmol			
Valine		10–2000 pmol			
Isoleucine (serum, / urine)		20–2000 pmol	5.6–9.7%		
Leucine (serum, / urine)		20–2000 pmol			
Valine (serum, / urine)		20–2000 pmol			
L-lactate dehydrogenase	O-alkylated nylon coils + glutaraldehyde	3–2000 U/L (25°C) 1–5000 U/L (37°C)	5.3–9.8%	—	66
L-malate	CNBr Sepharose-4B	1 μM to 10 mM	≤5%	—	67
Malate dehydrogenase		0.03–3 pmol	≤5%		
Glucose-6-phosphate		10 nM to 10 μM	≤5%		
Glucose-6-phosphate dehydrogenase		0.015–3 pmol	≤5%		
Oxalate	CNBr Sepharose-6MB	0.6 μM	3.8–8.4%	3000–4000 assays	68
Ammonia	CNBr Sepharose-4B	1.2–92.5 μM	3.5–8.6%	—	69
Adenosine deaminase		1–200 U/L			

Table 6 Characteristics and Performance of Chemiluminescence Flow Systems with Immobilized Reagents

Analyte	Support and activation technique	Immobilized compound	Chemiluminescent reagent	Linearity or detection limit	Reference
β-D-glucose	Sepharose	Glucose oxidase	Luminol/ferricyanide	10 nM to 0.1 mM	70
H_2O_2				10 nM to 10 μM	
β-D-glucose	Controlled pore glass	Glucose oxidase	TCPO/perylene	25–250 mg/L	71
H_2O_2				70 nM to 1 mM	
Formaldehyde	TiO_2-coated quartz beads	Aldehyde oxidase	TCPO/9, 10-diphenylanthracene	6 nM to 30 μM	72
Formic acid		Aldehyde oxidase		6 nM to 30 μM	
Oxalic acid	TiO_2-coated quartz beads	Oxalate oxidase	TCPO/9, 10-diphenylanthracene	20 nM to 0.8 mM	73
Uric acid		Urate oxidase		20 nM to 0.8 mM	
L-amino acids	Porous glass beads + glutaraldehyde	L-amino acid oxidase	TCPO/9, 10-diphenylanthracene	20 nM to 30 μM	74
β-D-glucose	Alkylamine glass beads + glutaraldehyde	Glucose oxidase	Luminol/ferricyanide	0.5–4 g/L	75
Uric acid		Urate oxidase		0.01–0.1 g/L	
H_2O_2	Controlled pore glass	Solid TCPO and 3-aminofluoranthene	TCPO/3-aminofluoranthene	15 μM	76
Cholesterol	Alkylamine controlled pore glass + glutaraldehyde	Cholesterol oxidase	Luminol/ferricyanide	5 mg/L	77
Creatinine		Creatininase + creatinase + sarcosine oxidase		0.3 mM	
β-D-glucose		Glucose oxidase		0.3 mg/L	
L-lactate		Lactate oxidase		0.3 mg/L	
Acetylcholine	Silica gel + glutaraldehyde	Acetylcholine esterase + choline oxidase	TCPO or CPPO/perylene	10 pmol to 10 nmol	78
Choline		Choline oxidase		10 pmol to 10 nmol	
β-D-glucose	Ion exchanger or controlled pore glass + glutaraldehyde	Glucose oxidase	TCPO/3-aminofluoranthene	50 nM	79
		Solid TCPO and 3-aminofluoranthene			
Sucrose	Controlled pore glass	Invertase + mutarotase	Luminol/peroxydase	5 μM to 1 mM	80

NH_3	Alkylamine glass beads + glutaraldehyde	Glutamate dehydrogenase + glutamate oxidase	Luminol/ferricyanide	50 μM to 1 mM	81
H_2O_2	Controlled pore class or silica	Luminol	Luminol/hemin	1 μM to 1mM	82
Glucose	Alkylamine glass beads + glutaraldehyde	Glucose oxidase	Luminol/ferricyanide	0.6–4.2 g/L	83
Lactate		Lactate oxidase		80–220 mg/L	
Blood urea nitrogen		Glutamate dehydrogenase + glutamate oxidase		30–900 mg/L	
Uric acid		Uricase		10–130 mg/L	
β-D-glucosides Salicin Phenylglucoside *p*-Nitrophenyl-glucoside	Controlled pore glass + glutaraldehyde	β-Glucosidase + glucose oxidase	Luminol/hemin/ peroxidase	≈0.1 μM	84
H_2O_2	Controlled pore glass + glutaraldehyde or silica gel + glutaraldehyde or adsorption on Ambersorb	Luminol or	Luminol/hemin/ peroxidase	0.15 μM	85
Luminol		peroxidase		0.1 nM	
Glucose	Controlled pore glass + glutaraldehyde	Glucose oxidase/	Luminol/hemin/ peroxidase	0.5 μM to 1 mM	86
Fructose		mutarotase and glucose isomerase		3 μM to 1 mM	
Lactose		or lactase		1 μM to 1 mM	
Maltose		or amyloglucosidase		0.5 μM to 1 mM	
Sucrose		or invertase/mutarotase		0.5 μM to 1 mM	
H_2O_2	Silica gel + glutaraldehyde	Luminol + hemoglobin + peroxidase	Luminol/hemoglobin/ peroxidase	40–600 μM 4–600 μM (electrode)	87
urea	Porous alkylamine glass + glutaraldehyde	Urease, L-glutamate oxidase, L-glutamate dehydrogenase	Luminol/ferricyanide	0.5–30 mM	88
Inorganic phosphate	Aminopropyl controlled pore glass beads + glutaraldehyde	Purine nucleoside phosphorylase, xanthine oxidase, urate oxidase	Luminol/microperoxidase	0.1–5 μM	89

multiplier tube, and a shutter was interposed between the fiberoptic sensor and the photomultiplier to protect it from excess light when the sensing tip was removed from the reaction medium (Fig. 3). When the probe was inserted in the reaction medium containing luminol in excess and hydrogen peroxide, light was generated in the enzyme phase and transmitted through the fiberoptic probe to the detector. Both pH and luminol concentration affected the chemiluminescence intensity. The time required to reach steady state, that is, the response time, varied with the hydrogen peroxide concentration. Under optimum conditions (1×10^{-3} M luminol and pH 9) a steady state was reached in about 4 sec at 1×10^{-5} M H_2O_2, whereas at 1×10^{-4} M the response time was close to 30 sec. The minimum detectable concentration was 1×10^{-6} M hydrogen peroxide, and the assay was linear up to 1×10^{-4} M. However, the fiberoptic biosensor showed a response that was second order in hydrogen peroxide. The precision was limited by variations in stirring and the RSD was typically 10% between replicate measurements of the same solution. To reduce the effect of irregular stirring a dialysis membrane was interposed between the polyacrylamide gel and the solution. Under these conditions, at 1×10^{-4} M H_2O_2, the precision improved from 9 to 5% RSD, but the signal decreased by a factor of 0.62 and the response time increased.

Instead of an enzymatic catalyst the combination of optical fibers and of a platinum electrode has been applied to electrogeneration of luminol chemiluminescence (91). The probe incorporated optical fibers and a "cage" platinum working electrode. The sensor was designed for flow injection analysis as well as for stationary solution measurement. Using the positively biased electrode in place of a catalyst, luminol was oxidized and in the presence of hydrogen peroxide chemiluminescence was generated. The flow system was operated at a flow rate of 3.3 ml/min, and luminol was used at a final concentration of 0.2 mM. With the injection of 500 μl H_2O_2 in the 1×10^{-3}–1×10^{-6} M range,

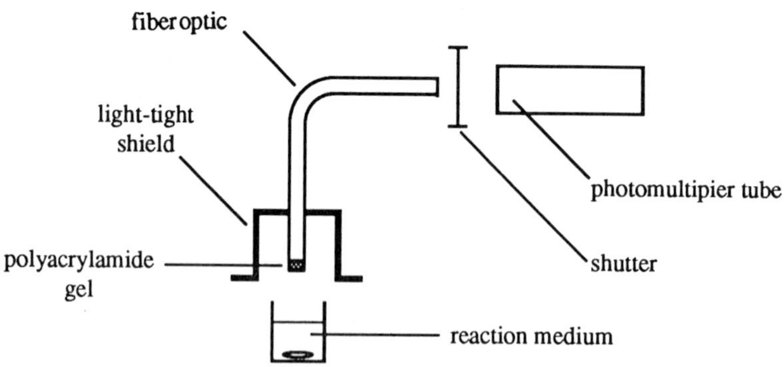

Figure 3 Chemiluminescence fiberoptic probe for hydrogen peroxide based on the luminol reaction. [After Freeman and Seitz (90).]

well-defined peaks were recorded, with a peak width at one-half maximum height of about 0.6 min. The detection limit was 1×10^{-6} M hydrogen peroxide, but the calibration curve was nonlinear with the luminescence signal and the response leveled off near 1×10^{-3} M. For two replicate measurements the coefficient of variation was 20 and 7.8% at 1×10^{-6} and 1×10^{-3} M H_2O_2, respectively. This was equal to 6.9 and 2.7% for four replicate measurements at 1×10^{-5} and 1×10^{-4} M H_2O_2, respectively.

A fiberoptic sensor for the determination of glucose using micellar enhanced chemiluminescence of the peroxyoxalate reaction was described by Abdel-Latif and Guilbault (92). The design of the cell compartment, constructed in such a way that ambient light does not interfere, is shown in Figure 4. The end of one of the two arms of a bifurcated fiberoptic bundle was fixed to the cell cover, the other end was fixed to the side of the cell (disposable glass vial, 15 × 45 mm). The common end of the fiber bundle was attached to a Lumac Biocounter, Model 2010, *via* an adapter. Stirring was accomplished by an egg-shaped stirring bar with the same diameter as the cell. Glucose oxidase was immobilized on a polyamide membrane (1 × 2.5 cm) mounted around a 2 cm long glass spacer held at the end of the arm of the fiberoptic bundle fixed to the cell cover. Cetyltrimethylammonium bromide (CTAB) was used as a surfactant to enhance the chemiluminescence generated from the reaction of TCPO with hydrogen peroxide in the presence of perylene. In addition, the use of micelles eliminated the problems of dissolution encountered in the peroxyoxalate chemiluminescence system. The effect of various experimental parameters on the chemilumines-

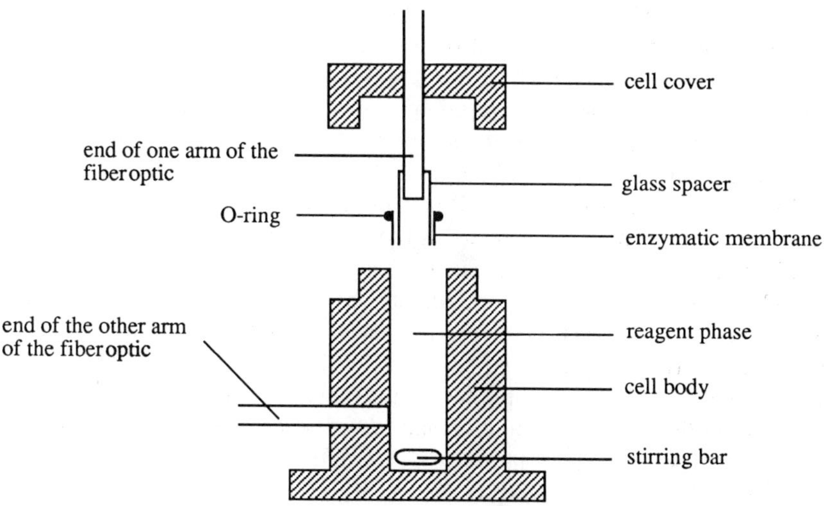

Figure 4 Design of the cell compartment of the fiberoptic sensor for the determination of glucose using micellar enhanced chemiluminescence of the peroxyoxalate reaction. [After Abdel-Latif and Guilbault (92).]

cence intensity was studied. The optimum concentrations of reagents were found to be 6.5 mg/ml for TCPO, 3 mg/ml for perylene, and 0.06 M for CTAB. After investigation of the effects of buffer concentration and pH, a 0.03 M phosphate buffer, pH 7.2, was selected. For the determination of H_2O_2 a 30 μl sample were introduced in the cuvette containing the chemiluminescence reagent and the buffer in a final volume of 1 ml. After quickly shaking the cuvette for 2 sec the light intensity was recorded at the maximum. The response was linear in the range of 8×10^{-9}–8×10^{-4} M hydrogen peroxide. The time required to reach the maximum light intensity depended slightly on the peroxide concentration and was between 4 and 10 s. At 1×10^{-7} M H_2O_2 the coefficient of variation for five replicate measurements was 0.3%. For glucose determination a sample of 100 μl was added to a cuvette containing 3 ml buffer. The cell was closed and the contents were stirred for 3 min. Then, 100 μl of the chemiluminescence reagent was added. The cell contents were stirred again and the maximum light intensity was recorded. The calibration plot for glucose was linear in the range of 3×10^{-6}–3×10^{-2} M, with a detection limit equal to 6×10^{-7} M. The maximum light intensity was also obtained in 4–10 sec, and the coefficient of variation for five measurements of 1×10^{-4} M glucose was 0.3%. The method has been used to analyze for glucose in soft drinks and compared very well with a classic colorimetric assay using soluble glucose oxidase/peroxidase and a dye.

Using the same fiberoptic setup, Abdel-Latif and Guilbault also described a fiberoptic sensor based on micellar-mediated chemiluminescence reaction of luminol (93). Soluble peroxidase was used as the catalyst of the chemiluminescence reaction, and for the quantification of glucose immobilized glucose oxidase was used. Although peroxidase allows the luminol chemiluminescence reaction to proceed at near neutral pH values, the reaction is most efficient at alkaline pH in the range 10–11. A problem then arises when coupling this light-emitting reaction to enzymatic reactions at neutral pH. To overcome the pH mismatch therefore, micelles were used by the authors. Several surfactants were tested, and all of them enhanced the luminescence intensity. However, the best results were obtained with CTAB. The analytical procedures used for H_2O_2 and glucose measurements were similar to those described for micellar enhanced chemiluminescence of the peroxyoxalate reaction with incorporation of fiberoptics (92). However, when using the luminol chemiluminescence reaction for the quantification of hydrogen peroxide, the reagents used were a 0.005 M phosphate buffer, pH 7.25, containing 1.6×10^{-2} M CTAB, 3×10^{-4} M luminol, and peroxidase. The time required for the maximum light intensity to be reached depended on the H_2O_2 concentration and varied from a few seconds to 4 min. The response was linear in the range 2.4×10^{-8}–1.2×10^{-4} M. For five replicates at 2.5×10^{-5} M the coefficient of variation was 2.6%. The glucose determination was performed in 2.5 ml of 0.005 M phosphate buffer, pH 7.25, containing 1.6×10^{-2} M CTAB, 0.1 ml peroxidase, and 0.1 ml glucose standard solution. The contents were stirred for 3 min and then 0.25 ml luminol,

3×10^{-3} M, was added. The maximum light intensity was proportional to the glucose concentration in the range 3×10^{-7}–3×10^{-4} M, and the detection limit was 1×10^{-7} M glucose. The coefficient of variation for five replicate measurements at 5×10^{-5} M was 3.8%. The determination of glucose in serum samples involved the same procedure except that serum samples of 2–10μl were used in place of the standard glucose solution. Serum controls of low and high glucose content were tested. The results obtained with this method were in excellent agreement with those obtained with a standard method. For eight replicates of both low- and high-glucose controls, the coefficients of variation were 5.5 and 2.9%, respectively.

Bioluminescence-Based Fiberoptic Sensors

Monofunctional biosensors

Batch analysis. New fiberoptic biosensors based on bioluminescence reactions have been developed in our laboratory for the quantification of ATP and NADH (94,95). The fiberoptic setup, primarily designed for batch analysis, is shown in Figure 5. A glass fiber bundle 1 m long and 8 mm in diameter was used. One end of the bundle was connected *via* an adapter to the photomultiplier tube of a Berthold Biolumat LB 9500 luminometer. On the other end an enzymatic membrane was maintained in close contact with the surface of the bundle by a screwcap. The biosensor was immersed in a stirred and thermostated reaction medium of 4.5 ml. To avoid ambient light interference the reaction vessel was surrounded by a light-tight polyvinyl chloride jacket. The sensor was fixed in the removable cell cover, also made of polyvinyl chloride. Samples were injected in the reaction medium by a syringe through a light-tight septum. The analysis of ATP was performed with the immobilized firefly luciferase from *P. pyralis*. For NADH measurements the immobilized luciferase and oxidoreductase from either *V. fischeri* or *V. harveyi* were used. In addition, the use of the chemiluminescence reaction of luminol catalyzed by immobilized peroxidase was investigated for the measurement of hydrogen peroxide with the fiberoptic biosensor.

The enzymes were covalently immobilized on polyamide membranes (Biodyne ImmunoAffinity, later called Immunodyne), supplied in a preactivated form from Pall Corporation, according to the general method originated in our laboratory to prepare enzymatic membranes with a very fast and convenient procedure (96) and now widely used by other authors. The binding reaction was conducted at ambient temperature simply by applying 10 μl of a suitable enzymatic solution on a disk 10 mm in diameter cut out of a preactivated polyamide sheet. The immobilization of enzymes was complete after 1 min.

The analytical procedure used with the fiberoptic biosensor was very simple and identical for each of the light-emitting system used, except the reaction medium content. The biosensor was immersed in the reaction medium containing

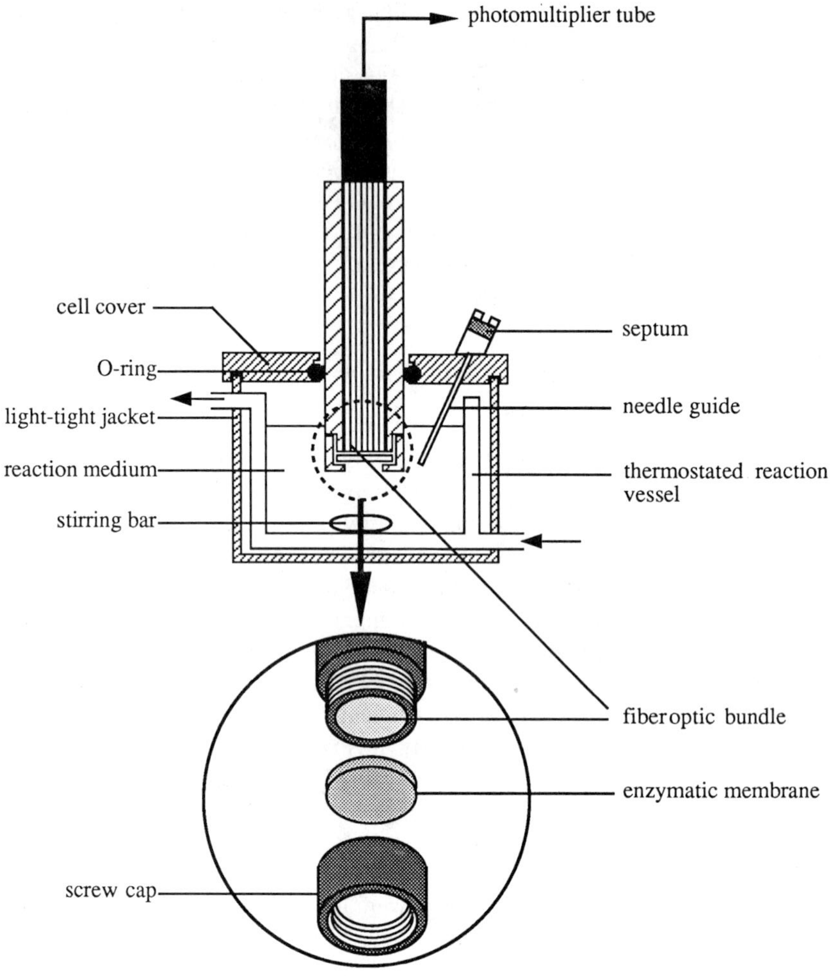

Figure 5 The bioluminescence-based fiberoptic biosensor setup.

the suitable reagents according to the analysis to be performed. The background light emission was recorded and then a sample was injected. Measurements of ATP were performed in a 0.05 M Tris-acetate buffer at pH 7.75 containing 5 mM $MgCl_2$, 2 mM dithiothreitol (DTT), and 0.2 mM luciferin. The light emission was linearly related to the ATP concentration in the range 2.8×10^{-10}–1.4×10^{-6} M. The time required to reach a steady-state luminescence depended mainly on the ATP concentration and varied from 1 min at 2.8×10^{-10} M to 6–7 min at 3×10^{-6} M, and then the light emission was stable for several minutes.

For hydrogen peroxide analysis the reaction medium was a 0.1 M Tris-HCl buffer, pH 8.5, with luminol at a final concentration of 0.05 mM. Under these conditions the standard curve was linear from 2×10^{-8} to 2×10^{-5} M H_2O_2. The maximum light intensity was reached within 1 min, but the luminescent signal decreased after a 20–30 sec stable light emission. Because of the strong interest in the microdetermination of NADH in clinical analysis (97) much attention has been paid to the NADH fiberoptic biosensor. We first studied the analytical performance of this biosensor equipped with immobilized enzymes from *V. fischeri* (98). The optimum conditions for the expression of maximum enzyme activities were determined, and the operational and storage stabilities of this immobilized bioluminescence system were carefully studied. The optimum conditions selected were a 0.1 M phosphate buffer, pH 7.0, and a final concentration of FMN and decanal equal to 2×10^{-5} M and 0.005%, respectively. In addition, the reaction medium contained 0.2% bovine serum albumin (BSA) and 2 mM DDT. Measurements could be then performed in the linear range 1×10^{-9}–3×10^{-6} M with a detection limit of 3×10^{-10} M. For 10 replicate measurements of 4×10^{-8} M NADH the coefficient of variation was 4.8%. The time required to reach the maximum light intensity varied with NADH concentration from 1 to 3 min, and the luminescent signal was stable for 1–2 min. When stored at −20°C with 20% glycerol the activity of the bioactive membrane was higher than that measured before freezing and then remained stable for several months. However, concerning the operational stability, a recalibration had to be performed every 10 assays because of the gradual thermal inactivation of the bacterial luciferase from *V. fischeri*.

The potential of the NADH fiberoptic biosensor was extended by co-immobilizing suitable dehydrogenases (DH) with the bienzymatic system from *V. fischeri*. Microdetermination of sorbitol, ethanol, and oxaloacetate was then performed at the nanomolar level (99). The dehydrogenase reactions involved are as follows:

$$\text{D-sorbitol} + \text{NAD}^+ \xrightarrow{\text{sorbitol DH}} \text{D-fructose} + \text{NADH} + \text{H}^+ \quad (12)$$

$$\text{Alcohol} + \text{NAD}^+ \xrightarrow{\text{alcohol DH}} \text{aldehyde} + \text{NADH} + \text{H}^+ \quad (13)$$

$$\text{Oxaloacetate} + \text{NADH} + \text{H}^+ \xrightarrow{\text{malate DH}} \text{NAD}^+ + \text{L-malate} \quad (14)$$

Analysis of sorbitol and ethanol was performed in a one-step procedure, with NAD^+ in excess, since the enzymatic oxidation of these two substrates leads to the production of NADH, which was then directly monitored by the bacterial bioluminescence system. For assay of sorbitol the selected reaction conditions were a 0.1 M phosphate buffer at pH 7.3 containing in addition to the reagents required for the bioluminescence reaction 1 mM NAD^+. A linear relationship was obtained in the range 2×10^{-8}–2×10^{-5} M by measuring either the maximum light intensity or the rate of increase of light intensity. The response time at the steady state was 6–7 min over the range of concentrations tested, and the coefficient of variation for 10 replicate measurements at 4.4×10^{-7} M sorbitol was 6%. The optimum conditions for ethanol measurements were a 0.1 M phosphate buffer, pH 7.5, with a NAD^+ concentration equal to 5.5×10^{-3} M. The calibration curve was linear from the detection limit, equal to 1×10^{-7} M to 7×10^{-5} M. For 10 replicate measurements at 5.7×10^{-6} M ethanol the coefficient of variation was 5.4%. Measurements could be performed by determining either the rate of increase in the light intensity or the maximum light intensity. In the latter case, however, the time necessary to reach the steady state depended on the ethanol concentration and varied from 1 min at 4×10^{-7} M to 8 min at 7×10^{-5} M. The selectivity of alcohol dehydrogenase is not absolute, and therefore some potentially interfering primary and secondary alcohols were tested. At pH 7.5 the relative rates of oxidation for ethanol, 1-propanol, and 1-butanol were 100, 16, and 4, respectively. The detection limit was 2.7×10^{-4} M for methanol and 2.6×10^{-3} M for 2-propanol. These concentrations were 4 and 37 times respectively, higher than the maximum measurable ethanol concentration (7×10^{-5} M), and consequently the risk of interference was weak. The assay of oxaloacetate involves the simultaneous consumption of NADH by malate dehydrogenase and bacterial oxidoreductase and was therefore conducted in two steps. Thus, NADH was first injected (1×10^{-6} M final concentration), and after a constant light emission was obtained the oxaloacetate sample was injected, leading to a light decrease. Either the variation in light intensity or the rate of light decrease could be linearly related to the oxaloacetate concentration in the range 3×10^{-9}–2×10^{-6} M. The detection limit was 1×10^{-9} M. For oxaloacetate concentrations below 1×10^{-6} M, however, the time required to reach the steady state was at least 10 min. The measurement of light decrease was then chosen to speed up the analytic procedure. Under these conditions for 10 replicates of 5.5×10^{-8} M oxaloacetate the coefficient of variation was 5.1%.

The stability of the bioactive membrane under working conditions remains a key point for designing an operational biosensor of practical use. Thus particular attention has been paid to the study of the operational stability of the immobilized bacterial bioluminescence system. When using immobilized luciferase and oxidoreductase from *V. fischeri*, an excellent storage stability was obtained when the membranes were frozen at −20°C with glycerol. Nevertheless, the operational stability was not sufficient to allow an intensive use of the biosensor without several recalibrations per day. We therefore investigated the potential of using a commercial mixture of luciferase and oxidoreductases from *V. harveyi* as the sensing element for the NADH fiberoptic biosensor (100). Whatever the origin of the bioluminescent enzymatic system, the microdetermination of NADH could be performed at the nanomolar level with detection limits of 2×10^{-9} and 3×10^{-10} M with the systems from *V. harveyi* and *V. fischeri*, respectively. Although we found that the use of polyethylene glycol improved the operational stability of the bienzymatic system from *V. fischeri*, it appeared necessary to proceed with recalibrations every 10 measurements. In contrast, when using the immobilized system from *V. harveyi* no loss in activity was observed after more than 100 assays performed within a day. Moreover, after carrying out 300 NADH measurements within a 10 day period, with storage at 4°C between each set of experiments, the residual activity was still 50% of the initial one. The use of polyethylene glycol or other preservatives was thus not needed. Despite the lower activity exhibited by the immobilized system from *V. harveyi* compared with the immobilized system from *V. fischeri*, the operational stability was much better, needing only one calibration per day, and this appeared as a decisive advantage for reliable determinations. Thus the excellent operational stability of the polyamide-bound enzymatic system from *V. harveyi* permitted the design of a really usable and operational fiberoptic biosensor.

Continuous-flow Analysis. Immobilized bioluminescent reagents in association with continuous-flow manifolds would facilitate automation. We thus developed a continuous-flow bioluminescent method for NADH (101) using a specially designed flow cell (Fig. 6) adapted to the fiberoptic biosensor. The bioluminescence enzymes from *V. harveyi* were immobilized on preactivated Immunodyne membranes. The flow cell was made of black polyvinyl chloride to avoid ambient light interference and included a minute reaction chamber of 250 μl volume. The fiberoptic bundle, similar to that used for batch analysis (94), was fixed in the flow cell, and the enzymatic membrane, in close contact with the

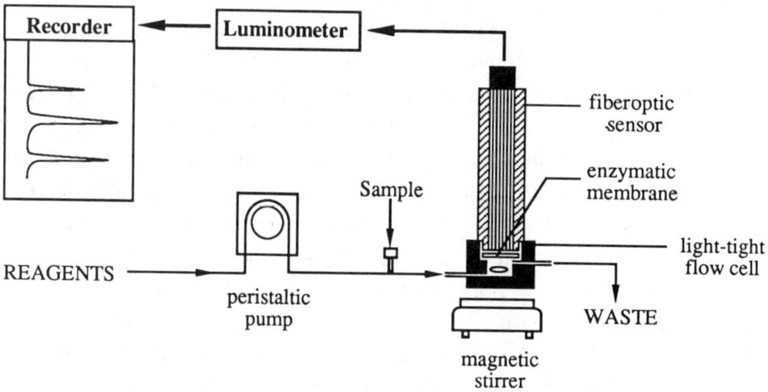

Figure 6 Flow diagram of the fiberoptic continuous-flow system for bioluminescence measurements.

surface of the bundle, faced the reaction chamber. Stirring was accomplished in the flow cell by a small magnetic bar (3 × 6 mm). Samples were injected into a continuously flowing stream by a syringe through a light-proof septum placed 3.5 cm before the flow cell. The reagent solution was delivered by a one-channel peristaltic pump including silicone rubber pump tubing (1 mm ID). Otherwise, polyethylene tubing (1 mm ID), was used throughout the system, and to prevent ambient light interferences, it was surrounded by black Iso-versinic tubing. The shape of the peak response depended mainly on the flow rate and on the stirring conditions in the flow cell. The optimum conditions determined were a flow rate of 0.74 ml/min and a stirring speed of 1200 rpm. The assay was then linear from 2×10^{-12} to 1×10^{-9} mol NADH with a coefficient of variation of 3.4% for 10 replicate measurements at 1×10^{-10} mol. At 5×10^{-12} mol NADH the cycle time, defined as the time for a complete cycle to be carried out, that is back to the baseline, was 1.5 min. No carryover was observed between successive samples of high and low NADH content with an analytic throughput of 25 samples per hour. When used over a 3 day period with storage at 4°C between sets of analysis, 150 assays could be performed without any decrease in the sensitivity of the analytic system.

Multifunctional Biosensors

When considering the needs in the field of biosensors, specificity and sensitivity are sometimes met but versatility is seldom encountered in the same device. Only monofunctional optical biosensors have been described, that is, biosensors specific for only one analyte, which required, using the same transducing system, changing the sensing element for the measurement of a chemically different analyte. It is thus obvious that the possibility of specifically determining different compounds with a single probe, without changing the sensing element, is particularly attractive.

In a new approach we have thus designed a multifunctional fiberoptic biosensor that involves two different enzymatic systems coimmobilized on the same membrane, allowing the preselected determination of different analytes depending only on the reaction conditions. The firefly luciferase from *P. pyralis* was coimmobilized on the same support with the bioluminescence bacterial enzymes either from *V. fischeri* or from *V. harveyi,* and these multienzyme membranes were used with the fiberoptic sensor for batch and continuous-flow analysis (102,103). To our knowledge such an approach has not been reported thus far. Thus ATP or NADH could be measured with the same biosensor without changing the sensing element. However, one of the main problems to be overcome with the use of a solid-phase multienzyme system deals with the reaction medium conditions, which must be compatible with both the stability of the different systems and the expression of their optimal activity. As a matter of fact, the firefly luciferase and the bacterial bioluminescence enzymes have specific needs in terms of reactants, optimum pH, and nature of the buffer. When using these two different systems coimmobilized on the same membrane, the substrates and reaction products of one system may interact with the other enzymatic system. In fact, when making NADH measurements the firefly luciferase, although not in use, is in contact with the bacterial enzyme substrates and reaction products and also with the specific buffer used. Conversely, when ATP measurements are made the bacterial enzymes are in contact with the firefly enzyme reagents. Therefore, all these effects have been studied to determine the optimum reaction conditions that maintain both the activities and the stabilities of the different enzymes involved.

In the presence of the bacterial enzyme substrates a 20–30% decrease in firefly luciferase activity was observed. However, the inhibition was completely reversible since the enzyme recovered its full activity when reassayed in the absence of the bacterial enzyme substrates. Similarly, when the activity of the bacterial bioluminescence systems was measured in the presence of the firefly luciferase substrates, a slight inhibition occurred that was also reversible since a 100% recovery was observed when the immobilized bacterial enzymes were assayed again in the absence of the firefly luciferase substrates.

As for the buffer used, the different thorough studies we performed showed that the use of a mixed 0.05 M Tris and −0.05 M phosphate buffer appeared the best compromise, allowing the use of a unique buffered solution to assay alternatively either ATP or NADH without impairing the stability of the different enzymatic systems coimmobilized on the same membrane. Nevertheless, utilization of the Tris-phosphate buffer, although it slightly affected the sensitivity of ATP analysis, resulted in a significant decrease in NADH sensitivity. For assays of NADH, therefore, the optimum conditions previously determined using the systems from *V. fischeri* (98) and from *V. harveyi* (100) were used.

The performance of the multifunctional fiberoptic sensor, used in batch or continuous-flow analysis, for the alternate determination of ATP or NADH is presented in Table 7. For continuous-flow bioluminescent ATP analysis the optimum conditions were found to be a flow rate of 0.54 ml/min and a stirring speed of 1200 rpm in the flow cell.

The storage and operational stabilities of the coimmobilized bioluminescent systems were also evaluated. For the bacterial bioluminescence systems coimmobilized with the firefly luciferase, the results were identical to those obtained with the systems immobilized alone (98,100). Concerning the coimmobilized firefly luciferase, the enzyme lost 30% of its initial activity during the first month of storage and then remained stable at this level for several months. After at least 35 consecutive continuous-flow measurements of ATP and NADH, performed within half a day with the same bioluminescence multienzymatic membrane, no decrease in the activity of the bacterial system was observed but a 10–20% decrease in firefly luciferase activity occurred. After washing the enzymatic membrane, however, the immobilized firefly luciferase recovered its full activity. This slight decrease in activity was probably a result of the presence, in the

Table 7 Performance of the Multifunction Fiberoptic Biosensor for the Alternate Determination of ATP or NADH

	Coimmobilized multienzymatic system					
	Luciferase/oxidoreductase from *Vibrio fischeri* + luciferase from *Photinus pyralis*, batch analysis		Luciferase/oxidoreductase from *Vibrio harveyi* + luciferase from *Photinus pyralis*			
			Batch analysis		Flow analysis	
	ATP	NADH	ATP	NADH	ATP	NADH
Detection limit, mol	5×10^{-13}	5×10^{-12}	5×10^{-13}	2.5×10^{-11}	2.5×10^{-13}	5×10^{-12}
Linear range, mol	$5 \times 10^{-13} - 5 \times 10^{-9}$	$5 \times 10^{-12} - 5 \times 10^{-9}$	$5 \times 10^{-13} - 5 \times 10^{-9}$	$2.5 \times 10^{-11} - 5 \times 10^{-9}$	$2.5 \times 10^{-13} - 3 \times 10^{-9}$	$5 \times 10^{-12} - 1 \times 10^{-9}$
RSD (for 10 measurements)	5% at 4×10^{-11} mol	6% at 2×10^{-10} mol	7.4% at 4×10^{-11} mol	6.2% at 2×10^{-10} mol	4.5% at 1×10^{-11} mol	4% at 5.5×10^{-10} mol

immediate vicinity of the immobilized firefly luciferase, of bacterial reactant traces not completely removed by the flow stream of buffer.

TRENDS AND PROSPECTS

The field of biosensors is expanding rapidly with a constant search for new transducing systems associated with stable biosensing elements. The ultra-sensitivity of bio- and chemiluminescence techniques, together with the convenience of immobilized compounds in combination with fiberoptics, constitutes an attractive opportunity for designing novel biosensors. The validity of such an approach is now widely proved, as the encouraging results reported in the recent papers cited here attest. In addition to the advantages of fiberoptic-based sensors, that is, the possibility of miniaturization and of remote sensing, bio- and chemiluminescence-based sensors require simpler instrumentation than those based on other spectroscopic techniques. The coupling of auxiliary enzymes to bio- and chemiluminescence-based fiberoptic sensors allows us to extend the range of compounds that can be monitored at the trace level, including enzyme activities, and new developments based on this concept can be reasonably expected. Bio- and chemiluminescence-based fiberoptic sensors in association with continuous-flow manifolds also offer a great analytic potential that is far from being completely exploited.

REFERENCES

1. DeLuca, M. (1976). Firefly luciferase, *Adv. Enzymol., 44*: 37.
2. DeLuca, M., and McElroy, W. D. (1978). Purification and properties of firefly luciferase, *Methods Enzymol., 57*:3.
3. McElroy, W. D., and DeLuca, M. (1985). Firefly luminescence. In: *Chemi- and Bioluminescence,* (J. G. Burr, Ed), Marcel Dekker, New York, p. 387.
4. White, E. H., McCapra, F., Field, G. F., and McElroy, W. D. (1961). The structure and synthesis of firefly luciferin, *J. Am. Chem. Soc., 83*: 2402.
5. Hastings, J. W., Baldwin, O. T., and Nicoli, M. Z. (1978). Bacterial luciferase: Assay, purification, and properties, *Methods Enzymol., 57*: 135.
6. Nealson, K. H. (1978). Isolation, identification, and manipulation of luminous bacteria, *Methods Enzymol., 57*: 153.
7. Puget, K. and Michelson, A. M. (1972). Studies in bioluminescence. VII. Bacterial NADH: flavin mononucleotide oxidoreductase, *Biochimie, 54*: 1197.
8. Seitz, W. R. (1978). Chemiluminescence detection of enzymically generated peroxide, *Methods Enzymol., 57*: 445.
9. Cormier, M. J., and Prichard P. M. (1968). An investigation of the luminescent peroxidation of luminol by stopped flow techniques, *J. Biol. Chem., 243*: 4706.
10. Kuwana, T. (1963). Electro-oxidation followed by light emission, *J. Electroanal. Chem., 6*: 164.
11. Epstein, B., and Kuwana, T. (1967). Electrooxidation of phthalhydrazides, *J. Electroanal. Chem., 15*: 386.

12. Thorpe, G. H. G., Kricka, L. J., and Carter, T. J. N. (1982). Chemiluminescence. In: *Clinical and Biochemical Luminescence* (L. J. Kricka, and T. J. N. Carter, Eds), Marcel Dekker, New York, p. 21.
13. Lee, R. T., Denburg, J. L., and McElroy, W. D. (1970). Substrate-binding properties of firefly luciferase. II. ATP-binding site, *Arch. Biochem. Biophys.*, *141*: 38.
14. Meighen, E. A., and MacKenzie, R. E. (1973). Flavin specificity of enzyme-substrate intermediates in the bacterial bioluminescent reaction. Structural requirements of the flavin side chain, *Biochemistry, 12*: 1482.
15. Duane, W., and Hastings, J. W. (1975). Flavin mononucleotide reductase of luminous bacteria, *Mol. Cell. Biochem.*, *6*: 53.
16. Gerlo, E., and Charlier, J. (1975). Identification of NADH-specific and NADPH-specific FMN reductases in *Beneckea harveyi, Eur. J. Biochem.*, *57*: 461.
17. Lundin, A. (1982). Analytical applications of bioluminescence: The firefly system. In: *Clinical and Biochemical Luminescence* (L. J. Kricka and T. J. N. Carter, Eds), Marcel Dekker, New York, p. 43.
18. Gould, S. J., and Subramani, S. (1988). Firefly luciferase as a tool in molecular and cell biology, *Anal. Biochem.*, *175*: 5.
19. Jablonski, E. G., and DeLuca, M. (1982). Analytical applications of bioluminescence: Marine bacterial system. In: *Clinical and Biochemical Luminescence* (L. J. Kricka, and T. J. N. Carter, Eds), Marcel Dekker, New York, p. 75.
20. Ulitzur, S., and Kuhn, J. (1987). Introduction of lux genes into bacteria, a new approach for specific determination of bacteria and their antibiotic susceptibility. In: *Bioluminescence and Chemiluminescence: New Perspectives* (J. Schölmerich, R. Andreesen, A. Kapp, and W. G. Woods, Eds), Wiley, Chichester, p. 463.
21. Erlanger, B. F., Isambert, M. F., and Michelson, A. M. (1970). Insoluble bacterial luciferases: A new approach to some problems in bioluminescence, *Biochem., Biophys. Res. Commun.*, *40*: 70.
22. Lee, Y., Jablonski, I., and DeLuca, M. (1977). Immobilization of firefly luciferase on glass rods. Properties of immobilized enzyme, *Anal. Biochem.*, *80*: 496.
23. Brovko, L. Y., Ugarova, N. N., Vasileva, T. E., Dombrovski, V. A., and Berezin, I. V. (1978). Use of immobilized firefly luciferase for quantitative determination of ATP and enzymes that synthesize and destroy ATP, *Biochem. USSR, 43*: 633.
24. Ugarova, N. N., Brovko, L. Y., and Berezin, I. V. (1980). Immobilized firefly luciferase and its use in analysis, *Anal. Lett.*, *13*: 881.
25. Wienhausen, G. K., Kricka, L. J., Hinkley, J. E., and DeLuca, M. (1982). Properties of bacterial luciferase/NADH: FMN oxidoreductase and firefly luciferase immobilized onto sepharose, *Appl. Biochem. Biotech.*, *7*: 463.
26. Ugarova, N. N., Brovko, L. Y., and Beliaieva, E. I. (1983). Immobilization of luciferase from the firefly *Luciola mingrelica:* Catalytic properties and thermostability of the enzyme immobilized on cellulose films, *Enzyme Microb. Technol.*, *5*: 60.
27. Blum, L. J., Coulet, P. R., and Gautheron, D. C. (1985). Collagen strip with immobilized luciferase for ATP bioluminescent determination, *Biotechnol. Bioeng.*, *27*: 232.
28. Ivanova, L. V., Brovko, L. Y., Shekhovtsova, T. N., Ugarova, N. N., and Dolmanova, I. F. (1986). Bioluminescent method of determining the creatine

phosphokinase activity using immobilized firefly luciferase, *J. Anal. Chem., USSR, 41*: 593.
29. Ugarova, N. N., Brovko, L. Y., Ivanova, L. V., Shekhovtsova, T. N., and Dolmanova, I. F. (1986). Bioluminescent assay of creatine kinase activity using immobilized firefly extract, *Anal. Biochem., 158*: 1.
30. Jablonski, E., and DeLuca, M. (1976). Immobilization of bacterial luciferase and FMN reductase on glass rods, *Proc. Natl. Acad. Sci, USA, 73*: 3848.
31. Haggerty, C., Jablonski, E., Stav, L., and DeLuca, M. (1978). Continuous monitoring of reactions that produce NADH and NADPH using immobilized luciferase and oxidoreductases from *Beneckea harveyi, Anal. Biochem., 88*: 162.
32. Jablonski, E., and DeLuca, M. (1979). Properties and uses of immobilized light-emitting enzyme systems from *Beneckea harveyi, Clin. Chem., 25*: 1622.
33. Ford, J., and DeLuca, M. (1981). A new assay for picomole levels of androsterone and testosterone using co-immobilized luciferase, oxidoreductase and steroid dehydrogenase, *Anal. Biochem., 110*: 43.
34. Saga, K., Kobayashi, T., and Shimizu, S. (1981). Hydrogen peroxide sensor using immobilized luminous bacterium, *Agri. Biol. Chem., 45*: 2895.
35. Rodriguez, O., and Guilbault, G. G. (1981). Immobilized bacterial luciferase for microscale analysis of creatine kinase activity. *Enzyme Microb. Technol., 3*: 69.
36. Roda, A., Kricka, L. J., DeLuca, M., and Hofmann, A. F. (1982). Bioluminescence measurement of primary bile acids using immobilized 7α-hydroxysteroid dehydrogenase: Application to serum bile acids, *J. Lipid Res., 23*: 1354.
37. Wienhausen, G., and DeLuca, M. (1982). Bioluminescent assays of picomole levels of various metabolites using immobilized enzymes, *Anal. Biochem., 127*: 380.
38. Schoelmerich, J., Hinkley, J. E., McDonald, I. A., Hofmann, A. F., and DeLuca, M. (1983). A bioluminescent assay for 12-α-hydroxy bile acids using immobilized enzymes, *Anal. Biochem., 133*: 224.
39. Schoelmerich, J., van Berge Henegouwen, G. P., Hofmann, A. F., and DeLuca, M. (1984). A bioluminescence assay for total 3-α-hydroxy bile acids in serum using immobilized enzymes, *Clin. Chim. Acta, 137*: 21.
40. Blum, L. J., and Coulet, P. R. (1984). Bioluminescent determination of reduced nicotineamide adenine dinucleotide with immobilized bacterial luciferase and flavin mononucleotide oxidoreductase on collagen film, *Anal. Chim., Acta, 161*: 355.
41. Rossi, S. S., Clayton, L. M., and Hofmann, A. A. (1986). Determination of chenodiol bioequivalence using an immobilized multi-enzyme bioluminescence technique, *J. Pharm. Sci., 75*: 288.
42. Lebedeva, O. V., Frumkina, I. G., and Ugarova, N. N. (1986). Determination of microquantities of pyridine nucleotides NADH, NAD and formate using coimmobilized bioluminescent systems, *J. Anal. Chem. USSR, 41*: 943.
43. Ugarova, N. N., Lebedeva, O. V., and Frumkina, I. G. (1988). Bioluminescent microassay of various metabolites using bacterial luciferase co-immobilized with multienzyme systems, *Anal. Biochem., 173*: 221.
44. Branchini, B. R., Salituro, F. G., Hermes, J. D., and Post, N. J. (1980). Highly sensitive assays for proteinases using immobilized luminogenic substrates, *Biochem., Biophys., Res. Commun., 97*: 334.

45. Lippman, R. D. (1980). Sensitive solid-phase chemiluminescence microassay of thiols, *Anal. Chim. Acta, 116*: 181.
46. Kobayashi, T., Saga, K., Shimizu, S., and Goto, T. (1981). The application of chemiluminescence of a *Cypridina* luciferin analog to immobilized enzyme sensors, *Agri. Biol. Chem., 45*: 1403.
47. Carter, T. J. N., Whitehead, T. P., and Kricka, L. J. (1982). Investigation of a novel solid-phase chemiluminescent analytical system, incorporating photographic detection for the measurement of glucose, *Talanta, 29*: 529.
48. Aizawa, M., Ikariyama, Y., and Kuno, H. (1984). Photovoltaic determination of hydrogen peroxide with a biophotodiode, *Anal. Lett., 17*: 555.
49. Blum, L. J., Plaza, J. M., and Coulet, P. R. (1987). Chemiluminescent analyte microdetection based on the luminol-H_2O_2 reaction using peroxidase immobilized on new synthetic membranes, *Anal. Lett., 20*: 317.
50. Demura, M., Asakura, T., Nakamura, E., and Tamura, H. (1989). Immobilization of peroxidase with a *Bombyx mori* silk fibroin membrane and its applications to biophotosensors, *J. Biotechnol., 10*: 113.
51. Kricka, L. J., Wienhausen, G. K., Hinkley, Y. J. E., and Deluca, M. (1983). Automated bioluminescent assays for NADH, glucose-6-phosphate, primary bile acids, and ATP, *Anal. Biochem., 129*: 392.
52. Vellom, D. C., and Kricka, L. J. (1986). Continuous-flow bioluminescent assays employing Sepharose-immobilized enzymes, *Methods Enzymol., 133*: 229.
53. Worsfold, P. J., and Nabi, A. (1986). Bioluminescent assays with immobilized firefly luciferase based on flow injection analysis, *Anal. Chim. Acta, 179*: 307.
54. Carrea, G., Bovara, R., Mazzola, G., Girotti, S., Roda, A., and Ghini, S. (1986). Bioluminescent continuous-flow assay of adenosine 5'-triphosphate using firefly luciferase immobilized on nylon tubes, *Anal. Chem., 58*: 331.
55. Carrea, G., Bovara, R., Girotti, S., Ferri, E., Ghini, S. and Roda, A. (1989). Continuous-flow bioluminescent determination of ATP in platelets using firefly luciferase immobilized on epoxy methacrylate, *J. Biolumin. Chemilumin, 3*: 7.
56. Girotti, S., Cascione, M. L., Ghini, S., Carrea, G., Bovara, R., Roda, A., Motta, R., and Petilino, R. (1989). Bioluminescence flow sensor for determination of creatine kinase activity in blood, *Anal. Chim. Acta, 227*: 29.
57. Kurkijärvi, K., Raunio, R., and Korpela, T. (1982). Packed-bed reactor of immobilized bacterial bioluminescence enzymes: A potential high-sensitive detector for automated analysis, *Anal. Biochem., 125*: 415.
58. Girotti, S., Roda, A., Ghini, S., Grigolo, B., Carrea, G., and Bovara, R. (1984). Continuous flow analyses of NADH using bacterial bioluminescent enzymes immobilized on nylon, *Anal. Lett., 17*: 1.
59. Roda, A., Girotti, S., Ghini, S., Grigolo, B., Carrea, G., and Bovara, R. (1984). Continuous-flow determination of primary bile acids, by bioluminescence, with use of nylon-immobilized bacterial enzymes, *Clin. Chem., 30*: 206.
60. Oda, K., Yoshida, S., Hirose, S., and Takeda, T. (1984). Photocounting determination of ultratrace levels of NAD, reduced form (NADH) by use of immobilized luciferase, *Chem. Pharm. Bull., 32*: 185.
61. Roda, A., Girotti, S., and Carrea, G. (1986). Flow systems utilizing nylon-immobilized enzymes. *Methods Enzymol., 133*: 238.

62. Nabi, A., and Worsfold, P. J. (1986). Bioluminescence assays with immobilized bacterial luciferase using flow injection analysis, *Analyst, 111*: 1321.
63. Girotti, S., Roda, A., Piazzi, S., Carrea, G., Piacentini, A., Angellotti, M., Bovara, R., and Ghini, S. (1987). Bioluminescent flow sensors: L-alanine determination in serum and urine, *Anal. Lett., 20*: 1315.
64. Nabi, A., and Worsfold, P. J. (1987). Flow injection procedure for the determination of ethanol and alcohol dehydrogenase using co-immobilized bacterial luciferase and oxidoreductase, *Analyst, 112*: 531.
65. Girotti, S., Roda, A., Angellotti, M. A., Ghini, S., Carrea, G., Bovara, R., Piazzi, S., and Merighi, R. (1988). Bioluminescence flow system for determination of branched-chain L-amino acids in serum and urine, *Anal. Chim., Acta, 205*: 229.
66. Girotti, S., Bassoli, C., Cascione, M. L., Ghini. S., Carrea, G., Bovara, R., Roda, A., Motta, R., and Petilino, R. (1989). Bioluminescence flow sensors: L-lactate dehydrogenase activity determination in serum, *J. Biolumin. Chemilumin., 3*: 41.
67. Nabi, A., and Worsfold, P. J. (1989). Indirect flow-injection assays for glucose-6-phosphate dehydrogenase/glucose-6-phosphate and malate dehydrogenase/L-malate using immobilized bacterial luciferase, *Anal. Lett., 22*: 1861.
68. Parkinson, I. S., Channon, S. M., Tomson, C. R. V., Adonai, L. R., Ward, M. K. and Laker, M. F. (1989). The determination of plasma oxalate concentrations using an enzyme/bioluminescent assay. 2. Co-immobilisation of bioluminescent enzymes and studies of in vitro oxalogenesis, *Clin. Chim. Acta, 179*: 97.
69. Oda, K., Yoshida, S., Hirose, S., and Takeda, T. (1989). Bioluminescent assay for serum adenosine deaminase with immobilized bacterial luciferase, *Clin. Chim. Acta, 185*: 17.
70. Bostick, D. T., and Hercules, D. M. (1975). Quantitative determination of blood glucose using enzyme induced chemiluminescence of luminol. *Anal. Chem., 47*: 447.
71. Williams, D. C. III, Huff, G. F., and Seitz, W. R. (1976). Evaluation of peroxyoxalate chemiluminescence for determination of enzyme generated peroxide. *Anal. Chem., 48*: 1003.
72. Rigin, V. I. (1981). Determination of formaldehyde and formic acid in natural waters by applying an immobilized enzyme and a chemiluminescence finish. *J. Anal. Chem. USSR, 36*: 1111.
73. Rigin, V. I. (1982). Chemiluminescence determination of microamounts of oxalate and urate by means of flow-through columns containing immobilized enzymes. *J. Anal. Chem. USSR, 37*: 1302.
74. Rigin, V. I. (1983). Determination of microamounts of L-amino-acids by enzymatic oxidation reaction. *J. Anal. Chem. USSR, 38*: 1328.
75. Tabata, M., Fukunaga, C., Ohyabu, M., and Murachi, T. (1984). Highly sensitive flow injection analysis of glucose and uric acid in serum using an immobilized enzyme column and chemiluminescence. *J. Appl. Biochem., 6*: 251.
76. Van Zoonen, P., Kamminga, D. A., Gooijer, C., Velthorst, N. H., and Frei, R. W. (1985). A solid-state chemiluminescence detector for hydrogen peroxide based on an immobilized luminophore. Application to rain water. *Anal. Chim. Acta, 174*: 151.
77. Petersson, B. A., Hansen, E. H., and Ruzicka, J. (1986). Enzymatic assay by flow injection analysis with detection by chemiluminescence: Determination of glucose,

creatinine, free cholesterol and lactic acid using an integrated FIA microconduit. *Anal. Lett., 19*: 649.
78. Honda, K., Miyaguchi, K., Nishino, H., Tanaka, H., Yao, T., and Imai, K. (1986). High-performance liquid-chromatography followed by peroxyoxalate chemiluminescence detection of acetylcholine and choline utilizing immobilized enzymes. *Anal. Biochem., 153*: 50.
79. Van Zoonen, P., De Herder, I., Gooijer, C., Velthorst, N. H., and Frei, R. W. (1986). Detection of oxidase generated hydrogen peroxide by a solid state peroxyoxalate chemiluminescence detector. *Anal. Lett., 19*: 1949.
80. Koerner, C. A., and Nieman, T. A. (1986). Chemiluminescence flow injection analysis determination of sucrose using enzymatic conversion and a microporous membrane flow cell. *Anal. Chem., 58*: 116.
81. Murachi, T., and Tabata, M. (1987). Use of a bioreactor consisting of sequentially aligned L-glutamate dehydrogenase and L-glutamate oxidase for the determination of ammonia by chemiluminescence. *Biotechnol. Appl. Biochem., 9*: 303.
82. Hool, K., and Nieman, T. A. (1987). Chemiluminescence analysis in flowing stream with luminol immobilized on silica and controlled-pore glass. *Anal. Chem., 59*: 869.
83. Ohyabu, M., Fujimura, M., Tanimizu, K., Okuno, Y., Tabata, M., Totani, M., and Murachi, T. (1987). "One-shot" flow-injection analysis with immobilized enzyme columns: Clinical applications, *Anal. Sci., 3*: 277.
84. Koerner, P. J., Jr., and Nieman, T. A. (1988). High-performance liquid chromatographic determination of glucosides (glucose conjugates) with post-column reaction detection combining immobilized enzyme reactors and luminol chemiluminescence, *J. Chromatogr., 449*: 217.
85. Nieman, T. A. (1988). Immobilized and solid-state reagent systems for luminol chemiluminescence in flow systems, *Mikrochim. Acta, III*: 239.
86. (Koerner) Swindlehurst, C. A., and Nieman, T. A. (1988). Flow-injection determination of sugars with immobilized enzyme reactors and chemiluminescence detection, *Anal. Chim. Acta, 205*: 195.
87. Hool, K., and Nieman, T. A. (1988). Immobilized luminol chemiluminescence reagent system for hydrogen peroxide determinations in flowing streams, *Anal. Chem., 60*: 834.
88. Tabata, M., and Murachi, T. (1988). A chemiluminometric method for the determination of urea in serum using a three-enzyme bioreactor, *J. Biolumin. Chemilumin., 2*: 63.
89. Kawasaki, H., Sato, K., Ogawa, J., Hasegawa, Y., and Yuki, H. (1989). Determination of inorganic phosphate by flow injection method with immobilized enzymes and chemiluminescence detection, *Anal. Biochem. 182*: 366.
90. Freeman, T. M., and Seitz, W. R. (1978). Chemiluminescence fiber optic probe for hydrogen peroxide based on the luminol reaction, *Anal. Chem., 50*: 1242.
91. Vandyke, D. A., and Cheng, H. -Y. (1989). Electrochemical manipulation of fluorescence and chemiluminescence signals at fiber-optic probes, *Anal. Chem., 61*: 633.
92. Abdel-Latif, M. S., and Guilbault, G. G. (1988). Fiber-optic sensor for the determination of glucose using micellar enhanced chemiluminescence of the peroxyoxalate reaction, *Anal. Chem., 60*: 2671.

93. Abdel-Latif, M. S., and Guilbault, G. G. (1989). Peroxide optrode based on micellar-mediated chemiluminescence reaction of luminol, *Anal. Chim. Acta, 221*: 11.
94. Blum, L. J., Gautier, S. M., and Coulet, P. R. (1988). Luminescence fiber-optic biosensor, *Anal. Lett., 21*: 717.
95. Blum, L. J., Gautier, S. M., and Coulet, P. R. (1989). Design of luminescence photobiosensors, *J. Biolumin. Chemilumin, 4*: 543.
96. Assolant-Vinet, C. H., and Coulet, P. R. (1986). New immobilized enzyme membranes for tailor-made biosensors, *Anal. Lett., 19*: 875.
97. Coulet, P. R., Blum, L. J., and Gautier, S. M. (1990). Luminescence optical biosensor oriented to clinical analysis. In: *Automation and New Technology in the Clinical Laboratory* (K. Okuda, Ed), Blackwell Scientific, Oxford, p. 239.
98. Gautier, S. M., Blum, L. J., and Coulet, P. R. (1989). Fibre-optic sensor with co-immobilised bacterial bioluminescence enzymes, *Biosensors, 4*: 181.
99. Gautier, S. M., Blum, L. J., and Coulet, P. R. (1990). Fibre-optic biosensor based on luminescence and immobilized enzymes: Microdetermination of sorbitol, ethanol and oxaloacetate, *J. Biolumin. Chemilumin., 5*: 57.
100. Blum, L. J., Gautier, S. M., and Coulet, P. R. (1989). Highly stable bioluminescence-based fiber-optic sensor using immobilized enzymes from *Vibrio harveyi, Anal. Lett., 22*: 2211.
101. Blum, L. J., Gautier, S. M., and Coulet, P. R. (1989). Continuous flow bioluminescent assay of NADH using a fibre-optic sensor, *Anal. Chim. Acta, 226*: 331.
102. Gautier, S. M., Blum, L. J., and Coulet, P. R. (1989). Alternate determination of ATP and NADH with a single bioluminescence-based fiber-optic sensor, *Sensors & Actuators, B1*: 580.
103. Gautier, S. M., Blum, L. J., and Coulet, P. R. (1990). Multifunction fibre-optic sensor for the selected bioluminescent flow determination of ATP or NADH, *Anal. Chim. Acta*, 235: 243.

11

Immunosensors

Masuo Aizawa

Tokyo Institute of Technology
Tokyo, Japan

INTRODUCTION

Immunosensors take advantage of the high selectivity provided by the molecular recognition of antibodies. Because of significant differences in affinity constants antibodies may confer an extremely high sensitivity to immunosensors in comparison to enzyme sensors. Furthermore, antibodies may be obtained in principle against an unlimited number of determinants. Immunosensors are thus characterized by high selectivity, sensitivity, and versatility.

In the early 1970s Janata proposed an immunotype electrode, although no immunochemical reaction was involved, consisting of an electrode coated with a concanavalin A-entrapped polyvinyl chloride membrane (1). The immunoelectrode responded to the corresponding saccharide in solution by a potential variation. Aizawa et al. reported two different types of immunosensors. The first was based on the measurement of the transmembrane potential across a membrane that incorporates an antibody (or antigen) (2,3). The membrane-bound antibody (or antigen) specifically binds the corresponding antigen (or antibody) in solution, which results in a change in transmembrane potential. The other was designed on the basis of an enzyme-linked immunoassay (4,5). An oxygen electrode was covered with an antibody binding membrane. A catalase-labeled antigen was added to a solution containing the antigen to be determined. Both antigens competitively reacted with the membrane-bound antibody. The labeling enzyme was assayed by measuring the sensor output after the addition of hydrogen peroxide. This type of immunosensor exhibited an extremely high sensitivity provided by chemical amplification due to enzyme labeling.

There has been extensive research and development on immunosensors in the last two decades (6–8), concentrated in recent years, however, probably because of both the strong demand for immunoassays and the remarkable progress in immunochemical technology. It should also be noted that innovative technologies, such as optoelectronics, have recently been introduced into immunosensor technology.

A high sensitivity as well as a high selectivity and reproducibility are essential requisites for immunosensors. These requisites have made it difficult to launch immunosensors on the market. It should be noted, however, that immunosensors are developed by the integration of current sensor technology and biotechnology.

This chapter is concerned with the principles of various immunosensors that have been described, and the state of the art on the subject is presented here.

PRINCIPLES OF NONLABELED AND LABELED IMMUNOSENSORS

Immunosensors can be divided in principle into two categories: nonlabeled and labeled immunosensors (6). Nonlabeled immunosensors are designed so that the immunocomplex, that is, the antigen-antibody complex, is directly determined by measuring the physical changes induced by the formation of the complex. In contrast, in a labeled immunosensor a sensitively detectable label is incorporated. The immunocomplex is thus determined through measurement of the label.

Nonlabeled immunosensors are based on several principles (Fig. 1). Either the antibody or the antigen is immobilized on the solid matrix to form a sensing device. The solid matrix should be sensitive enough, in its surface characteristics, to detect the immunocomplex formation. Electrode, membrane, piezoelectric material, or optically active material surfaces may be used to construct nonlabeled immunosensors. The antigen or the antibody to be determined is dissolved in a solution and reacts with the complementary matrix-bound antibody or antigen to form an immunocomplex. This formation thus alters the physical properties of the surface, such as the electrode potential, the transmembrane potential, the intrinsic piezofrequency, or the optical properties.

Transmembrane potential is sensitive to immunocomplex formation on the antibody-binding surface according to the following mechanism. Transmembrane potential ($\Delta\phi_m$) is approximated by Equation (1) when both surfaces of a membrane are in contact with solutions of a uni-univalent electrolyte at concentrations C_1 and C_2(9). A concentration gradient then exists across the membrane.

$$\Delta\varphi_m = \frac{RT}{F}\left\{\frac{\ln\left(\frac{C_1}{C_2}\right) - \ln\left(-\theta + \sqrt{\theta^2 + 4C_1{}^2}\right)}{-\theta + \sqrt{\theta^2 + 4C_2^2}} + \frac{(1-2t)\ln\left[(1-2t)\theta + \sqrt{\theta^2 + 4C_1{}^2}\right]}{(1-2t)\theta + \sqrt{\theta^2 + 4C_2{}^2}}\right\} \tag{1}$$

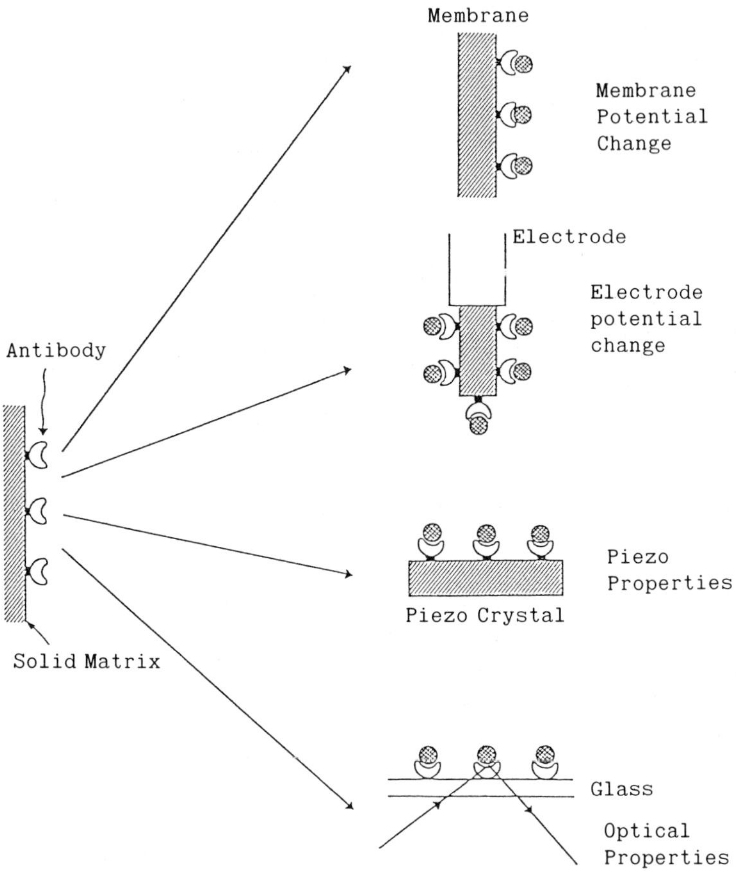

Figure 1 Principle of nonlabeled immunosensors.

In this equation R is the universal gas constant, T the temperature, F Faraday's constant, θ the charge density of the membrane phase, and t the transport number of anion. Transmembrane potential consists of diffusion potential in the first term and interfacial potential in the second term.

A sufficiently high selectivity may be obtained with nonlabeled immunosensors, although such problems as nonspecific adsorption onto the matrix-bound antibody surface remain unsolved. To meet the requirements of immunosensors intensive research has been conducted to enhance sensitivity as well as selectivity. Various labels have been incorporated into immunosensors to achieve high sensitivity, but radioisotopic labels have essentially been avoided. Several nonisotopic labels for immunosensors are listed in Table 1. These involve enzymes, catalysts, fluorophores, electrochemically active molecules, and liposomes. Among these labels, enzymes, catalysts, and liposomes provide chemical

Table 1 Nonisotopic Labels for Immunosensors

Label		Determination
Enzyme	Catalase	Amperometric determination of O_2
	Glucose oxidase	
	Peroxidase	Luminescent determination
	Luciferase	
	Urease	Potentiometric determination of phenol
Catalyst	Hemin	Luminescent determination
Fluorophore	FITC	Fluorescent determination
Electroactive substances	Ferrocene	Amperometric determination
	DTPA-In	
	Pyrene	Electrochemical luminescent determination
	Luminol	
Liposome	Ionic marker	Potentiometric determination of ionic marker

amplification. Highly sensitive immunosensors may be designed by incorporating these labels.

Labeled immunosensors are basically designed so that immunochemical complexation takes place on the surface of the sensor matrix. There are several variations of the procedure to form an immunocomplex on this matrix. In the final step, however, the label should be incorporated into the immunocomplex, which can thus be determined (Fig. 2).

When an enzyme is used as label it is covalently bound to either the antibody or the antigen, and catalase is one of the labels that has been successfully used for such a purpose (4,5). An oxygen electrode can be coupled with an antibody binding membrane to form an immunosensor. In the final step of the analytical procedure, both nonlabeled and labeled immunocomplexes are adsorbed on the surface of the antibody binding membrane.

NONLABELED IMMUNOSENSORS

Potentiometric Immunosensors

Three types of potentiometric immunosensors have been proposed (Fig. 1). The first is based on the determination of transmembrane potential across an antibody (or antigen) membrane that specifically binds the corresponding antigen (or antibody) in solution. Concentrations of either the target antigen or antibody can be determined by measuring a change in the transmembrane potential occurring when the immunocomplex forms on the membrane surface. The second type is based on determination of the electrode potential. The surface of an electrode is modified by an antibody or antigen able to bind specifically the corresponding antigen or antibody. Immunocomplex formation causes the electrode potential to

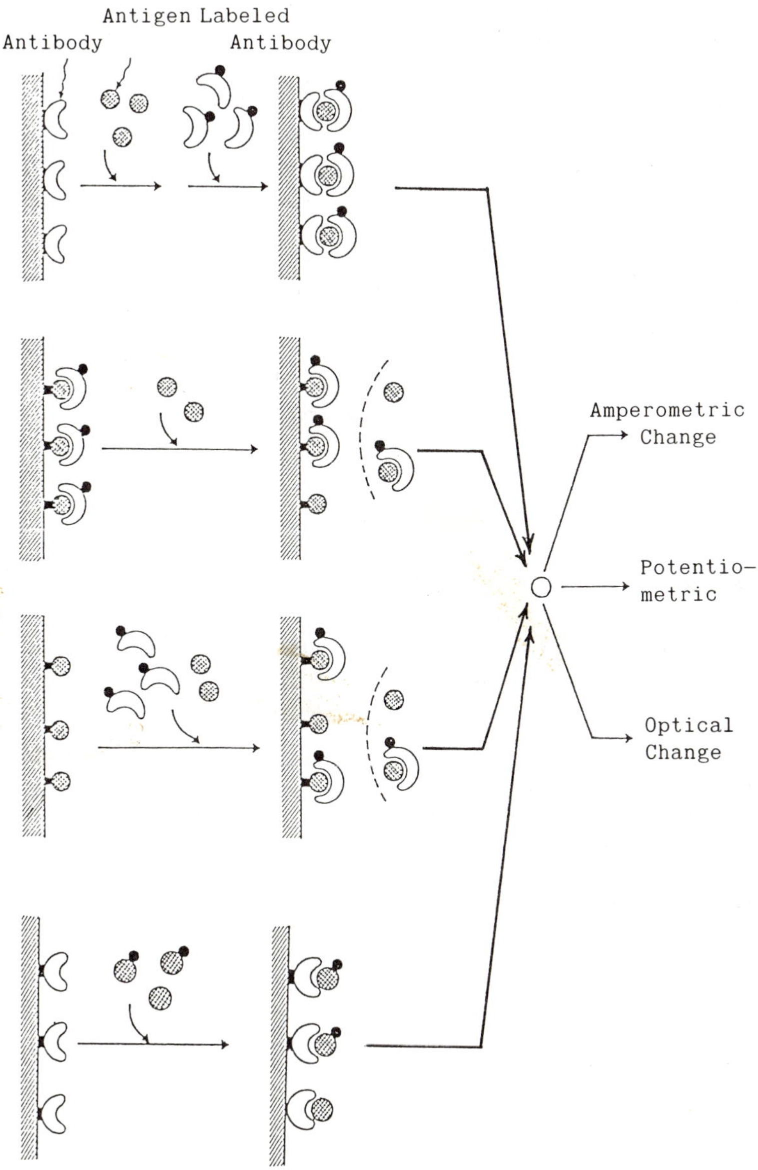

Figure 2 Principle of labeled immunosensors.

vary, primarily as a result of a change in surface charge related to the concentration of the analyte in solution. The third type is based on the determination of surface potential of the gate of a field effect transistor (FET) covered by a thin antibody binding membrane. The surface potential of the FET gate may vary with the concentration of the corresponding antigen in solution.

Results for potentiometric immunosensors for syphilis and blood typing have been reported by Aizawa et al. (2,3,10). They are based on the determination of the transmembrane potential across an immunoresponsive membrane. For blood typing, for example, the immunoresponsive membrane incorporates blood group substances and a pair of reference electrodes for measuring the transmembrane potential (Fig. 3). A, B, and O blood samples have been typed using this potentiometric immunosensor.

Yamamoto et al. (11) proposed a potentiometric immunosensor with an antibody against human chorionic gonadotropin (HCG) hormone, covalently bound to the surface of an electrode (Fig. 4). The electrode is reported to respond to HCG in solution.

Using an antibody-modified FET Janata et al. described the possible determination of the corresponding antigen. No reproducible data, however, have yet been published.

Although these potentiometric immunosensors are characterized by their simplicity of operation, they suffer from insufficient sensitivity.

Optical Immunosensors (12)

If the solid surface is sensitive enough to allow changes in its optical properties with immunocomplex formation, optical immunosensors without label may be constructed. Surface plasmon resonance (SPR) is so sensitive that an immu-

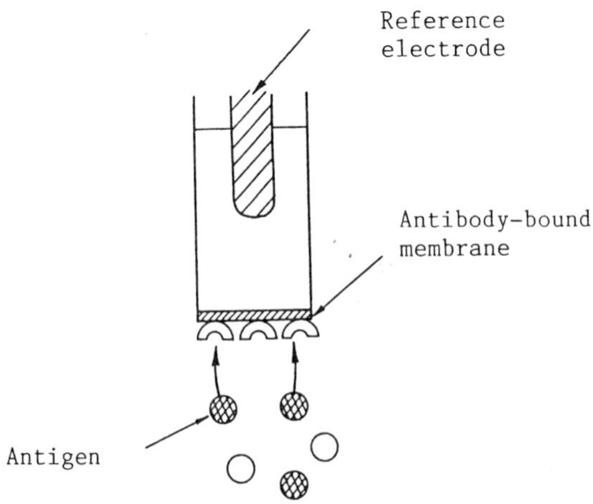

Figure 3 Principle of an immunosensor based on membrane potential.

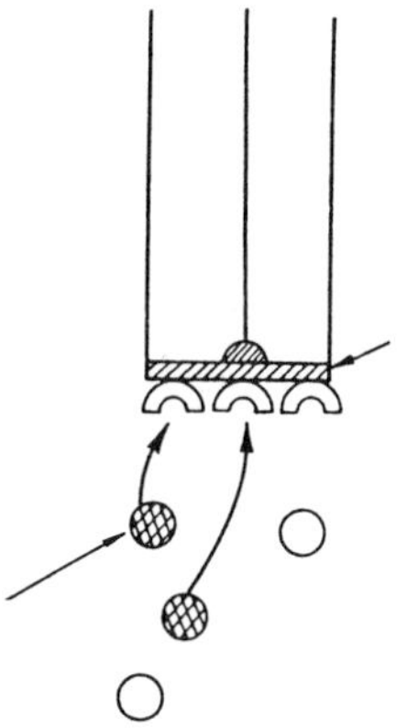

Figure 4 Principle of an immunosensor based on electrode potential.

nocomplex may be detected on the surface of a silver-coated solid. However, it seems difficult to obtain reproducible data in highly diluted samples.

Acoustic Immunosensors

Acoustic immunosensors have attracted many researchers because of their simplicity of operation. The surface of a piezoelectric material is coated with antibodies or antigens. The intrinsic frequency of the piezoelectric material is expected to shift when responding to the corresponding antigen or antibody in solution. Although several papers have been published on the subject, further confirmation is required because of the uncertainty due to numerous artifacts.

IMMUNOSENSOR LABELS FOR HETEROGENEOUS IMMUNOASSAYS

Enzyme Labels

A variety of labeled immunosensors have been proposed. Enzyme labels can enhance the sensitivity of immunosensors owing to chemical amplification. Since an enzyme rapidly converts a substrate into a product, the product that accumulates can be detected with various electronic and optoelectronic devices.

Some enzyme labels are electrochemically detected with a high sensitivity. These include catalase, glucose oxidase, and peroxidase. Both catalase and glucose oxidase may be associated with an oxygen electrode.

When catalase, which catalyzes the decomposition of hydrogen peroxide into oxygen according to the following reaction, is the labeling enzyme for an antigen, an enzyme immunosensor can be constructed by assembling an antibody-bound membrane and an oxygen-sensing electrode.

$$H_2O_2 \longrightarrow H_2O + 1/2O_2$$

In heterogeneous enzyme immunoassays the labeling enzyme activity is measured by amperometry with the oxygen-sensing device. Because only a short time is required for measuring the labeling enzyme a rapid and highly sensitive enzyme immunoassay can be achieved.

An enzyme immunosensor was prepared by attaching the antibody-bound membrane to a Clark type oxygen electrode involving an oxygen-permeable synthetic (e.g., Teflon) membrane on the cathode surface (Fig. 5).

An antibody membrane with anti-α-fetoprotein (AFP) was attached to an oxygen electrode to form an enzyme immunosensor for AFP (13). The sensor is immersed in a test solution to which a known amount of catalase-labeled AFP was added. After a thorough washing the sensor was placed in phosphate buffer. After a background current due to dissolved oxygen was obtained, hydrogen peroxide was injected. The sensor responded very rapidly to the generated oxygen, and a steady-state current was obtained within 30 sec.

A calibration curve for competitive enzyme immunoassay for AFP obtained with this sensor is presented in Figure 6. It shows that AFP can be determined in the range 5×10^{-11}–5×10^{-8} g/ml. The standard deviation for 25 assays of 10^{-9} g AFP is 15%.

Ochratoxin A (OTA), a secondary metabolite of *Aspergillus ochraceus, Penicillium viridicatum*, and strains of some other species of both genera, causes nephropathy in animals and probably also in humans. Recently carcinogenicity was also demonstrated with feeding experiments, and a simplified and rapid analysis for OTA is urgently needed.

An enzyme immunosensor for OTA was recently developed. The competitive immunoassay of OTA is carried out by using the toxin sensor, which consists of an amperometric oxygen electrode and an OTA-bound membrane. OTA is determined under a constant concentration of catalase-labeled antibody. The standard curve shifts to the lower concentration range with a decrease in antibody concentration (Fig. 7). The detection limit is of the order of 10^{-10} g/ml when the concentration of labeled antibody is fixed at 20 μg/ml.

Enzyme labels are efficiently detected with optoelectronic devices in a manner similar to that for electrochemical devices. Peroxidase and luciferase, for instance, catalyze luminescent reactions of luminol and luciferin, respectively, to generate photons. These enzymes may be incorporated as labels to form optical enzyme immunosensors (14,15).

An optical enzyme immunosensor consists of an optical fiber having one end modified with either an antibody or an antigen (Fig. 8). According to this principle an optical enzyme immunosensor for insulin was designed on the basis of subtle differences in antibody bioaffinity for the corresponding antigen

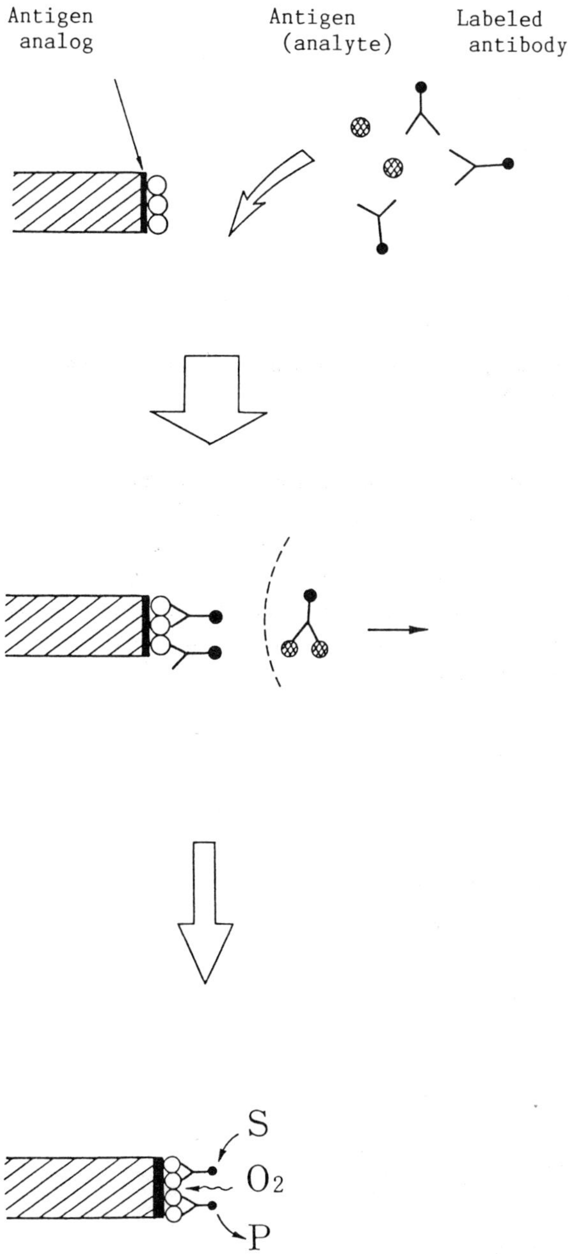

Figure 5 Principle of an amperometric enzyme immunosensor.

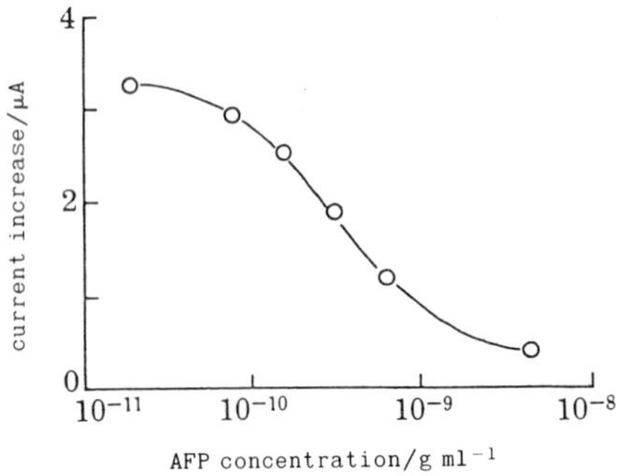

Figure 6 Calibration curve for competitive enzyme immunoassay for AFP with an amperometric enzyme immunosensor.

(bovine insulin) and its analog compound (porcine insulin) (16). Porcine insulin is bound on the surface of a transparent plastic plate and fixed to the tip of an optical fiber. Anti-bovine insulin antibody is labeled with peroxidase and immunochemically complexed with plate-bound porcine insulin. The optical fiber sensor is immersed in a solution containing the free bovine insulin to be determined, and then the anti-bovine insulin antibody-porcine insulin complex may be dissociated. The dissociated enzyme-labeled antibody then forms a stable

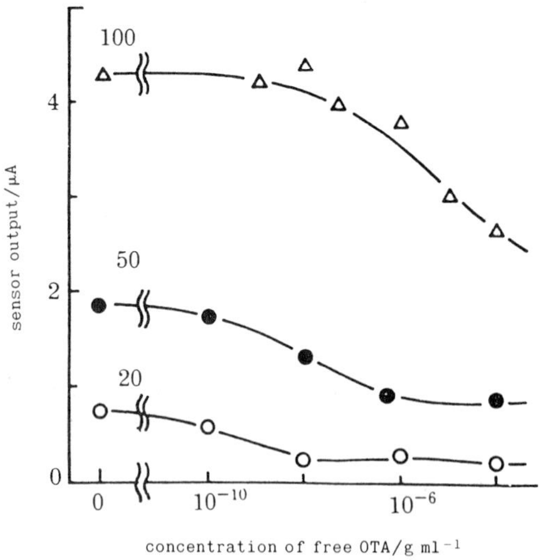

Figure 7 Standard curves for competitive enzyme immunoassay for OTA at different concentrations of labeled antibody (20, 50, and 100 mg/L).

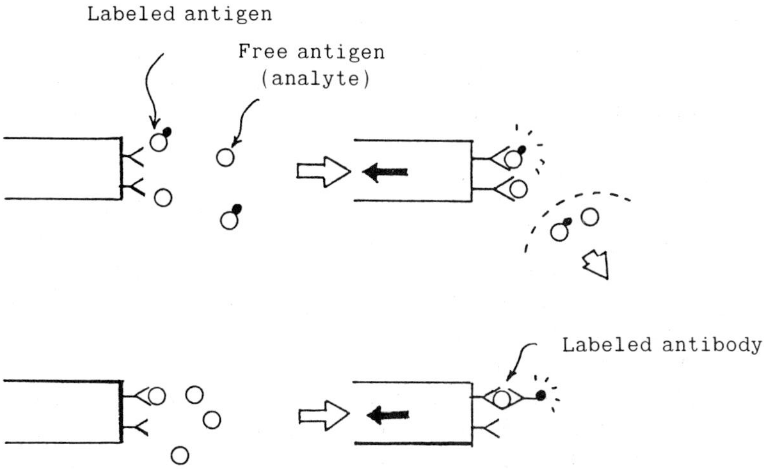

Figure 8 Principle of an optical enzyme immunosensor: (a) competitive and (b) sandwich immunoassays.

complex with bovine insulin in solution. Insulin is thus determined by measuring the peroxidase retained on an immunoplate. The luminol-H_2O_2 system is employed to detect peroxidase activity. The emitted light is then transmitted to a photomultiplier through the optical fiber.

Electrochemically Active Labels

Several redox substances can be used as labels for electrochemical immunoassays (17). These may include amperometrically detected ferrocene and In^{2+}.

A voltammetric immunoassay for morphine and codeine was explored in conjunction with the continuous-flow amperometric detection of ferrocene. Morphine acts as an electroinactive antigen competing for antibody binding sites. Competition between the two molecules resulted in the release of morphine from the antibody complex with a subsequent increase in the signal current due to unbound morphine.

Estriol is also determined in a similar manner with the use of Hg-labeled estriol. Estriol and Hg-labeled estriol compete for the limited binding site on the estriol antibody.

Ferrocene-mediated amperometric immunoassay was modified in such a way that ferrocene and its derivatives are used as electron acceptors for glucose oxidase (GOD). The binding of ferrocene-antigen complex by antibody inhibits its ability to act as a mediator in the GOD-catalyzed reaction, and thus the catalytic current is greatly decreased. This can be reversed by adding a nonlabeled antigen (i.e., the analyte) that competes for the available antibody binding sites. Thus the catalytic current produced in the reaction depends on the concentration of the analyte in solution. Lidocaine in plasma was determined over the concentration range 5–50 μM with a relative standard deviation of 3–6%.

An indicator metal can be used in metal chelate labels. With human serum albumin (HSA) chosen as a model antigen, diethylenetriaminepentaacetic acid (DTPA) is coupled to the amino residues and acts as a site-specific chelating agent for metal ions. Indium (In^{3+}) is very tightly bound by DTPA, making the labeled protein complex stable until the In^{3+} is purposely released by acidification. The heterogeneous voltammetric immunoassay is based on the competition between labeled and native antigen (HSA) for a limited number of antibody binding sites that have been insolubilized on an immunoadsorbent. After equilibration the antibody "bound" complexes are released by acidification and are detected separately by differential pulse anodic stripping. A detection limit for HSA of approximately 5 μg/ml is obtained.

Liposome Labels

Liposomes may work as chemically amplifying labels if marker substances are encapsulated.

IMMUNOSENSOR LABELS FOR HOMOGENEOUS IMMUNOASSAYS

Fluorescence Labels

Since homogeneous immunoassays require no bound/free species separation process, they may be advantageous over heterogeneous immunoassays (Fig. 9). Although intensive efforts have been concentrated on the development of homogeneous immunosensors, a very limited number of papers has been published on the subject.

An optical fiber structure has been proposed for designing an homogeneous immunosensor (Fig. 10). An antibody is immobilized on the surface of an optical fiber core, and a fluorescence compound is used as label. Both labeled antigen and free antigen to be determined react competitively with bound antibody to form an immunocomplex on the core surface.

The surface-bound label can be excited by an evanescent wave that passes through the optical fiber core. On the other hand, fluorescence labels in the bulk solution cannot be excited even if the excitation beam comes through the optical fiber core. Labels attached to the surface-bound immunocomplex are thus discriminated from labels in solution. Although the details have not been reported, the principle is elegant and homogeneous immunoassay may be possible.

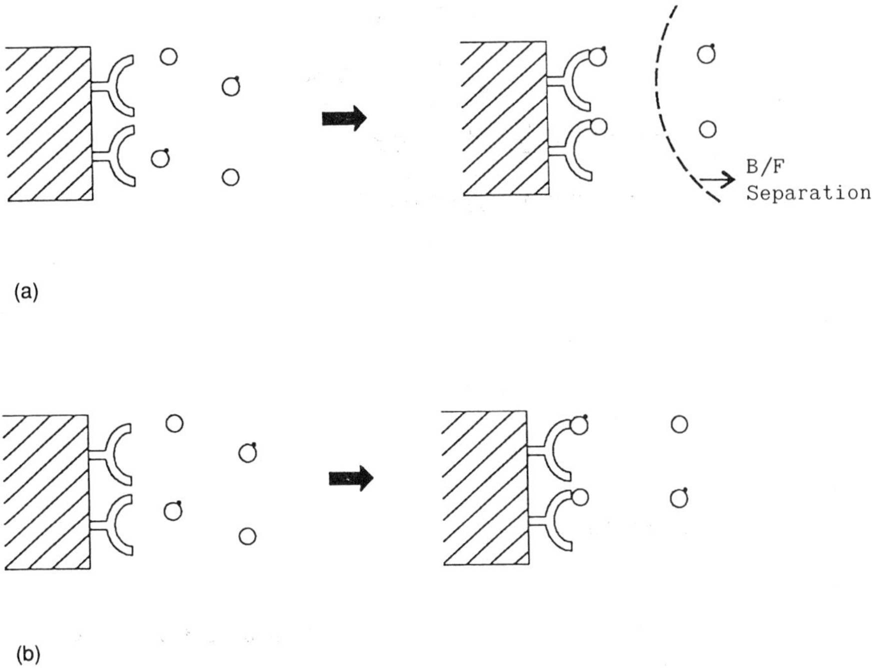

Figure 9 Heterogeneous (a) and homogeneous (b) immunoassays.

Electrochemiluminescence Labels

Some electrochemically active substances able to generate photons on an electrode surface can be used as labels for homogeneous immunoassays. A labeled antigen exhibits an electrochemical reactivity and generates luminescence, but when it is immunochemically complexed the labeled antigen loses its electrochemiluminescence property (Fig. 11).

An optical immunosensor for homogeneous immunoassay was assembled by spattering platinum on the end surface of an optical fiber (Fig. 10). Spattered platinum maintains optical transparency and works as an electrode. An optical fiber electrode efficiently collects photons generated on the surface of the transparent electrode.

Luminol as label offers excellent characteristics for designing a homogeneous immunosensor and generates luminescence by two different types of electrochemical excitation (Fig. 12) (18–20). One is based on a two-step electrochemical excitation, cathodic excitation followed by anodic excitation. The cathodic excitation produces hydrogen peroxide that causes anodically generated luminol radical to emit photons. The other is a single-step electrochemical excitation in

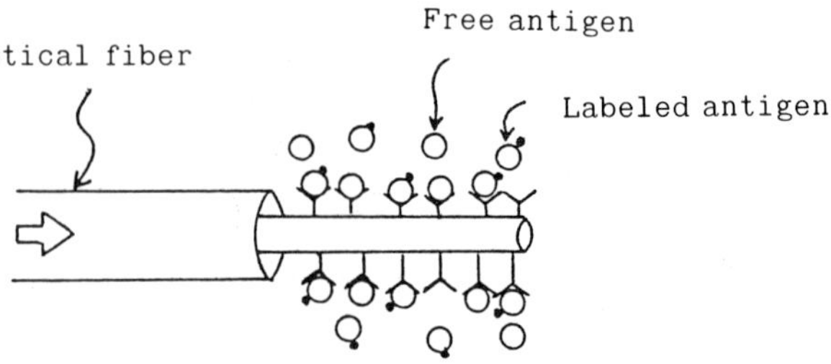

(a)

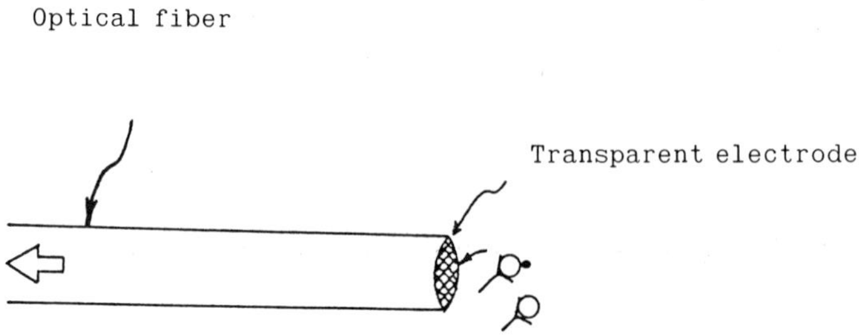

(b)

Figure 10 Optical immunosensors for homogeneous immunoassay. (a) Optical fiber immunosensor based on evanescent waves; (b) optical fiber electrode for electrochemical luminescence-based immunoassay.

the presence of hydrogen peroxide. Luminol is simply oxidized by anodic excitation to generate radicals, followed by emission of photons. The two-step electrochemical excitation provides a very high sensitivity with a limit of detection equal to 10^{-13} mol/L of luminol.

Homogeneous immunoassay with immunoglobulin G (lgG) as a model antigen, labeled with luminol, was thus performed by the two-step electrochemical excitation using an optical fiber electrode. Similarly to free luminol, labeled lgG generates electrochemical luminescence in the presence of hydrogen peroxide by anodic oxidation. Electrochemical luminescence sharply decreases by immunocomplexation with anti-lgG antibody. The addition of 10^{-13} g/L of antibody

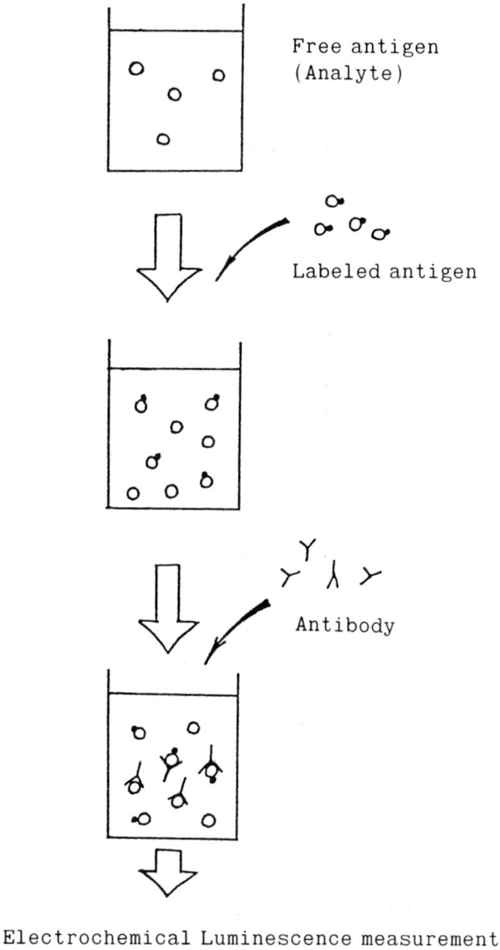

Figure 11 Homogeneous immunoassay with an optical fiber electrode.

results in an appreciable suppression of luminescence. The lower limit of detection may then fall in the range of 10^{-13} g/L of antibody.

CONCLUDING REMARKS

Immunosensors have been designed by a combination of immobilized antibody (or antigen) with an electronic device. A high selectivity may be attained with both nonlabeled and labeled immunosensors as a result of the specific affinity of antigen for the corresponding antibody. However, the immunochemical reaction

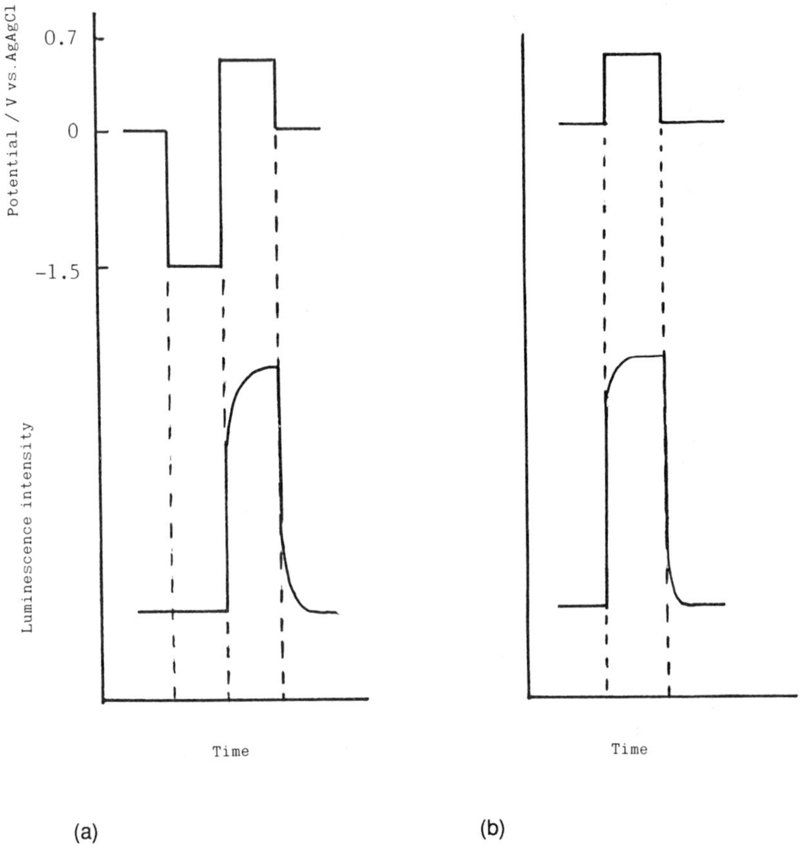

Figure 12 Electrochemical excitation of luminol to generate luminescence: (a) two-step excitation, and (b) single-step excitation.

of matrix-bound antibody (or antigen) is influenced by nonspecific adsorption of coexisting substances onto the matrix. Several methods have been proposed to prevent nonspecific binding, although they are not satisfactory. It is desirable that further research be concentrated on the development of improved materials for immunosensors.

Nonlabeled immunosensors are characterized by a high selectivity and simplicity of operation. Most of these sensors are based on potentiometric detection. Further investigations are required to seek novel nonlabeled immunosensors with both high selectivity and sensitivity.

In contrast to nonlabeled immunosensors, various enzyme labels have been successfully incorporated into labeled immunosensors to work as effective chemical amplifiers. These labels are sensitively detected with electrochemical and optoelectronic devices. The lower limit of detection may be around 10^{-10} g/ml, and further improvements in sensitivity are expected.

Several immunosensors for homogeneous immunoassays have been developed. These include an optical fiber immunosensor using evanescent wave and an optical fiber electrode based on electrochemical luminescence.

REFERENCES

1. Janata, J. (1975). Immunoelectrode, *J. Am. Chem. Soc., 97*: 2915
2. Aizawa, M., Suzuki, S., Nagamura, Y., Shinohara, R., and Ishiguro, I. (1977). An immunosensor for specific protein, *Chem. Lett., 1977*: 779.
3. Aizawa, M., Kato, S., and Suzuki, S. (1977). Immunoresponsive membrane, *J. Membr. Sci., 2*: 125.
4. Aizawa, M., Morioka, A., Matsuoka, H., Suzuki, S., Nagamura, Y., Shinohara, R., and Ishiguro, I. (1976). An enzyme immunosensor for lgG, *J. Solid-Phase Biochem., 1*: 319.
5. Aizawa, M., Suzuki, S., Nagamura, Y. (1979). An enzyme immunosensors. III. *Anal. Biochem., 94*: 22.
6. Aizawa, M. (1977). Immunosensors, *Philos. Trans. Roy. Soc. Lond. Biol., 316*: 121.
7. Ngo, T. T. (Ed). (1987). *Electrochemical Sensors in Immunological Analysis*, Plenum Press, New York.
8. Aizawa, M. (1990). Optical fiber electrodes for electrochemical luminescence-based homogeneous immuno-assay. In: *Biosensor Technology* (R. P. Buck, W. E. Hatfield, M. Umana, and E. F. Bowden, Eds), Marcel Dekker, New York, pp. 209–218.
9. Kobatake, Y., Takeguchi, N., Toyoshima, Y., and Fujita, H. (1965). Studies of membrane phenomena, *J. Phys. Chem., 69*: 3981.
10. Aizawa, M., Kato, S., and Suzuki, S. (1980). Electrochemical typing of blood using affinity membranes, *J. Membr. Sci., 7*: 1.
11. Yamamoto, N., Nagasawa, Y., Sawai, M., Suda, T., and Tsubomura, H. (1978). Potentiometric investigations of antigen-antibody and enzyme-enzyme inhibitor reactions using chemically modified metal electrodes. *J. Immunol. Methods, 22*: 309.
12. Arwin, H., and Lundstrom, I. (1988). Surface oriented optical methods for biomedical analysis, *Methods Enzymol., 137*: 366–381.
13. Aizawa, M., Morioka, A., and Suzuki, S. (1980). An enzymc immunosensor for the electrochemical determination of the tumor antigen α-fetoprotein, *Anal. Chim. Acta., 115*: 61.
14. Aizawa, M., Suzuki, S., Kato, T., Fujiwara, T., and Fujita, Y. (1980). Solid-phase luminescent enzyme immunoassay of IgG and anti-IgG using a transparent and nonporous antibody-bound plate. *J. Appl. Biochem., 2*: 190.
15. Ikariyama, Y., Suzuki, S., and Aizawa, M. (1982). Luminescence immunoassay of human serum albumin with hemin as labelling catalyst. *Anal. Chem., 54*: 1126.
16. Ikariyama, Y., and Aizawa, M. (1983). Bioaffinity sensor for insulin with luminescence detection, In: *Proceedings 3rd Sensors Symposium*, pp. 17–20.
17. Nakamura, K., Aizawa, M., and Miyawaki, (1988). Electroenzymology. In: *Coenzyme Regeneration*, Springer-Verlag, Berlin, pp. 48–52.

18. Ikariyama, Y., Kunoh, H., and Aizawa, M. (1985). Electrochemical luminescence based homogeneous immunoassay. *Biochem. Biophys. Res. Commun., 128*: 987.
19. Aizawa, M., Tanaka, M., Ikariyama, Y., and Shinohara, H. (1989). Luminescence biosensors, *J. Biolum. Chemilum., 4*: 535.
20. Aizawa, M., Tanaka, M., and Ikariyama, Y. (1989). A fiber-optic electrode for opto-electrochemical biosensors. *ACS Symp. Ser., 403*: 129.

12

Microbial Biosensors

I. Karube and M. E. SangMok Chang

Research Center for Advanced Science and Technology, University of Tokyo, Tokyo, Japan

INTRODUCTION

The industrial applications of biochemical and microbiological processes in such fields as the production of pharmaceuticals, food manufacturing, wastewater treatment, and energy production are increasing. Fermentation reactions play a very important role in such biotechnological processes. Precise control of these systems is essential if complex and changeable reaction mixtures and broths are to be used on the industrial scale. Without such control substrates, products, catalysts, pH, and other variables cannot be kept at optimal levels.

Most of these materials in broth can be determined by spectrophotometric methods. When natural materials are used as raw materials in fermentations, however, the broths are not optically clear and may therefore be unsuitable for on-line measurement.

Rapid and sensitive on-line monitoring and control of the variable factors already mentioned calls for sensors that are not only specific to the substrates and products of fermentation but can also measure the numbers of viable whole cells present in the fermentation broth.

Biosensor design and technology play a prominent role in fulfilling these criteria because they have distinct advantages: for example, test samples do not need to be optically clear and can be measured over a wide range of concentrations without pretreatment. Biosensors also offer the possibility of real-time analysis, which is particularly important for the rapid measurement of analytes in industry, for example in process monitoring and control, in which there is a demand for the in situ determination of flow rates, levels of contaminants, and so on.

Recently several kinds of biosensors have been developed for the determination of organic compounds (1–6). These methods have predominantly utilized enzyme-catalyzed reactions and microbially-catalyzed reactions. Enzyme sensors may be highly specific for the substrates of interest, but the enzymes employed are generally expensive and unstable. Microbial sensors are therefore considered more suitable for on-line control of biochemical processes and are being developed by our group.

The advantages of microbial biosensors are as follows (7,8): (1) they are less sensitive to inhibition by solutes and are more tolerant of suboptimal pH and temperature values than enzyme electrodes; (2) they have a longer lifetime than enzyme electrodes; and (3) they are cheaper because an active enzyme need not be isolated.

However, they may also have disadvantages, which are as follows: (1) some have a long response time than enzyme electrodes; (2) they need more time to return to the baseline signal after use; and (3) cells contain many enzymes and precautions must be taken to ensure a degree of selectivity, for example by optimizing storage conditions or adding specific enzyme reactions. Mutant microorganisms lacking certain enzymes can also be used. The advantages of microbial biosensors must therefore be balanced against their known disadvantages.

Microbial sensors developed by the authors involve the assimilation of organic compounds by microorganisms, changes in respiration activity, or the production of electroactive metabolites. These parameters can be monitored directly by an electrochemical device (5,9,10).

This chapter describes the principles, the construction, and the measurement of microbial biosensors, largely developed by our group, together with discussion of some other current topics in this broad field.

PRINCIPLES OF MICROBIAL BIOSENSORS

Microbial biosensors are devices incorporating a biological sensing element (a microorganism) that can specifically recognize species of interest, either intimately connected to or integrated within a suitable transducing system. The transducer functions to quantitatively convert the biochemical signal into an electronic signal that can be suitably processed and output. In this way analytes of interest may be measured by using the assimilation capacity of the microorganism as an index of the respiration activity or of the metabolic activity. The transducer can take many forms, including photodetectors, potentiometric or amperometric electrodes, and thermistors. Many aspects of biosensors based on these different transducing systems have been reviewed in the literature (1–3).

Electrochemical Microbial Sensor

Microbial biosensors of this kind consist of a combination of microorganisms immobilized on a membrane in close contact with an electrochemical transducer (4,5). Microbial biosensors can be classified as measuring either respiratory activity or electrochemically active metabolites.

With those that measure respiration activity, changes in respiratory activity of microorganisms caused by assimilation are detected by an oxygen electrode. Substrate concentrations can be estimated from these changes. Aerobic microorganisms are used in these sensors. A microbial biosensor is dipped into a buffer solution saturated with oxygen. Substrate concentration can be measured from the decrease in oxygen.

The other type of microbial biosensor can detect electrochemically active metabolites, such as H_2, CO_2, NH_3, and organic acids, that are released from microorganisms. This type of microbial biosensor is not limited to aerobic but can also employ anaerobic microorganisms.

However most of the microbial biosensors developed are of the type that measure respiratory activity.

Photomicrobial Sensors

The luminescence intensity of photobacteria (luminobacteria) is dependent on metabolic activity. It is therefore possible to construct high-sensitivity microbial biosensors by combining these bacteria with a photodetector (e.g., a photomultiplier or a photodiode) (11).

The in vivo luminescence of photobacteria occurs by the following enzyme reactions (12,13) (for details see Chap. 10):

$$\mathrm{NAD(P)H} + \mathrm{H}^+ + \mathrm{FMN} \xrightarrow{\text{NAD(P)-FMN oxidoreductase}} \mathrm{NAD(P)} + \mathrm{FMNH_2}$$

$$\mathrm{E(luciferase)} + \mathrm{FMNH_2} \longrightarrow \mathrm{E\text{-}FMNH_2}$$

$$\mathrm{E\text{-}FMNH_2} + \mathrm{RCHO} + \mathrm{O_2} \longrightarrow \mathrm{FMN} + \mathrm{RCOOH} + \mathrm{H_2O} + h\nu$$

where RCHO = aliphatic aldehyde and RCOOH = fatty acid.

Such luminescence is strongly affected by changes in the bacteria external conditions, which can subsequently lead to the transient accumulation of intracellular concentrations of NAD(P)H, FMN (flavine mononucleotide) H_2, ATP, and aldehyde (the enzymatic synthesis of aldehyde requires ATP as a cofactor). Using this principle the nutrients of these microorganisms (e.g., glucose and amino acids) and their inhibitors (e.g., toxins and heavy metals) can be detected.

Generally luminescence intensity is a more sensitive parameter for metabolic activity than for respiratory activity or heat generation.

Microbial Thermistor

Microbial biosensors can be constructed by placing immobilized microorganisms in close proximity to a thermistor that measures the metabolic heat they produce (14–16). This type of sensor is based on the general principle of measuring the enthalpy change. Because many metabolic reactions are accompanied by a considerable evolution of heat, calorimetry may be applied to measure a large variety of analytes. The lack of specificity due to the general detection principle is adequately compensated by the use of specific, immobilized biocatalysts. Unfortunately, however, the major part of the heat evolved in the metabolic reaction may be lost in the surrounding solution without being detected by the thermistor, which may limit the sensitivity of this technique.

Although many applications of calorimetry in biochemical analysis have been reported, calorimetry has not gained widespread use in routine bioanalysis. This may be attributed to the lack of sensitivity and the high cost and complexity of the instrumentation.

TRANSDUCERS OF MICROBIAL BIOSENSORS

Various kinds of transducers are available for biosensors, including photodetectors, thermistors, and potentiometric and amperometric electrodes. In this section we focus largely on recent studies using electrochemical devices. A general discussion of other transducers can be found in other chapters in this volume and in recent reviews (1–3,8).

Oxygen Electrode

A commercially available conventional dissolved oxygen electrode is the most general and suitable transducer for a microbial biosensor. The membrane electrode as originally designed by Clark is still widely used (17). The geometrical configuration of this electrode is very important; in particular the thickness of the electrolyte layer between the cathode and the membrane must be narrow to ensure good linearity of response and a low background current. Clark electrodes may be classified as either polarographic electrodes or galvanic electrodes.

Polarographic electrodes consist of a platinum cathode and a silver anode, both immersed in the same solution of saturated potassium chloride. A suitable polarization voltage (typically -650 mV versus Ag/AgCl) between the anode and the cathode selectively reduces oxygen at the cathode:

$$O_2 + 2H_2O + 4e^- \longrightarrow 4OH^- \qquad \text{cathode}$$

$$4Ag + 4Cl^- \longrightarrow 4AgCl + 4e^- \qquad \text{anode}$$

These redox reactions result in the production of a current that is quantitatively proportional to the concentration of dissolved oxygen in the reaction vessel.

The galvanic electrode has a lead anode and a silver (or platinum) cathode. It is a self-driven electrode and does not need an externally supplied voltage. The reactions are as follows:

$$0_2 + 2H_2O + 4e^- \longrightarrow 40H^- \quad \text{cathode}$$

$$Pb + 4OH^- \longrightarrow PbO_2 + 2H_2O + 4e^- \quad \text{anode}$$

This type of electrode is very simple and economical but has the disadvantages of slower response times and worse stability than a polarographic electrode

Other Electrodes

Ion-sensitive and gas-sensing electrodes can be used as transducers for microbial biosensors. Fuel cell electrodes (H_2 detection) (18), CO_2 electrodes (19), NH_3 electrodes (20), and pH electrodes (including ISFET) (21), can also be used as transducers for microbial biosensors. Most of these electrodes, except for the fuel cells, are based on potentiometry. When using an electrode to make potentiometric measurements of the activity of a given ion in solution it is important to remember that the device is affected by the activity of the ion. Therefore species that may complex the ion of interest and lower its activity must be removed. In addition it is often necessary to use a buffer solution to control ionic strength and pH. Although these sensors have a wide measurable range, they may respond to other contaminants and have a relatively poor detection limit.

The gas-sensing electrodes are prepared by using a gas-permeable membrane, a housing containing a pH electrode, and an internal solution. Electrodes of this type are available for ammonia, carbon dioxide, and hydrogen sulfide sensing systems.

IMMOBILIZATION OF MICROORGANISMS

It is vital that leakage of the biologically active components does not occur to any extent during use of the biosensor. It is therefore necessary to ensure that the method of immobilization of the microorgamism works very well. Many methods have been employed to this end (22). Among the techniques reviewed, adsorption, physical entrapment, and covalent attachment appear to be the most suitable to microbial biosensor construction. Care must be taken of the following points. (1) In the case of oxygen electrode-based sensors, the microorganism-immobilized membrane must remain permeable to gas. (2) If the sensor is based on the functions of living cells, a very gentle method for microbe immobilization must be selected so that the viability of the organism is not compromised. It is important to test whether these biological components retain activity and stability when attached to the support matrix in the microbial biosensor.

In early works on microbial biosensors gel entrapment methods (e.g., polyacrylamide and collagen) were mainly employed. More recently most microbial biosensors are constructed using membrane entrapment methods.

Membrane Entrapment

Membrane entrapment is probably the least severe and simplest technique to immobilize microorganisms and is carried out by methods as described here (23).

(1) Culture the appropriate microorganism to a cell density corresponding to an absorbance at 526 nm of 8.6. (2) Take a porous cellulose acetate membrane and drop 3 ml of the microbial culture onto it while applying gentle suction with a water pump. (3) Air dry the membrane containing the entrapped cells, and store it at room temperature. (*Note*: A Millipore type HA membrane with a pore size of 0.45 μm, a diameter of 47 mm, and a thickness of 150 μm is suitable.)

Cellulose acetate or nitrocellulose membranes may both be used, as illustrated in Figure 1. If vinyl spacers (14 mm OD, 6 mm ID, and thickness 50 μm) are attached to the cellulose acetate membrane, several microbe-immobilized membranes may be made at one time. The membranes should be stored in buffered saline solutions.

Gel Entrapping

Entrapment in a gel matrix is a favored method of immobilization, particularly for microbes that have small substrates, which diffuse into the matrix more readily than large substrates (18). Numerous matrices have been employed, but recently the most suitable for microbial biosensors have been found to be either polyacrylamide or agar.

Polyacrylamide Entrapment of Cells

(1) Grow the appropriate organisms, and harvest by centrifugation. (2) Wash the cells by resuspending them in physiological saline solution, and then recentrifuge

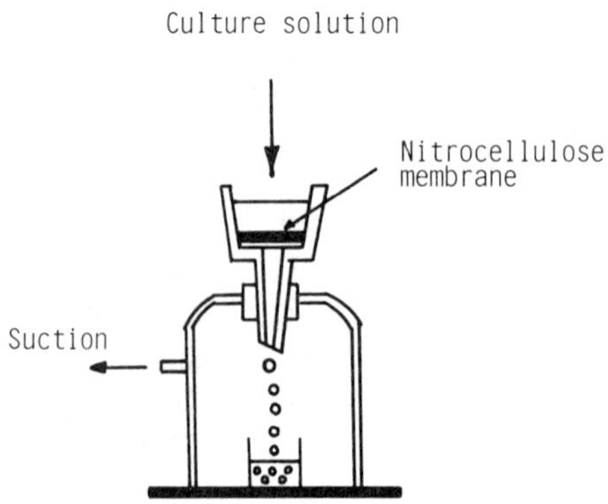

Figure 1 Membrane entrapping of microorganism.

then. (3) Place a 20 mesh nylon net (1.3 cm diameter and 200 μm thick) on a clean glass plate. (4) Suspend 100 mg of the washed cells, 90 mg acrylamide, and 10 mg *N,N'*-methylenebisacrylamide in 1 mL of physiological saline solution at 0°C. (5) Saturate the suspension of cells and acrylamide with nitrogen. (6) Initiate the polymerization by adding 30 μL of 10% (wt/vol) dimethylaminopropionitrile and 0.1 mg potassium persulfate. (7) Cast 250 μl of the suspension over the nylon net, and place the whole assembly in an oxygen-free atmosphere at 37°C for 30 min. (8) Remove the membrane, and store in 0.1 M phosphate buffer at pH 7.

Agar Entrapment of Microbial Cells

(1) Grow and wash the cells as described in steps 1 and 2 of the polyacrylamide procedure. (2) Suspend 100 mg washed cells in 0.1 mL physiological saline solution. (3) Dissolve 20 mg agar in 0.9 ml physiological saline at 100°C, and then cool the solution to 50°C. (4) Mix the agar solution with the cell suspension, and cast 250 μL of the mixture onto a nylon net placed on a glass plate. Cool to 5°C. (5) After the agar has set store the membrane in 0.1 M phosphate buffer, pH 7.

Unfortunately this method suffers from two major drawbacks; First, large diffusional barriers to the transport of substrate and product lead to a retardation of the reaction, particularly with high-molecular-weight substrates, such as ribonuclease, trypsin, and dextranase. Second, there is a continuous loss of activity due to the escape of microorganisms through the large pores.

BASIC CONSTRUCTION AND MEASUREMENT

To construct microbial biosensors we must understand the mechanism of the sensing system and consider the factors that affect the stability, dynamic range, and response time to optimize the performances. The operation of a microbial biosensor can be seen as a five-step process: (1) the substrate is transported to the surface of the sensing system in the vessel, (2) the substrate diffuses through the membrane to the immobilized microorganism, (3) the reaction occurs at the immobilized microorganism, (4) the products are formed in the microorganism and are transported through the membrane to the surface of the detector, and (5) the products are measured at the detector surface.

Culture of Microorganisms

Culture procedures of microorganisms for microbial biosensors are the same as for normal microbial culture. In general, logarithmic-phase growing cells are suitable for microbial biosensors.

Preparation of Microbe-Immobilized Membrane

Microorganism-immobilized membrane is prepared by one of the methods described in the previous section.

Construction of a Microbial Biosensor

The scheme of a microbial biosensor is illustrated in Figure 2. A Clark oxygen electrode consists of a Teflon membrane, a platinum cathode, an aluminum anode, and a saturated potassium chloride electrolyte. The microorganism-immobilized membrane is placed on the Teflon membrane and fixed in place using either a bored cap or a dialyzed membrane O-ring. The preparation of a typical microbial electrode is as follows: (1) Take the base sensor, and turn it upside down. (2) Cover the sensor with a piece of wet microorganism-immobilized membrane. (3) Cover with a piece of dialysis membrane about twice the diameter of the size of the electrode sensor. (4) The membranes are held in place with a rubber O-ring with a diameter that fits the electrode body snugly. Gently push the O-ring onto the electrode body so that microbe forms a nicely uniform layer on top of the electrode surface. Place the electrode in buffer solution overnight to allow penetration of buffer into the microbe and permit the loss of entrapped air. Store the electrode in buffer in a refrigerator between uses.

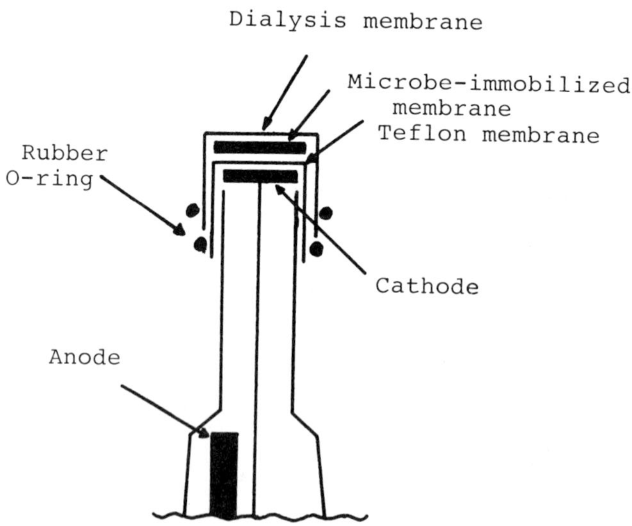

Figure 2 Construction of a microbe electrode.

Measurement System and Procedure

Batch Measurement

Figure 3 is a schematic diagram of a batch measurement system (24). A microbial sensor consisting of immobilized whole cells of *Pseudomonas fluorescens* in conjunction with an oxygen electrode was investigated. The electrode consisted of an alkaline electrolyte, platinum cathode, and lead anode with a double membrane, one layer of which was a bacteria-collagen membrane (40 μm thick) and the other an oxygen-permeable Teflon membrane (27 μm thick). The membranes were in direct contact with the platinum cathode and were tightly secured with rubber O-rings. The sensor was placed in a thermostatically controlled, circulating water jacket. The temperature was maintained at 30°C, and the cell contained a total volume of 50 ml of 0.05 M phosphate buffer at pH 7.0. The buffer was saturated with dissolved oxygen (aeration rate 200 mL/min) and stirred magnetically during the measurement. The current output of the oxygen electrode was measured with a digital multimeter (e.g., Takeda Riken, model TR6843) and with an electronic polyrecorder (e.g., TOA Electronics, model EPR-200A). When the output current became stable, 100 μL of sample solution was injected into the sensor cell.

Typical response curves are shown in Figure 4. The current at time zero was that obtained in a sample solution saturated with dissolved oxygen. The bacteria

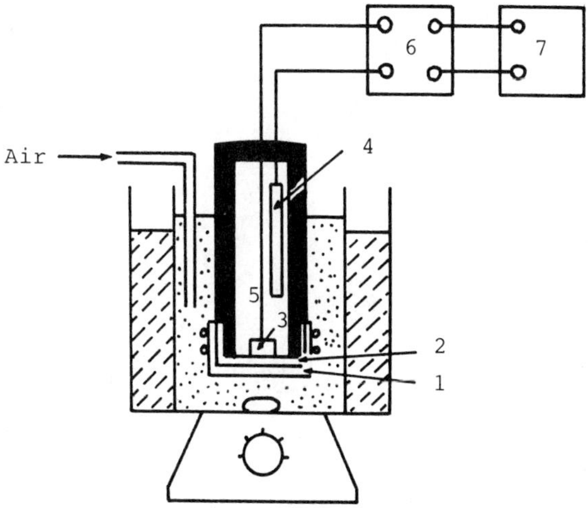

Figure 3 The microbial sensor for glucose: (1) bacteria-collagen membrane: (2) Teflon membrane; (3) platinum cathode; (4) lead anode; (5) electrolyte (KOH); (6) multimeter; (7) recorder. [Redrawn from Karube, Mitsuda, and Suzuki (1979), courtesy of Springer-Verlag, Heidelberg.]

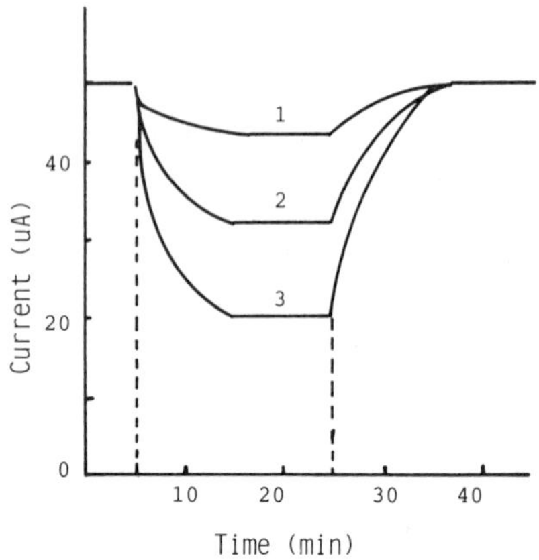

Figure 4 Response curves of the microbial sensor shown in Figure 3. Glucose concentrations were (1) 3–6 mg/L; (2) 10 mg/L. [Redrawn from Karube, Mitsuda, and Suzuki (1979), courtesy of Springer-Verlag, Heidelberg.]

began to utilize glucose in the reaction vessel when the sensor was placed in it, and as a result there was a decrease in the dissolved oxygen at the membrane. Consequently the current produced by the sensor decreased markedly until a steady state was reached, indicating that the consumption of oxygen by the bacteria and the diffusion of oxygen from the solution to the membrane were in equilibrium; this occurred within 10 min at 30°C. The level of the steady-state current depended on the concentration of glucose. When the sensor was removed from the sample and placed in a glucose-free solution the current gradually increased, returning to the initial level within 15 min at 30°C. (Hereafter, ''current'' means the steady-state current.)

Figure 5 shows the calibration curve of the microbial sensor. Using the steady-state method already described, a linear relationship was observed between the current and the concentration of glucose below 20 mg/L. The minimum concentration for determination was 2 mg/L. The current was reproducible with a coefficient of variation of ±6% when a sample solution containing 10 mg/L of glucose was used. The standard deviation was 0.6 mg/L in 20 experiments.

The microbial sensor was able to determine glucose over the concentration range 10^{-4}–10^{-5} mol/L compared with ordinary enzyme (GOD) electrodes, which have a linear range of 10^{-2}–10^{-4} mol/L. The sensitivity of the microbial sensor was therefore better than that of the ordinary enzyme (GOD) electrodes (25).

The microbial sensor was used to determine the concentration of glucose in molasses and gave results that were comparable with those obtained by the

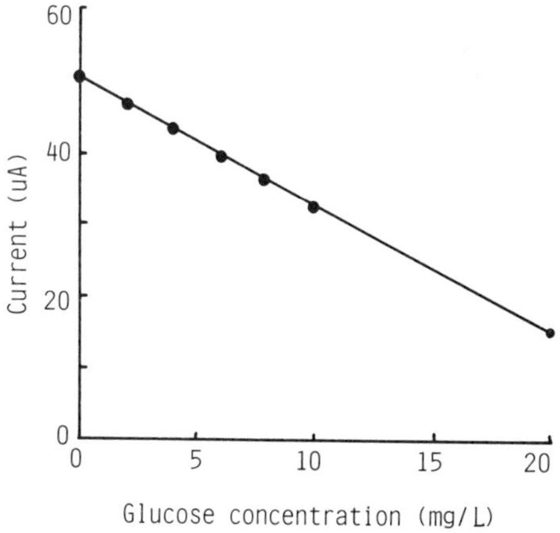

Figure 5 Calibration curve of the microbial sensor. Sample solutions (0.1 M phosphate buffer, pH 7.0) containing various amounts of glucose were employed for the experiments. [Redrawn from Karube, Mitsuda, and Suzuki (1979), courtesy of Springer-Verlag, Heidelberg.]

enzymatic method; the average coefficient of variation with the microbial sensor was ± 10%. Long-term stability tests revealed no decrease in current output over a 2 week period and 150 assays.

Flow Measurement

A flow system for a microbial biosensor consisted of a jacketed flow cell containing a microbial electrode, a peristaltic pump, an automatic sampler, and a recorder (Fig. 6) (26). The microbial sensor itself comprised an oxygen electrode consisting of a Teflon membrane (50 μm thick), a platinum cathode, an aluminum anode, and a saturated potassium chloride electrolyte. A porous membrane bearing the immobilized *Trichosporon brassicae* was fastened at the surface of the Teflon membrane and covered with a gas-permeable Teflon membrane: thus the microorganisms were trapped between the two porous membranes. The temperature of the jacketed flow cell was maintained at 30°C using a thermostated water bath. Phosphate buffer solution (0.01 M, pH 7.0) saturated with dissolved oxygen was transferred to the flow cell at 1 mL/min together with air at a flow rate of 250 ml/min. When the electrode response reached a steady state a sample was injected into the flow cell at a rate of 0.2 mL/min for 20 min at 60 min intervals. The principle of this microbial sensor was similar to that described previously. The sample was kept at a pH well below the p*K* value for acetic acid (4.75 at 30°C) because acetate ions cannot pass through the membrane. Acetic acid from the sample passed through the membrane and was assimilated by the microorganisms, so that the concentration of dissolved oxygen around the mem-

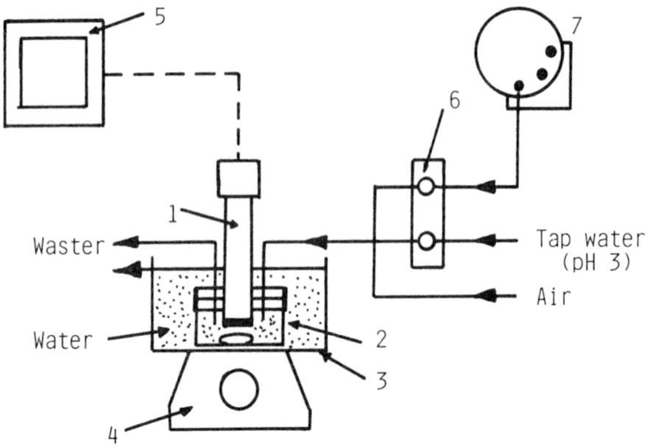

Figure 6 Acetic acid sensor system: (1) microbial electrode; (2) flow cell; (3) jacket; (4) magnetic stirrer; (5) recorder; (6) peristaltic pump; (7) sampler. [Redrawn from Hikuma, Kubo, Yasuda, Karube, and Suzuki (1979) by permission of the publishers, Elsevier Science Publishers B.V., Amsterdam.]

branes decreased and the current then decreased accordingly until a steady state was reached. The steady-state current depended on the concentration of acetic acid in the sample. Because of the long time taken to reach the steady state, however, samples were passed into the flow cell for only 3 min for the detection of acetic acid. The maximum current decrease was only 75% of that obtained by the steady-state method, but the total measurement time was less than 4 min. Figure 7 shows the response curves obtained for concentration of acetic acid between 18 and 72 mg/L using this method. The total time required for an assay of acetic acid was 30 min using the steady-state measurement and 15 min for the shorter rate method, which was therefore employed for the assay.

The calibration graphs obtained showed linear relationships between the current decrease and the concentration of acetic acid up to 72 mg/L by the rate method. The minimum concentration that could be determined was 5 mg/L. The current variation was reproducible within $\pm 6\%$ for an acetic acid sample containing 54 mg/L, and the standard deviation was 1.6 mg/L in 20 experiments.

The microbial sensor for acetic acid did not respond to volatile compounds, such as formic acid and methanol, or to nonvolatile nutrients, such as glucose and phosphate ions. Although *T. brassicae* utilizes propionic acid, *n*-butyric acid, and ethanol, these are not generally present in fermentations for glutamic acid or are present in concentrations too low to affect the measurement of acetic acid. The sensor was therefore considered specific for acetic acid.

The concentration of acetic acid in fermentation broth that was producing glutamic acid was determined by the microbial sensor and also by gas chromatography: good agreement was achieved, the regression coefficient being 1.04

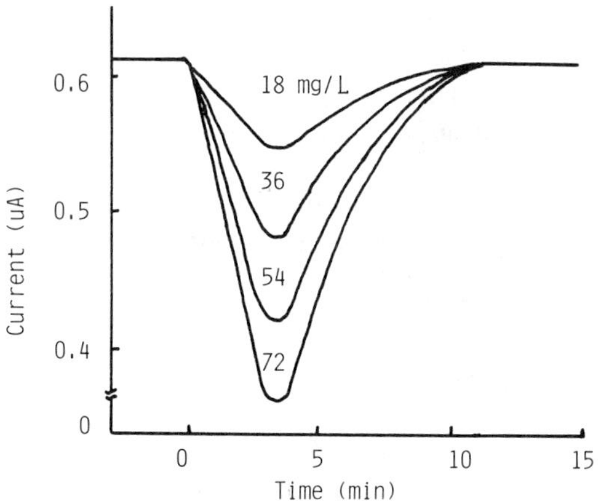

Figure 7 Response curve of the acetic acid sensor. Sample solution (2–4 ml) was passed into the flow cell for 3 min. [Redrawn from Hikuma, Kubo, Yasuda, Karube, and Suzuki (1979) by permission of the publishers, Elsevier Science Publishers B.V., Amsterdam.]

for 26 experiments. The current output (0.29–0.25 μA) of the sensor was constant (within ±10% of the original values) for more than 3 weeks and 1500 assays.

Photomicrobial Detector

A photomicrobial sensor may consist of a flow cell containing the immobilized *Photobacterium phosphoreum* MT10204, luminous bacteria, and a photometer equipped with a photomultiplier, a peristaltic pump, a water bath, and a recorder (11). The flow cell surface was positioned to face the photomultiplier tube of a photometer.

The temperature of the flow cell was maintained at 30°C using a thermostated water bath. The measurement was performed using flow injection analysis (FIA). Phosphate buffer (0.1 mM, pH 7.0) containing 3% wt/vol NaC1 was continuously transferred to the flow cell at a flow rate of 1.2 mL/min using a peristaltic pump. When the baseline of the luminescence intensity reached a constant value a sample was injected to the flow cell with a microsyringe, and the observed luminescence intensity change was monitored. The luminescence intensity change was expressed in arbitrary units.

The injection of a sample of 20 μl of 0.55 mM glucose solution increased the emission intensity immediately, and within 1 min a maximum response was observed. The sensor response returned to the baseline within 6 min at a flow rate of 1.2 mL/min.

The bioluminescence of photobacterium was catalyzed by the bacterial enzyme luciferase. This reaction requires several cofactors, such as NADH, ATP, and $FMNH_2$. As a result the bioluminescence is strongly affected by the regeneration of these cofactors. It was assumed that the addition of glucose resulted in cofactor regeneration and the subsequent increase in luminescence.

The time required for the recovery from maximum sensor response to baseline was more rapid compared to that with other types of microbial sensors. For example, a microbial glucose sensor based on respiratory change measurement required more than 10 min (24).

Microbe Thermistor

The microbe thermistor can be constructed using yeast gel grains packed in an insulated glass column with a thermistor at its outlet. An example of such a sensor involving immobilized whole cells, such as *Saccharomyces cerevisiae*, baker's yeast, can be used as a broad specificity sensor (16). The solution to be analyzed is first pumped through a heat exchanger in an ultrastable thermostatically controlled water bath and then through the column. Heat changes occurring in the gel bed are recorded. Glucose, fructose, and casein can be measured with this system. Additional examples of thermistor-based sensors are provided in Chapter 5.

Maintenance

Generally microbial biosensors have a long lifetime, but maintenance of these sensors is very important. The total activity of immobilized microorganisms should be kept constant. Normally microbial biosensors should be stored in phosphate buffer (pH 7) without nutrients at 4°C; otherwise microorganisms grow in the membrane. If the activity of the microorganisms decreases the sensor must be dipped into nutrient medium until the activity is recovered by the growth of new cells. Care to avoid contamination is much more important than for enzyme biosensors because antibacterial reagents, such as sodium azide or antibiotics, cannot be used. In general, contamination affects both the selectivity and the sensitivity of microbial biosensors.

APPLICATION OF MICROBIAL BIOSENSORS

Many kinds of microbial biosensors have been developed, some of which have already been used for environmental monitoring and in the fermentation and food industries. Characteristics of these microbial sensors are summarized in Table 1.

Glucose Sensor

The determination of glucose levels is an important prerequisite of fermentation process control. Enzyme electrodes have been used for the determination of glucose in both clinical and food analysis (1). Because enzymes are often

Table 1 Characteristics of Microbial Sensors

Sensor	Immobilized cells	Type of electrode	Range (mg/L)	Response (min)	Stability (days)
Amperometric method					
Glucose	*Pseudomonas fluorescens*	Oxygen	3–20	10	14
Assimilable sugars	*Brevibacterium lactofermentum*	Oxygen	20–200	10	
Acetic acid	*Trichosporon brassicae*	Oxygen	10–200	15	30
Ammonia	Nitrifying bacteria	Oxygen	3–45	5	20
Methanol	Unidentified bacteria	Oxygen	3–22	15	30
Ethanol	*T. brassicae*	Oxygen	3–30	15	30
Nystatin	*Saccharomyces cerevisiae*	Oxygen	1.2–800	60	—
Carcinogen	*Bacillus subtilis*	Oxygen	1–10	60	—
BOD	*Trichosporon cutaneum*	Oxygen	3–30	10	30
Cell population	—	Fuel cell	10^6–10^{11}*	15	60
Vitamin B_1	*(Lactobacillus fermenti)*	Fuel cell	10^{-3}–10^{-2}	360	60
Formic acid	*Clostridium butyricum*	Fuel cell	1–1000	10	30
Potentiometry method					
Cephalosporins	*Citrobacter freundii*	pH	60–500	10	7
Nicotinic acid	*Lactobacillus arabinosus*	pH	10^{-2}–5	60	30
Glutamic acid	*Escherichia coli*	CO_2	8–800	5	20
Lysine	*E. coli*	CO_2	10–100	5	20

unstable and expensive, however, and because they may be inhibited by compounds in the culture media, enzyme electrodes are not suitable for monitoring fermentation processes. The assimilation of glucose by microorganisms and the attendant respiratory activity can be monitored with an oxygen electrode, however (24). The general discussion of this type of sensor is presented in the preceding section.

Acetic Acid Sensor

When microorganisms are cultured with acetic acid as the carbon source, excess acetic acid inhibits growth and the optimal concentration must therefore be maintained by on-line monitoring. As gas chromatography is unsuitable for on-line measurements, a microbial sensor comprising immobilized yeasts (*T. brassicae*), a gas-permeable Teflon membrane, and an oxygen electrode was investigated for the continuous determination of acetic acid in fermentation broths (26). This microbial sensor for acetic acid is now commercialized in Japan. A general discussion of this type of sensor is given in the preceding section.

Assimilable Sugar Sensor

During the cultivation of microorganisms in cane molasses, which contains various sugars, determination of the total assimilable sugars in the broth is important for the control of the fermentation process. For example, catabolite repression occurs at a high sugar concentration, causing an inhibition of cell growth. Reduced sugars and sucrose in culture broths can be determined using a method based on the reduction of ferricyanide (27). This method, however, is not completely reliable because unassimilable substances can interfere with the determination.

Assimilation of organic compounds by microorganisms can be determined from their respiratory activity, which can be directly measured using an oxygen electrode. A microbial sensor consisting of immobilized living whole cells of *Brevibacterium lactofermentum* and an oxygen electrode was constructed for the continuous determination of total assimilable sugars (glucose, fructose, and sucrose) in a fermentation broth. *B. lactofermentum* was immobilized on a strip of nylon net (1 × 1 cm, 20 mesh) and attached to the oxygen probe. The consumption of oxygen by the immobilized microorganisms was measured (28). Addition of a glucose aliquot to the sensor system resulted in an increased oxygen consumption by the microorganisms. This lowered the dissolved oxygen concentration of the solution, causing the electrode current to decrease markedly with time until steady state was reached. The response time was 10 min if a steady-state current was measured, but this could be reduced to 1 min if the pulse method was used. A linear relationship was found to exist between the decrease in current and the concentration of glucose (1 mM), fructose (1 mM), and sucrose (0.8 mM), respectively. Sensitivity of the microbial sensor to glucose, fructose, and sucrose, normalized to glucose, was 1.00:0.80:0.92, respectively. The results were reproducible to within 2% of the relative standard deviation against a standard sample solution containing glucose (0.8 mM). Total assimilable sugar was calculated by a summation of the responses of glucose, fructose, and sucrose. The difference between the observed and calculated concentrations was within 8%. The microbial sensor was installed in a fermentor for the production of glutamic acid, where it operated reliably for more than 10 days and 960 assays.

Ammonia Gas Sensor

The determination of ammonia is important in clinical, environmental, and industrial process analyses. An ammonia gas electrode consisting of a combined glass electrode and a gas-permeable membrane is most often used for this purpose (29). In this case the determination must be performed under strongly alkaline conditions (above pH 11). This type of electrode is based on the potentiometric detection of ammonia. Since volatile compounds, such as amines, may interfere with the potentiometric electrode, an amperometric sensor is more desirable (30,31).

Nitrifying bacteria contain two genera of bacteria. The genus *Nitrosomonas* sp. utilizes ammonia as the sole source of energy, with 1.5 mol O_2 used per mol NH_3 to give $NO_2 + H_2O + H^+$. The other genus, *Nitrobacter* sp., oxidizes nitrite to nitrate, requiring 0.5 mol O_2 per mol NO_2. The oxidation of both NH_3 and NO_2 proceeds at a high rate, and oxygen uptake by the bacteria can be determined directly by an oxygen electrode in close contact with immobilized bacteria. The nitrifying bacteria were isolated from activated sludges and were immobilized onto a porous acetylcellulose membrane of 0.45 μm pore size. The porous membrane, retaining the immobilized bacteria, was fixed on the surface of the Teflon membrane of an oxygen electrode. The bacterial membrane was covered with a gas-permeable Teflon membrane of 0.5 μm pore size and fastened with rubber rings.

The microbial sensor was inserted into a glycine-buffered sample solution saturated with dissolved oxygen at pH 10.0. A current was measured at time zero, which corresponded to the endogenous respiratory level of the immobilized bacteria. When the ammonia solution was injected into the buffer solution the ammonium ion changed to ammonia gas. Ammonia gas permeated through the gas-permeable membrane and was utilized by the immobilized bacteria. Oxygen was then consumed by the bacteria, resulting in a decrease in the concentration of dissolved oxygen in proximity to the membrane. Consequently the current decreased until it reached a steady state, which indicated that the consumption of oxygen by the bacteria and the diffusion of oxygen from the sample solution to the bacteria membrane were in equilibrium. The response time for the determination of ammonia was within 4 min. The pH of the sample solution had to be kept sufficiently above the pK for ammonia (9.1 at 30°C); otherwise ammonium ions could not pass through the gas-permeable membrane. When the sensor was inserted in tap water the current of the sensor returned to its initial level within 5 min.

A linear relationship was observed between the current decrease (the current difference between the initial and the steady state) and the ammonia concentration below 42 mg/L. The minimum concentration for the determination of ammonia was 0.1 mg/L. The current decrease was reproducible within ±4% of the relative error when a sample solution containing 21 mg/L of ammonium hydroxide was used as a calibrant. The standard deviation was 0.7 mg/L over 20 experiments. Thus it was possible to determine ammonia using an amperometric microbial sensor. The sensitivity of the microbial sensor approached that of a glass electrode.

Further, the selectivity of the microbial electrode for ammonia was examined. The sensor did not respond to volatile compounds, such as acetic acid, ethanol, and amines (diethylamine, propylamine, and butylamine), or to nonvolatile nutrients, such as glucose, amino acids, and metal ions (potassium ion, calcium ion, and zinc ion). The long-term stability of the microbial sensor was investigated using a sample solution containing 33 mg/L of ammonia. The current output of the electrode was constant for more than 10 days and 200 assays.

This microbial sensor was applied to the determination of ammonia in human urine. The urine was diluted and the concentration of ammonia was determined by the electrochemical sensor and by a conventional spectrophotometric method. Good comparative results were obtained (correlation coefficient 0.9).

In conclusion, the microbial sensor appears to be quite promising and very attractive for the amperometric determination of ammonia.

Alcohol Sensor

When yeasts are cultured using sugar as a carbon source it is well known that ethanol as a by-product reduces the yield of whole cells per unit of sugar (32). Similarly, during the culture of microorganisms using methanol as a carbon source, the concentration of methanol must be maintained at an optimal level to avoid substrate inhibition (33). The concentration of methanol has previously been estimated by analysis of the exhaust gas from a fermentor with a flame ion detector (32). However, this method is not reliable because the values obtained are influenced by operational fermentor conditions, such as the aeration and agitation rates. Although many enzyme electrodes for the determination of alcohols have been described (1), a more stable biosensor is preferable.

With the knowledge that many microorganisms utilized alcohols as carbon sources and that this assimilation can be determined from their respiratory activity, a microbial sensor for alcohols has been devised using immobilized microorganisms (yeast or bacteria), a gas-permeable Teflon membrane, and an oxygen electrode (34). For the detection of ethanol, immobilized *T. brassicae* was used. The pH of the culture medium was 6.0–6.2. A linear relationship was observed between the current decrease when the sensor was placed in the alcohol-containing sample and the concentration of ethanol below 22.5 mg/L. The minimum concentration of alcohol that could be determined was 2 mg/L. The current difference was found to be reproducible within ±6% when a sample solution containing 16.5 mg/L of ethanol was used. The standard deviation was 0.5 mg/L in 40 experiments.

Under the test conditions used, with a pH above 6.0, the sensor did not respond to volatile compounds, such as methanol, formic acid, acetic acid (compare with the use of *T. brassicae* to determine acetic acid at pH 3.0), or propionic acid or other nutrients, such as carbohydrates, amino acids, or ionic compounds. Because the microbial sensor was covered with a gas-permeable membrane only volatile compounds could permeate the membrane. Since *T. brassicae* does not utilize methanol, the selectivity of the microbial sensor for ethanol was considered satisfactory.

When the concentration of ethanol in fermentation broths was determined using both a microbial sensor and gas chromatography, satisfactory comparative results were obtained, with a correlation coefficient of 0.98 for 20 experiments. When ethanol solutions (5.5–22.3 mg/L) were used to test the long-term stability of the sensor, the current output was found to be constant for more than 3 weeks and 2100 assays. An unidentified bacterium was used for the determination of methanol. As before, the microbial sensor consisted of immobilized bacteria (which used methanol in this case), a gas-permeable membrane, and an oxygen electrode. The determination was performed under the same conditions as those described for the ethanol sensor. A linear relationship was observed between the current decrease and the concentration of methanol, showing that the sensor may be of use for the determination of methanol.

BOD Sensor

The biochemical oxygen demand (BOD) is one of the most widely used tests in the measurement of organic pollution. However, the conventional BOD test requires a 5 day incubation period. A more rapid and reproducible method is therefore required for assessing BOD.

To this end *Trichosporon cutaneum*, which is used for wastewater treatment, was used for the BOD sensor. The sensor configuration was the same as previously described (9,23,35).

Phosphate buffer solution (0.01 M, pH 7) saturated with dissolved oxygen was transferred to the flow cell at a flow rate of 1 ml/min. When the current reached a steady-state value a sample was injected into the flow cell at a rate of 0.2 mL/min. The steady-state current was dependent on the BOD of the sample solution, and the current of the microbial sensor gradually returned to the initial level. The response time of the microbial sensor (time required for the current to reach steady state) depended on the nature of the sample solution used.

A linear relationship was observed between the current difference (between initial and steady-state levels) and the 5 day BOD assay of the standard solution (glucose and glutamate solution) for concentrations up to 60 mg/L. The minimum measurable BOD was 3 mg/L. The current was reproducible within ±6% of the relative error when a BOD standard of 40 mg/L was employed over 10 experiments.

The microbial sensor was applied to the estimation of a 5 day time course for BOD in untreated wastewater from a fermentation factory. The 5 day BOD of the wastewater was determined using the JIS method (Japanese Industrial Standard Committee). Good correlation was obtained between BOD estimated by the microbial sensor and those determined by the JIS method. The regression coefficient was 1.2 over 17 experiments, and the ratios of BOD estimated by the microbial sensor to 5 day BOD determined by the JIS method were within the range of 0.85–1.36. This variation may have been caused by a change in composition of the organic wastewater. The BOD of various kinds of untreated industrial wastewater were estimated by the sensor, and the response was found to depend on compounds present in the wastewater. The BOD sensor system shown in Figure 8 has now been commercialized in Japan (36).

Methane Sensor

Worldwide interest has arisen in the production of methane by fermentation of biomass. Methane is an attractive energy source and a main component of the natural gases used for fuel. Rapid methods for the detection and determination of methane in air are required, for example in the coal mining and industries, as it may form an explosive mixture with air (5–14%). A methane sensor consisting of immobilized methane-oxidizing bacteria and an oxygen electrode has been developed.

The system comprises two oxygen electrodes, two reactors, and an electrometer (Fig. 9) (37). Both reactors contain culture media, although only one contains the immobilized bacterium *Methylomonas flagellata*. The electrodes were fixed inside custom-made Teflon flow-through cells constructed with glass and Teflon tubing. Two vacuum pumps were used, one to evacuate the gas sample tube and the other to transport the sample gas through the system. The flow rate of the sample gas through the reactors was controlled using glass valves and equal to 80 mL/min. A cotton filter removed other microorganisms in the gas samples, preventing contamination of the reactors and gas lines. The latter were designed to maintain symmetry between the measuring and reference flows.

When sample gas containing methane was passed into the reactor it was assimilated by the immobilized microorganisms with a subsequent consumption of oxygen, causing the current from the oxygen electrode to decrease to a

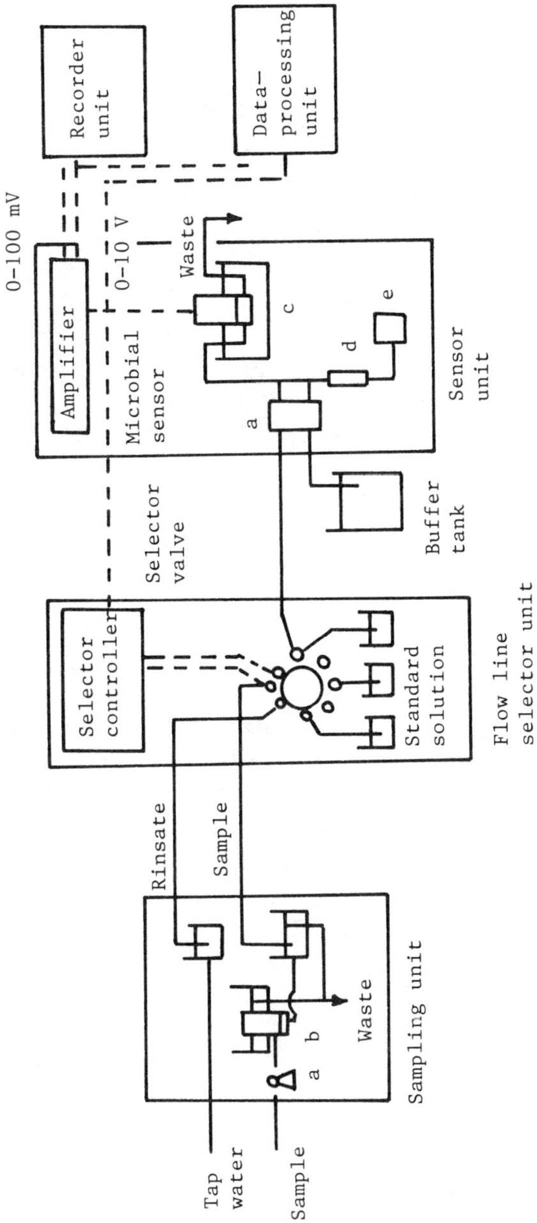

Figure 8 Continual measuring system for BOD: (a) pump; (b) filter; (c) incubator; (d) flow meter; (e) air pump.

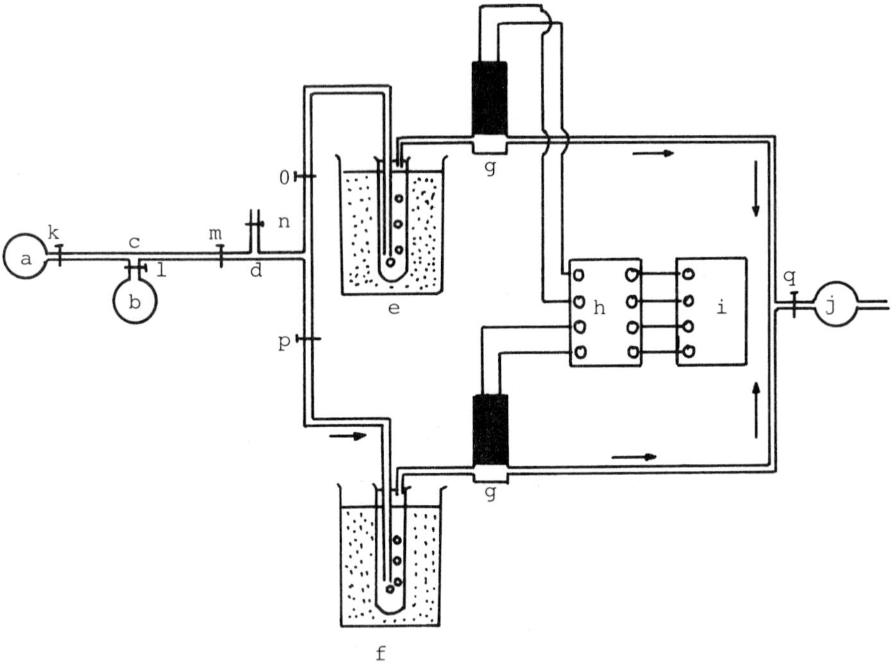

Figure 9 Microbial sensor system for methane: (a) vacuum pump; (b) sample gas bag; (c) gas sample line; (d) cotton filter; (e) control reactor; (f) methane-oxidizing bacteria reactor; (g) oxygen electrode; (h) amplifier; (i) recorder; (j) vacuum pump; (k–q) glass stopcocks. [Redrawn from Karube, Okada, and Suzuki (1982) by permission of the publishers, Elsevier Science Publishers B.V., Amsterdam.]

minimum steady state. Because the system contained two oxygen electrodes, the maximum difference between the currents depended on the concentration of methane in the sample gas. When pure air was again passed through the reactors the current returned to its initial level within 60 sec. The response time for the measurement of methane was less than 60 sec, and the overall time required for each methane assay was 2 min.

Calibration graphs for the system were linear for methane concentrations in the range of 0–6.6 mmol/L, and the minimum determinable concentration was 5 μmol.

Glutamic Acid Sensor

Glutamic acid is produced by a fermentation process and is used as a seasoning for foods. A rapid and automatic method is required for determination of its concentration. Glutamate decarboxylase catalyzes the decarboxylation of glutamic acid, producing carbon dioxide and amine. Enzyme-based autoanalyzers can therefore be used, but the enzyme is expensive and unstable. Certain microorganisms, however, contain glutamate decarboxylase.

Consequently a microbial sensor for glutamic acid was designed incorporating immobilized *Escherichia coli* (as a source of glutamate decarboxylase activity) in conjunction with a CO_2-sensing electrode (19). Preliminary experiments have shown that *E. coli* does not evolve carbon dioxide under anaerobic conditions in the absence of glutamic acid. Any carbon dioxide produced by these bacteria under such conditions results from the glutamate decarboxylase reaction. Nitrogen gas was passed through the flow cell to remove any dissolved oxygen from the buffer and sample solution. When a sample solution containing glutamic acid was injected into the system glutamic acid permeated through the membrane to the immobilized microorganisms and was metabolized to produce carbon dioxide. The enzyme reaction was carried out at pH 4.4, which was sufficiently below the p*Ka* value (6.34 at 25°C) of carbon dioxide. As a result the potential of the CO_2-sensing electrode increased with time. The assay could be performed using an injection period of 1–3 min and measuring the maximum potential, with little loss of sensitivity.

The selectivity of a microbial sensor to various amino acids was examined. The sensor responded to glutamic acid and glutamine and very slightly to some other amino acids. The response to glutamine could be decreased, if necessary, using acetone-treated *E. coli*. The microbial sensor did not respond under anaerobic conditions to such organic substances as glucose (7800 mg/L) and acetic acid (9200 mg/L), and the influence of inorganic ions on the response was negligible.

When this sensor was used to determine known concentrations of glutamic acid in a fermentation broth, satisfactory recovery data were obtained (99–103%) that were in good agreement with autoanalyzer determinations. The sensor was considered highly selective, stable, and reproducible.

A plot of the maximum potential versus the logarithm of the glutamic acid concentration was linear in the range 100–800 mg/L. When replicates of a glutamic acid solution (400 mg/L) were measured the standard deviation was 1.2 mg/L (20 experiments).

Cephalosporin Sensor

Antibiotics are usually determined by microbioassay based on a tunbidimetric or titrimetric method, but these methods require complicated procedures and are not suitable for rapid assay procedures.

It was found that the microorganism *Citrobacter freundii* produced an enzyme, cephalosporinase, that catalyzes the reaction of cephalosporin with consequent liberation of hydrogen ions. Cephalosporinase is very unstable, however, and as a result is difficult to utilize as a pure enzyme preparation. Immobilized whole cells of *C. freundii* were employed for the cephalosporin sensor, however, and were immobilized onto a collagen membrane. This bacteria-collagen membrane was then inserted into a membrane reactor.

The system used for continuous determination of cephalosporins is illustrated in Figure 10 (38). The reactor is a membrane type with a spacer located in the center. The pH change caused by the enzymatic reaction was measured using a combined glass electrode.

Sample solutions containing various amounts of cephalosporins were inserted into the reactor, causing the electrode potential difference to increase with time until a maximum was reached. The minimum response time depended on the flow rate and activity of the bacteria-collagen membrane. For a flow rate of 2 mL/min the maximum potential difference was reached after 10 min.

A linear relationship was obtained between the logarithm of the cephalosporin concentration and the maximum potential difference. Cephaloridine, 7-Phenylacetylamidodesacetoxysporanic acid (phenylacetyl-7ADCA), cephalothin, and cephalosporin C were all determined using this the cephalosporin sensor.

The stability of the microbial sensor was examined with a solution containing 125 μg/mL of phenylacetyl-7ADCA. The cephalosporin determination was carried out several times a day, giving no change in the observed potential difference response after 1 week.

The system was applied to the determination of cephalosporin C in a broth of *Cephalosporium acremonium* and was compared with a method based on high-performance liquid chromatography (HPLC). The relative error of the determination by the microbial system was 8%. Accordingly the method was considered suitable for continuous analysis of cephalosporins in fermentation broths.

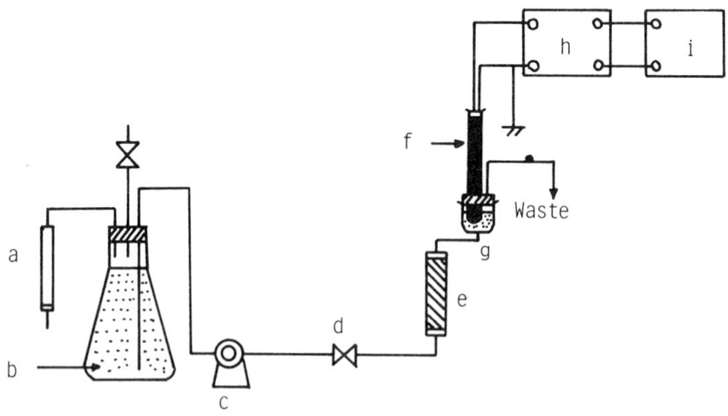

Figure 10 Immobilized whole-cell–based flow sensor for cephalosporins: (a) soda lime; (b) buffer reservoir; (c) peristaltic pump; (d) sample inlet; (e) immobilized whole-cell reactor; (f) combined glass electrode; (g) sensing chamber; (h) amplifier; (i) recorder. [Redrawn from Matsumoto, Seijo, Watanabe, Karube, Satoh, and Suzuki (1979) by permission of the publishers, Elsevier Science Publishers B.V., Amsterdam.]

Hybrid Biosensor

By combining cells with an immobilized enzyme membrane one of the disadvantages of microbial biosensors, which is low selectivity, can be improved. As such an example, an urea sensor is discussed here.

The determination of urea is important in various fields, such as environmental and industrial process analyses. Furthermore, the determination of urea in blood or urine is also important for diagnostic tests routinely carried out in clinical fields. Because of the importance of urea measurements, urease, a catalyst for the hydrolysis of urea, has been intensively studied in recent years (1,29,39–41).

At present the determination of urea is performed by spectrophotometric methods. These methods for the determination of urea are mostly based either on a direct color reaction (e.g., the Fearom reaction method) (42,43) or on the measurement of ammonia produced by the enzyme reaction, for example (44). The enzyme-based methods are generally more specific and more precise than the direct colorimetric procedures (45). However, both these methods require a long reaction time and additional reagents. Spectrophotometric methods are not suitable for the determination of urea in the colored samples or biologic fluids.

On the other hand, electrochemical monitoring systems for urea have been developed. These electrode systems consist of immobilized urease (immobilized in a gel layer and fixed on the surface of an ammonia electrode) and an ammonium ion-sensing electrode (29) or an ammonia gas-sensing electrode. However, ions or volatile compounds, such as amines, sometimes interfere with the determination of ammonia and ammonium ion.

A microbial sensor consisting of immobilized nitrifying bacteria and an oxygen electrode has been developed for the amperometric determination of ammonia by the present authors (30,31). However, these electrodes required a sample pH of 8.5 or higher (pH 11–12): such a high pH range is not suitable for enzymatic reaction (optimum pH of urease 7.0).

Therefore, an improved ammonium ion sensor has been developed and applied to the determination of ammonium ion in urine. The microbial sensor can be used at a neutral pH range and is suitable for the determination of ammonium ion in biological fluids, such as urine and blood (46,47).

Figure 11 shows the principle of the sensor for urea. When the sensor is inserted into a sample solution containing urea, urea is hydrolyzed to the ammonium ion and carbon dioxide by the immobilized urease on the collagen membrane. Ammonium ion then passes through the cation-exchange membrane and is converted to ammonia gas in the alkaline bed (pH 10.0). It permeates through the gas-permeable membrane and is utilized as the sole source of energy by nitrifying bacteria. The nitrifying bacteria require dissolved oxygen for nitrification, and the consumption of oxygen is determined by the oxygen electrode. The concentration of urea can subsequently be determined indirectly from the current decrease in the oxygen electrode.

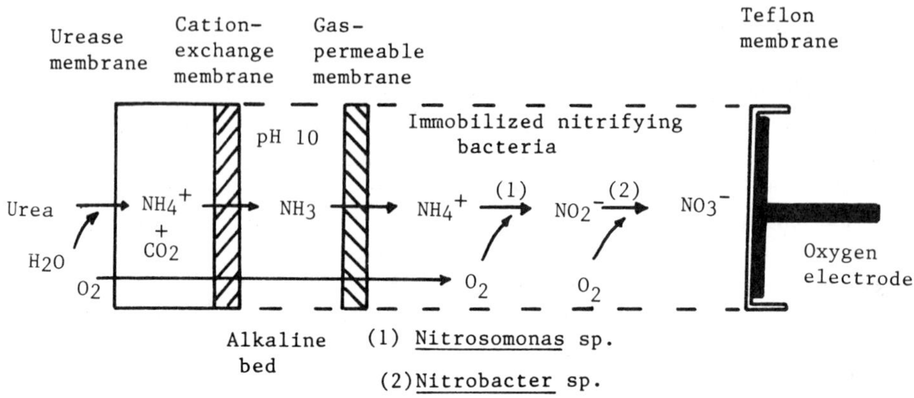

Figure 11 Principle of the hybridized urea sensor. [Redrawn from Okada, Karube, and Suzuki (1982), courtesy of Springer-Verlag, Heidelberg.]

The initial steady-state level of the output current was obtained and corresponded to the endogenous respiratory activity of the immobilized bacteria in the phosphate buffer (pH 7.0). When the sensor was placed in a sample solution containing urea the output current decreased as a result of oxygen consumption by the immobilized nitrifying bacteria. This output current reached a steady-state level within 7 min, which was dependent on the concentration of urea in the sample. When the sensor was inserted in tap water the output current of the sensor returned to its initial level within 2 min. Thus the overall time required for each urea assay was 9 min.

Inhibition of the nitrifying bacteria with nitrite formed by the microorganisms was not observed during these experiments during 150 assays. Nitrifying bacteria isolated from activated sludges could therefore be used for the determination of urea. The response of the sensor for urea was not dependent on the harvesting time of nitrifying bacteria. The microbial sensor was applied to the determination of urea in human urine. Urine was diluted with the phosphate buffer and employed for experiments. The concentration of urea was determined using both the biosensor and the conventional method (Fearon reaction method) as a reference (42). Good agreement was obtained between the sensor and the conventional method (correlation coefficient 0.97).

In addition, the long-term stability of the sensor was investigated using a sample solution containing 150 mM urea. The output current of the sensor was constant for more than 10 days and 150 assays.

The selectivity of the sensor for urea was also examined and was considered satisfactory. The sensor did not respond to some amines (diethylamine, propylamine, or butylamine) and nonvolatile compounds, such as glucose, amino acids, and metal ions (sodium ion, potassium ion, and calcium ion). The sensor proposed gives an economical, selective, and reliable method for the assay of urea in biological fluids that would be reliable for at least 1 week.

Mutagen Sensor

Long-term carcinogenicity tests using laboratory mammals are not only time consuming but are also resource demanding. The mutagenic activity of carcinogens has recently been confirmed in great number of cases. The existence of a high correlation between the mutagenicity and carcinogenicity of chemicals is now evident, and thus the use of microbial systems is important to investigate with mutagenic chemicals. Recently a number of microbial methods for detecting the various types of mutagens have been developed, for example microbial reversion assays using *Salmonella typhimurium* (48) or *E. coli* have been employed for screening tests of chemical carcinogens (49). A method called the rec$^-$ assay utilizing *Bacillus subtilis* has also been proposed for screening chemical mutagens and carcinogens. These methods are more rapid and simple than the carcinogen test using animals. However, the microbial reversion assays and the rec$^-$ assay still require a lengthy incubation of bacteria and complicated procedures.

A sensor consisting of aerobic recombination-deficient bacteria and the oxygen electrode can be applied to the preliminary screening of chemical mutagens and carcinogens. The recombination-deficient strain *B. subtilis* M45 Rec$^-$ and the wild strain *B. subtilis* H17 Rec$^+$ were immobilized onto a porous acetylcellulose membranes. The bacteria membrane was fixed on a Teflon membrane of the oxygen electrode (Fig. 12). The electrode system consisted of two microbial electrodes: the electrode of *B. subtilis* Rec$^-$ (Rec$^-$ electrode) and the electrode of *B. subtilis* Rec$^+$ (Rec$^+$ electrode). Each electrode was composed of immobilized bacteria and an oxygen electrode.

When the Rec$^-$ and Rec$^+$ electrodes were inserted into the glucose buffer solution (0.3 g/L of glucose), steady-state currents were obtained. When the mutagen AF-2 was added to the solution and after 20–40 min the current of the Rec$^-$ electrode began to increase, giving a sigmoidal response. However, the current of the Rec$^+$ electrode did not increase. The rate of current increase of the Rec$^-$ electrode is a measure of the concentration of mutagen and is most easily measured as the linear slope at the midpoint of the sigmoidal curve. Table 2 summarizes the response of the electrode system to various typical chemical mutagens. When such chemical mutagens as AF-2, mitomycin, captan, 4NQQ, *N*-methyl-*N*-nitro-*N*-nitrosobuanidine, and aflatoxin B were added to the glucose buffer solution, the rates of the current increase at the Rec$^-$ and Rec$^+$ electrodes were measured. An increase in current at the Rec$^-$ electrode was indicative that the reagents were mutagens and formed the basis by which the mutagenicity of chemicals could be estimated with an electrochemical system.

In particular, the relationships between the rate of current increase in the Rec$^-$ electrode and the concentration of AF-2 and mitomycin were examined.

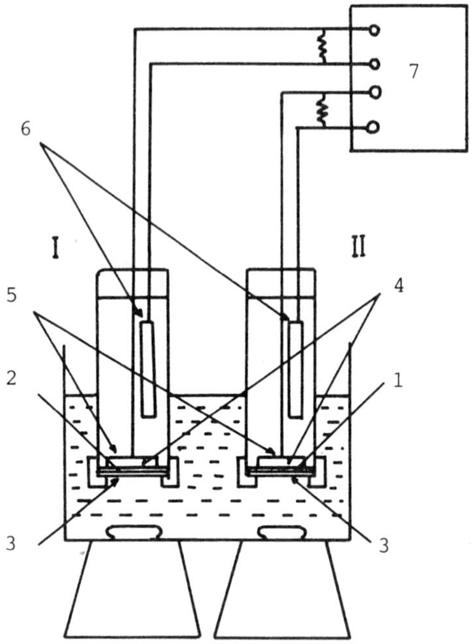

Figure 12 The electrode system for rapid detection of chemical mutagen: (I) Rec$^-$ electrode, (II) Rec$^+$ electrode, (1) *Bacillus subtilis* Rec$^+$, (2) *B. subtilis* Rec$^-$, (3) membrane filter, (4) Teflon membrane, (5) Pt cathode, (6) Pb anode, (7) recorder. [Redrawn from Karube, Matsunaga, Nakahara, Suzuki, and Kada (1981) by permission of the publishers, Publications Division, ACS, Washington, D.C.]

Linear relationships were obtained in the range over 1.6–2.8 μg/mL for AF-2 and 2.4–7.3 μg/mL for mitomycin. The sensor did not response to antibiotics, respiration inhibitors, or bactericide.

The electrochemical method is very different in principle from the conventional streak method. In the latter the inhibition of growth is measured by the conventional Rec$^-$ assay, whereas in the electrochemical method, the decrease in the cell number is measured from the respiratory activity of immobilized bacteria. Since the respiration of bacterial cells is directly and immediately converted to an electrical signal, the preliminary screening of mutagens is possible within an hour. Moreover, the microbial sensor system employs a homogeneous suspension of cells, and consequently, the sensitivity of the microbial sensor is higher than in the Rec$^-$ assay and the Ames test. The minimum measurable mutagen concentration was estimated as 1.6 μg/mL by the microbial sensor compared with 5.0 μg/mL by the rec$^-$ assay and 10 μg/mL by the Ames test for AF-2.

Table 2 Response to Various Mutagens

Mutagen	Concentration (μg/mL)	Rate of current increase (μA/hr)	
		Rec$^-$ electrode	Rec$^+$ electrode
AF-2	0.5	0	0
	1.6	2.7	0
	2.8	4.8	4.0
Mitomycin	0.9	0	0
	7.2	3.6	0.5
	14.4	12	11
Captan	0.5	0	0
	2.0	14	0
4NQO	5.0	0	0
	16.0	11	0
N-Methyl= *N*-nitro=	5	1.0	0.5
N-nitrosogu anidine	20	22	1.0
Aflatoxin	0.8	1	0
B_1	12	12	1

Recombinant-deficient (Rec$^-$) bacteria are killed with mutagens. Therefore, a new bacteria membrane is needed for each analysis although the preparation and exchange of a bacteria membrane is very easy and replacement takes only 1 min. Moreover, the bacteria can be stored for 6 months at −20°C. The electrode current was reproducible within 5% in the buffer solution containing 1 g/L of glucose when the bacteria membrane used for analysis was replaced with a new membrane prepared from the stock bacteria.

The microbial electrode system appears promising and attractive for use in the routine preliminary screening of mutagens and carcinogens.

Sodium Nitrite Sensor

The principal gaseous oxides of nitrogen (NO and NO_2) are of great importance in air pollution sampling and analysis. During the combustion of all types of fossil fuels at flame temperature up to about 0.5% nitric oxide may be formed, along with much smaller amounts of nitrogen dioxide. When discharged to the atmosphere nitric oxide oxidizes at a measurable rate to nitrogen dioxide. Nitrogen dioxide is the most reactive of the gaseous oxides of nitrogen and is a primary absorber of sunlight in atomospheric reactions that produce photochemical smog (50). The determination of nitrogen dioxide is therefore important in environmental and industrial process analyses and is at present performed by a spectrophotometric method (Japanese Standards Association, 1976). Various types of autoanalyzers have been developed for the measurement of nitrogen dioxide in the ambient atmosphere. Earlier developments in such analyzers, using colorimetric estimation of an azo dye-forming reaction, have been described (51). However, these methods require a long reaction time and additional reagents.

A new approach involves the use of infrared lasers and infrared spectroscopy for monitoring gaseous air pollutants (52). Nevertheless, these methods require a large and expensive instrumentation.

Increasing attention is being paid in environmental evaluation to the use of biochemical sensing systems. Such an approach using hemoglobin for measurement of nitrogen dioxide has been developed in our laboratory (53). However, this monitoring system was unstable and not reusable.

We have developed a second, more successful method using a microbial biosensor based on *Nitrobacter* sp. As previously reported, *Nitrobacter* sp. utilize nitrite as the sole source of energy and oxygen is consumed by respiration as follows:

$$2NO_2^- + O_2 \longrightarrow 2NO_3^-$$

Oxygen uptake by the immobilized bacteria can be directly determined by an oxygen electrode, forming the basis for measurement of NO_2 generated in the buffer (pH 2.0).

Figure 13 shows the principle of the sensor for sodium nitrite (54). When the sample solution (sodium nitrite solution) is introduced into the flow cell nitrous ions change to nitrogen dioxide gas at pH 2.0. The nitrogen dioxide passes

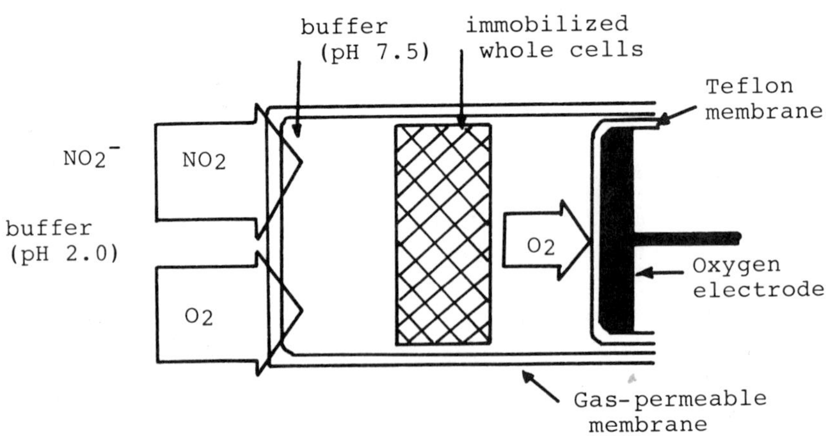

Figure 13 Principle of sodium nitrite sensor. [Redrawn from Karube, Okada, Suzuki, Suzuki, Hikuma, and Yasuda (1982), courtesy of Springer-Verlag, Heidelberg.]

through the gas-permeable membrane and changes to nitrous ions in the bacterial layer, which are then utilized by *Nitrobacter* sp. as their sole source of energy. The consumption of oxygen around the membrane is determined by the oxygen electrode. As a result the concentration of sodium nitrate can be indirectly determined from the current decrease at the oxygen electrode.

The background current at time zero was obtained with the buffer solution saturated with dissolved oxygen and showed the endogenous respiratory level of the immobilized bacteria. When the sample solution containing sodium nitrite was injected into the system for 2 min nitrogen dioxide was produced in the flow cell and permeated through the gas-permeable membrane. Nitrite was formed in the bacterial layer and assimilated by the immobilized bacteria. The consumption of oxygen by the bacteria began and caused a decrease in dissolved oxygen around the membrane. As a result the current of the electrode decreased markedly with time until a steady state was reached, usually within 10 min. At this time the consumption of oxygen by the bacteria and the diffusing species from the sample solution to the membrane were in equilibrium.

When a sufficient quantity of the bacteria is immobilized at the sensor the current of the sensor for a sodium nitrite solution depends mainly on the rate of diffusion of nitrite from the sample solution to the immobilized bacteria. The O_2 content in sample solutions was checked using a differential sensor system, with a sensor without an immobilized bacterial layer as reference to negate the influence of oxygen in the sample solution so that the steady-state current depended on the concentration of sodium nitrite alone. The difference between the initial and steady-state currents was directly proportional to the concentration of sodium nitrite. When only the buffer solution was transferred to the flow cell the current of the microbial sensor returned to its initial level within 12 min. The next sample could then be determined by continuously using the same sensor system.

The long-term stability of the microbial sensor was examined with a sample solution containing 0.25 mM sodium nitrite. The current output of the sensor was constant for more than 21 days and 400 assays. In the same experiments the concentration of sodium nitrite was determined by both the sensor proposed and the conventional method (dimethylnaphthylamine method) as reference. The results obtained for sodium nitrite concentrations determined by the two methods showed a good correlation (correlation coefficient 0.99).

Other Microbial Sensors

Various other microbial sensors have been developed by our group (55–57), and new approaches using genetic engineering are also being conducted to achieve more stable microbial sensors (58). For example, they aim to construct *E. coli* strains capable of producing stable photochemical reaction. Light production in *E. coli* can be introduced by cloning the genes encoding bacterial luciferase from *Vibrio harveyi*.

CONCLUSION

The use of microorganisms, which are highly sensitive to changes in their environment, seems a very good strategy for the rapid and precise monitoring of the various factors involved in fermentation reactions and the environment. This chapter has touched upon some of the ways in which microbial sensors might be employed to monitor levels of fermentation substrates, reaction products, nutrients, and antibiotics.

This account has dealt almost entirely with those microbial sensors that are being investigated and developed in Japan for application to the monitoring of fermentation processes and environmental materials. Here, as in research laboratories around the world, work on these devices continues and much remains to be done. Sensors that work well under laboratory conditions do not always succeed when placed in industrial media, and so basic research on sensors using not only microorganisms now available, but also possibly new mutants with improved properties, must be backed up by developmental modification and refinement. Nevertheless, these and other types of biosensor that can be used in on-line systems continuously to record changes in biochemical and microbiological processes with maximum accuracy and minimum delay appear to be indispensable for the efficient operation of numerous biotechnological processes and should therefore amply repay further investigation.

REFERENCES

1. Guilbault, G. G. (1976). *Handbook of Enzymatic Methods of Analysis*, Marcel Dekker, New York.
2. Guilbault, G. G. (1980). Enzyme electrode probes, *Enzyme Microb. Technol. 2*: 258–264.
3. Guilbault, G. G. (1982). Ion selective electrodes applied to enzyme systems, *Ion Selective Electrode Rev. 4*: 187–231.
4. Suzuki, S., and Karube, I. (1981). Bioelectrochemical sensors based on immobilized enzymes, whole cells and proteins, *Appl. Biochem. Bioeng., 3*: 145–174.
5. Suzuki, S., Satoh, I., and Karube, I. (1982). Recent trends of biosensors in Japan, *Appl. Biochem. Biotechnol., 7*: 147–155.
6. Mosbach, K., Mandenius, C. F., and Danielsson, B. (1983). New biosensor devices. In: *Biotech 83: Proceedings of the International Conference on the Commercial Applications and Implications of Biotechnology*, Online Publications, Ltd., Northwood, UK, pp. 665–678.
7. Wilson, K., and Goulding, K. H. (Eds) (1986). *A Biologist's Guide to Principles and Techniques of Practical Biochemistry*, 3rd ed., Edward Arnold, London.
8. Turner, A. P. F., Karube, I., and Wilson, G. S. (Eds) (1987). *Biosensors: Fundamentals and Applications*, Oxford University Press, Oxford.
9. Karube, I., Matsunaga, T., and Suzuki, S. (1977). A new microbial electrode for BOD estimation, *J. Solid-Phase Biochem., 2*: 97–104.
10. Karube, I., and Suzuki, S. (1983). Application of biosensor in fermentation pro-

cesses. In: *Annual Reports on Fermentation Process* (G. Tsao, Ed), Academic Press, New York, pp. 203–236.

11. Tamiya, E., Sode, K., Lee, S.-M., and Karube, I. (1989). Microbial sensor system using photobacterium. In: MRS Intl Mtg. on Adv. Mats., Vol. 14, pp. 241–243.
12. Lavi, J. T., Lövgren, T. N.-E., and Raunio, R. P. (1981). Comparison of luminous bacteria and their bioluminescence-linked enzyme activities, *FEMS Microbiol. Lett., 11*: 197–199.
13. Makiguchi, N., Arita, M., and Asai, Y. (1980). Immobilization of a luminous bacterium and light intensity of luminous materials, *J. Ferment. Technol., 58*(1): 17–21.
14. Mosbach, K., and Danielson, B. (1974). An enzyme thermistor, *Biochim. Biophys. Acta, 364*: 140–145.
15. Borgerud, A., and Scott, M. (1975). Determination of heat changes in the proximity of immobilized enzymes with an enzyme thermistor and its use for the assay of metabolites, *Biochim. Biophys. Acta, 403*: 256–265.
16. Mattiasson, B., Larsson, P. O., and Mosbach, K. (1977). The microbe thermistor, *Nature, 268*: 519–520.
17. Ingold Co. Operating instructions O_2 electrodes.
18. Matsunaga, T., Karube, I., and Suzuki, S. (1980). A specific microbial sensor for formic acid, *Eur. J. Appl. Microbiol. Biotechnol., 10*: 235–243.
19. Hikuma, M., Obana, H., Yasuda, T., Karube, I., and Suzuki, S. (1980). A potentiometric microbial sensor based on immobilized *Escherichia coli* for glutamic acid, *Anal. Chim. Acta, 116*: 61–67.
20. Rechnitz, G. A., Kobos, R. K., Riechel, S. J., and Gebauer, C. R. (1977). A bio-selective membrane electrode prepared with living bacterial cells, *Anal. Chim. Acta, 94*: 357–365.
21. Matsunaga, T., Karube, I., and Suzuki, S. (1979). Rapid determination of nicotinic acid by immobilized *Lactobacillus arabinosus, Anal. Chim. Acta, 99*: 233–239.
22. Chibata, I., and Wingard, L. B. Jr. (1983). *Applied Biochemistry and Bioengineering*, Vol. 4, *Immobilized Cells*, Academic Press, London.
23. Hikuma, M., Suzuki, H., Yasuda, T., Karube, I., and Suzuki, S. (1979). Amperometric estimation of BOD by using living immobilized yeasts, *Eur. J. Appl. Microbiol. Biotechnol., 8*: 289–297.
24. Karube, I., Mitsuda, S., and Suzuki, S. (1979). Glucose sensor using immobilized whole cells of *Pseudomonas fluorescens, Eur. J. Appl. Microbiol. Biotechnol., 7*: 343–350.
25. Weibel, M. K., Dritschilo, W., Bright, H. J., and Huuphrey, A. E. (1973). Immobilized enzymes. Prototype apparatus for oxidase enzymes in chemical analysis utilizing covalently bound glucose oxidase. *Anal. Biochem., 52*: 402–414.
26. Hikuma, M., Kubo, T., Yasuda, T., Karube, I., and Suzuki, S. (1979). Amperometric determination of acetic acid with immobilized *Trichosporon brassicae, Anal. Chim. Acta, 109*: 33–38.
27. Technicon Industrial Systems, No. 142-71A (1972).
28. Hikuma, M., Obana, H., Yasuda, T., Karube, I., and Suzuki, S. (1980). Amperometric determination of total assimilable sugars in fermentation broths with use of immobilized whole cells. *Enzyme Microb. Technol., 2*: 234–238.
29. Bailey, P. L. (1978). *Analysis with Ion-Selective Electrodes*, Heyden Spectrum House, London, P. 147.

30. Hikuma, M., Kubo, T., Yasuda, T., and Karube, I. (1980). Ammonia electrode with immobilized nitrifying bacteria, *Anal. Chem., 52*: 1020.
31. Karube, I., Okada, T., and Suzuki, S. (1981). Amperometric determination of ammonia gas with immobilized nitrifying bacteria, *Anal. Chem., 53*: 1852–1854.
32. Wang, H. Y., Cooney, C. L., and Wang, D. I. C. (1977). Computer-aided baker's yeast fermentations. *Biotechnol. Bioeng. 19*: 69–86.
33. Reuss, M., Grieser, J., Reng, H. G., and Wagner, F. (1975). Extended culture of *Candida boidinii* on methanol, *Eur. J. Appl. Microbiol., 1*: 295–305.
34. Hikuma, M., Kubo, T., Yasuda, T., Karube, I., and Suzuki, S. (1979). Microbial electrode sensor for alcohols. *Biotechnol. Bioeng., 21*: 1845–1853.
35. Karube, I., Mitsuda, S., Masunaga, T., and Suzuki, S. (1977). A rapid method for estimation of BOD by using immobilized microbial cells, *J. Ferment. Technol., 55*: 243–248.
36. Nissin Electric Co., Ltd., Kyoto, Japan.
37. Karube, I., Okada, T., and Suzuki, S. (1982). A methane gas sensor based on oxidizing bacteria. *Anal. Chim. Acta, 135*: 61–67.
38. Matsumoto, K., Seijo, H., Watanabe, T., Karube, I., Satoh, H., and Suzuki, S. (1979). Immobilized whole cell-based flow-type sensor for cephalosporins, *Anal. Chim. Acta, 105*: 429–432.
39. Rich, S. Ianniello, R. M., and Jaspersen, N. D. (1979). Development and application of a thermistor enzyme probe in the urea-urease system, *Anal. Chem., 51*: 204–206.
40. Ruzicka, J., Hansen, E. H., and Ghose, A. K. (1979). Enzymatic determination of urea in serum based on pH measurement with the flow injection method, *Anal. Chem., 51*: 199–203.
41. Sampson, E. J., Baird, M. A., Burtis, C. A., Smith, E. M., Witte, D. L., and Bayse, D. D. (1980). A coupled-enzyme equilibrium method for measuring urea in serum: Optimization and evaluation of the AACC study group on urea candidate reference method, *Clin. Chem., 26*: 816–826.
42. Le Mar, R. L., and Bootzin, D. (1957). Modification of the diacertyl determination of urea, *Anal. Chem., 29*: 1233–1236.
43. Rosenthal, H. L. (1955). Determination of urea in blood and urine with diacetyl monoxime, *Anal. Chem., 27*: 1980–1982.
44. Fawcett, J. K., and Scott, J. E. (1960). A rapid and precise method for the determination of urea, *J. Clin. Pathol., 13*: 156–159.
45. Llenado, R. A., and Rechnitz, G. A. (1974). Automated serum urea determination using membrane electrodes, *Anal. Chem., 46*: 1109–1112.
46. Kubo, I., Karube, I., and Suzuki, S. (1983). Amperometric determination of creatinine with a biosensor based on immobilized creatininase and nitrifying bacteria, *Anal. Chim. Acta, 151*: 371–376.
47. Okada, T., Karube, I., and Suzuki, S. (1982). Hybrid urea sensor, *Eur. J. Appl. Microbiol. Biotechnol., 14*: 149–154.
48. Karube, I., Nakahara, T., Matsunaga, T., and Suzuki, S. (1982). *Salmonella* electrode for screening mutagens, *Anal. Chem., 54*: 1725–1727.
49. Karube, I., Matsunaga, T., Nakahara, T., Suzuki, S., and Kada, T. (1981). Preliminary screening of mutagens with a microbial sensor, *Anal. Chem., 53*: 1024–1026.
50. Stern, A. C. (1976). *Air Pollution*, (3rd Ed.,) Academic Press, New York, pp. 259–272.

51. Moore, G. E., Cole, A. F. W., and Katz, M. (1957). Concurrent determination of sulfur dioxide and nitrogen dioxide in the atmosphere, *J. Air Pollut. Control Ass.*, *7*: 25–28.
52. Hanst, P. L. (1970). Infrared spectroscopy and infrared lasers in air pollution research and monitoring, *Appl. Spectrosc.*, *24*: 161–174.
53. Matsuoka, H., Aizawa, M., and Suzuki, S. (1979). Membrane biochemical indicator of NO_x exposure, *J. Membr. Sci.*, *5*: 265–273.
54. Karube, I., Okada, T., Suzuki, S., Suzuki, H., Hikuma, M., and Yasuda, T. (1982). Amperometric determination of sodium nitrite by a microbial sensor, *Eur. J. Appl. Microbiol. Biotechnol.*, *15*: 127–132.
55. Matsunaga, T., Karube, I., and Suzuki, S. (1978). Electrochemical microbioassay of vitamin B_1, *Anal. Chim. Acta*, *98*: 25–30.
56. Matsunaga, T., Karube, I., and Suzuki, S. (1979). Electrode system for the determination of microbial population, *Appl. Environ. Microbiol.*, *37*: 117–121.
57. Karube, I., Matsunaga, T., and Suzuki, S. (1979). Microbioassay of nystatin with a yeast electrode. *Anal. Chim. Acta*, *109*: 39–44.
58. Korpela, M., Mantsala, P., Lilius, E. M., and Karp, M. (1989). Stable-light-emitting *Escherichia coli* as a biosensor, *J. Biolumin. Chemilumin.*, *4*: 551–554.

13

In Vivo Biosensors

P. Vadgama and M. A. Desai

University of Manchester, Hope Hospital
Salford, England

INTRODUCTION

Although it is commonly asserted that the biological organism is characterized by internal biochemical stability in the face of a highly variable environment, this constitutes only an approximation of the truth. Certainly the absolute levels of various biochemical parameters remain within certain clearly defined bands (1), often referred to as the reference ranges, but these range from tightly controlled levels for, say, blood calcium to serum iron levels, which can show substantial (percentage) variations on a diurnal weekly and seasonal basis. The time scale for such changes and fluctuations, furthermore, may be relatively short, as exhibited by growth hormone and certain peptide hormones acting on target endocrine glands; these hormones are best described as secreted in a pulsatile manner. In disease, not only may mean levels slip out of the reference range band but there can be changes in the fluctuation dynamics of biochemical entities, as well as in the dynamic interactions between different species. The advantages of electrophysiological measurement on a real-time basis are already well illustrated by the enormous clinical benefits accruing from conventional ECG and EEG monitoring. Transducer-based, continuous measurement of physical parameters, such as pulse, arterial and central venous blood pressure, blood flow, temperature, and ventilation rate, has proved relatively easy to carry out, but few techniques are available for continuous in vivo monitoring of organic biochemical species, ions, or gases. The centralized hospital laboratory has powerful automated techniques for the rapid spectrophotometric assay of organic molecules of biochemical importance, but these techniques clearly do not adapt to decentralized monitoring.

Biosensors are an attractive technology for continuous patient monitoring (2) for reasons rather similar to those that have led to their prominence in discrete in vitro measurement (Table 1). For the purpose of in vivo monitoring, however, most of these listed attributes are a necessity, not an analytical luxury.

The challenge posed by in vivo monitoring with biosensors is considerably greater than for use in discrete measurements in vitro and also makes it obligatory for researchers in the field to exploit (and adapt) techniques that are more commonly regarded as mainstream to the biological, rather than physical sciences. In general it must be accepted that functional devices have seen relatively slow development. Unquestionably, however, real-time biochemical monitoring is set to have a major impact on the practice of a range of specialities, including intensive care medicine. The rapid return of biochemical information, especially in relation to the metabolically unstable patient, would not only permit early institution of effective therapy but enable its titration to precise levels, altering with medical needs.

In this chapter the broader definition of biosensors is used by which the device is a chemically responsive transducer directly in contact with a biological matrix or surface but does not itself necessarily incorporate a biological component. Although increasingly the use of a bioreagent will become commonplace, direct chemical transducers for in vivo monitoring share similar problems of stability, reliability, biocompatibility, and safety; as such they may be usefully grouped together. Also, in vivo frequently implies that the device is implanted in the body for selective location in a body compartment, namely the intravascular space, peritoneal cavity, or subcutaneous tissue. Much work has been done, however, on continuously sampled anticoagulated blood simply pumped through an electrode flow cell; a description of such extracorporeal monitoring is included in this chapter. Clinically important devices for transcutaneous gas monitoring are also described.

ELECTROCHEMICAL GAS SENSORS

Disease states leading to major organ failure, especially that of the lungs, heart, or kidney, may be associated with changes to normal arterial pO_2 and pCO_2. The degree of change provides an index of the severity of the condition, its nature,

Table 1 Advantages of Biosensors for In Vivo Use

Reagentless	Disposable
Miniaturized	Multiparameter assay possible
Implantable	Low materials costs
Direct electrical signal	IC technology usable
Stabilized bioreagent	Reversible response

and often the extent of the therapeutic response that is necessary. Real-time monitoring of blood gases may consequently provide important information that helps to avoid a fatal clinical outcome. A specific area in which monitoring has led to clinical benefit on a large scale has been in the management of the critically ill neonate. Here, during therapy for respiratory problems, it has proved possible to avoid not only the dangers of hypoxia but the damaging consequences of prolonged exposure to high pO_2 levels.

Oxygen Measurements

Catheter Sensors

The majority of sensors employ the polarographic principle of electrochemical O_2 reduction at a noble metal working electrode polarized at approximately -0.65 V versus Ag/AgCl.

The Clark design, in which an O_2-permeable plastic membrane separates a complete reference/pseudoreference electrochemical cell from the biological solution, has proved to be the mainstay of in vivo O_2 sensors (3) and has permitted the operation of such devices largely independent of sample stirring rate, at the same time protecting the metal working electrode surface from fouling with blood constituents. This bipolar electrode system has some limitations on the minimal external diameter that can be achieved (4) (typically 0.6 mm OD), but certainly umbilical artery catheterization has proved possible in the neonate (5) and peripheral artery cannulation in the adult (6). Detailed design considerations have been reviewed by Rolfe (7) and highlight the importance of membrane properties with respect to stable in vivo operation. Although polyvinyl chloride (PVC) was one of the earliest gas-permeable barriers to be used (8) and could relatively easily be bonded to the catheter structure, changes in its state of hydration as well as creep contributed to signal drift during use. Other membrane materials to have been explored include Teflon, polystyrene, and polyethylene, but often, although permeability has been optimized, it has proved necessary to employ some mechanical means of retaining membranes over the working electrode (9). Silver has been used, as well as platinum, for the working electrode, and bipolar silver electrodes have been reported (10,11). Continuous intravascular monitoring with such devices has provided important insight into the labile nature of blood pO_2 in certain disease states (Fig. 1).

Dry storage of bipolar electrodes is made difficult by the need to incorporate internal electrolyte between the covering gas-permeable membrane and the electrode elements; solid electrolyte crystals (e.g., KCl) deposited over the latter (10,11), however, rehydrate on exposure to blood, or to steam during sterilization, and allow reliable O_2-responsive currents to be generated. This reconstitution process, however, makes in vivo recalibration of such probes mandatory. The reproducible deposition of electrolyte as well as of covering membranes over integrated microfabricated electrodes may well furnish greater predictability in performance (12).

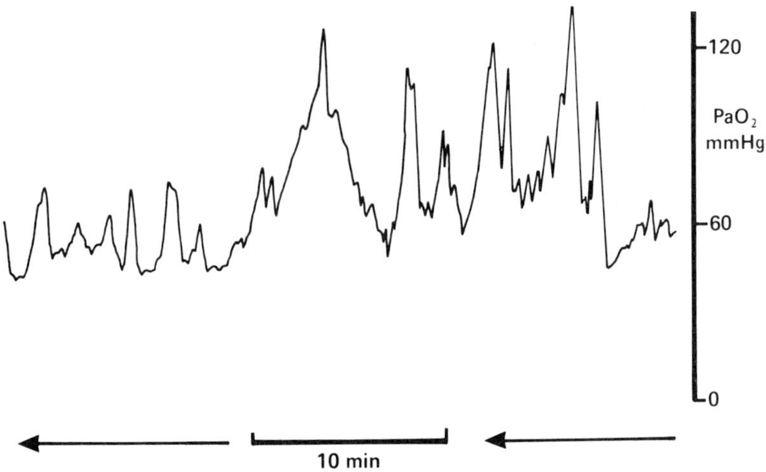

Figure 1 Continuous intraarterial recording of pO_2 in a spontaneously breathing premature infant recovering from lung hyaline membrane disease. [After Conway et al. (11).]

A galvanic cell requires no external applied voltage; within an alkaline internal electrolyte an anodic reaction takes place at a lead electrode that drives the cathodic reduction of O_2 at a silver electrode. The device therefore appears to have an instrumental advantage for long-term implantation; the intrinsic polarizing voltages achieved are suboptimal for O_2 reduction, however, and although a design has been reported for oxygen sensing (13) signal stability has been a problem. For the Clark amperometric device Hahn (14) has outlined the benefits of pulsed over constant polarization voltages. With sufficient short pulses the field of influence of the polarized electrode on local oxygen is reduced to within the membrane and not only is O_2 consumption minimized but signal size becomes less dependent on membrane properties and sample flow.

A constructionally simpler variant of the amperometric O_2 electrode is the monopolar sensor reported by Harris and Neugent (15). Here a single anodic wire coated with a hydrophilic (electrolyte-permeable) membrane of Hydron is sited intravascularly. Membrane characteristics are of additional importance, since electrical continuity is maintained with a skin surface anode through the membrane. Results have shown poor correlation with in vitro pO_2 estimations, and more seriously, the wire has tended to induce clots in the circulation. Although the incidence of complications were considered by some (16) to be no greater than with standard medical catheters, there remains the indeterminate electrolytic path between the two electrodes leading to uncertain polarizing voltages as well as a high vulnerability to biopotential effects (14). The nature of any sensor surface exposed to blood has a bearing on eventual in vivo electrode performance, and a hydrophobic gas-permeable membrane appears in this context to be superior. Specific surface modifications are also worth considering. Nilsson et al. (17) immobilized heparin over electrode membranes of Clark catheters. Their results for arterial pO_2 in nonanticoagulated dogs show good sensor stability and demonstrate further benefits of an engineered nonthrombogenic surface. Of interest from their work is the way an unstable in vivo electrode can give the impression of true biological fluctuations in a measured biochemical parameter (Fig. 2).

Provided the membrane design problems can be overcome, considerable information of pathophysiological relevance can be obtained with very rapid response oxygen sensors. An indication of the possibilities was given by Oeseburg et al. (18), who used a low polarizing voltage to avoid blood flow dependence at a bare, rapid-response O_2 catheter. These workers demonstrated pO_2 fluctuations that were of pathological significance, occurring within a single cardiac cycle, and also showed changes during respiratory exertions (Fig. 3).

Tissue Electrodes

Needle oxygen electrodes of <0.5 mm diameter have been reported for tissue oxygen monitoring. Silver (19) produced an early miniature Clark electrode, but the probe was fragile and difficult to construct. Aarnoudse et al. (20) devised a polystyrene-covered monopolar electrode, implanted subcutaneously, to follow fetal oxygenation during labor; electrode drift was <3.5%/hr, but membrane deformation was a possible cause of calibration shifts.

A tip design in which the cathode surface is recessed restricts the O_2 diffusion field to within the recess chamber, provided the recess length to cathode diameter ratio is >10; this design eliminates stirring artifacts and any local tissue pO_2 perturbations (21). Construction of microelectrodes (tip diameter < 5 μm) has

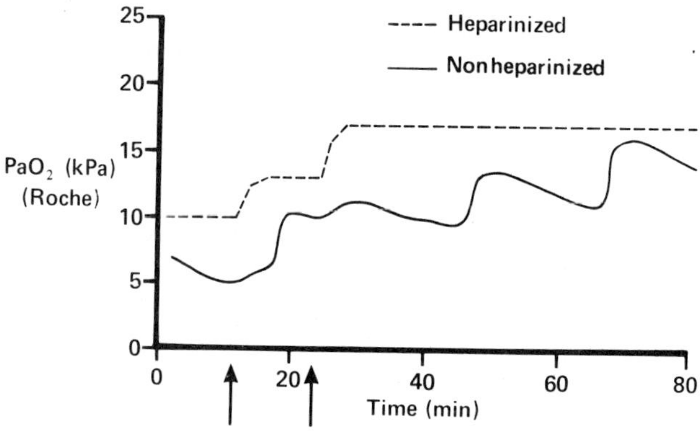

Figure 2 Femoral artery pO_2 traces in dogs using O_2 catheters. Constant inspired oxygen levels were used; arrows indicate increases in O_2 administration via respiration. [After Nilsson et al. (17).]

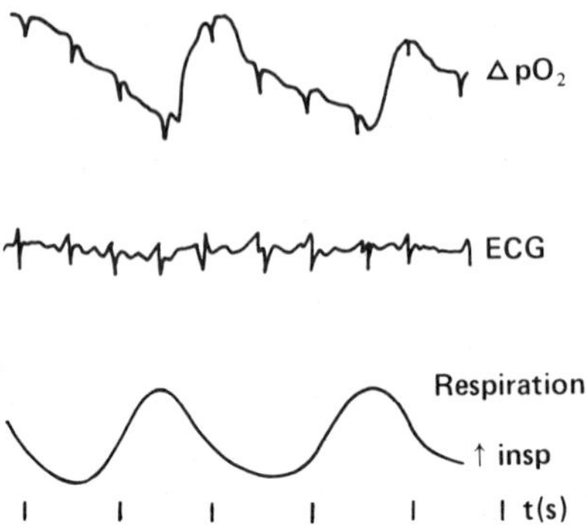

Figure 3 pO_2 tracing using a bare Pt electrode catheter sited near a septal defect in the right atrium of the heart. During expiration the venous return (low pO_2) to the atrium diminishes and pO_2 rises as a result of admixture of blood from the left atrium. [After Oeseburg et al. (18).]

been difficult, and Linsenmeier and Yancey (22) have reported a simplified construction mode in which the creation of a recess is with a low melting point metal alloy. Beyond specialist use of such sophisticated oxygen microelectrodes, van der Kleij and de Koning (23) have managed to observe the intramuscular effects of shock in experimental animals with larger, robust recessed tip cathode/anode combination electrodes. One development of such electrodes may be the fabrication of multicathode arrays (24) to enable measurement of pO_2 over a wider, more representative zone of tissue. Whatever probe design is selected tissue constitutes a heterogeneous medium, with localized regions of high and low oxygen, and this needs to be taken into account in interpreting data (25).

Transcutaneous Sensors

The closest approximation to the ideal of noninvasive monitoring is probably that evidenced by transcutaneous monitoring with heated Clark electrodes. Normally little oxygen gets through the skin, but vasodilation produced by heat (or other agents) results in the O_2 supply far exceeding the metabolic requirements of skin. The effects of increased tissue respiration on skin surface pO_2 are furthermore balanced by a shift to the right of the hemoglobin dissociation curve at elevated temperatures; this reduces retention of oxygen in the skin by hemoglobin. Although the electrode is sited outside the body, the net flux of O_2 to the electrode surface is affected by the variable, heterogeneous, and ill-defined compartment of respiring and nonrespiring cells (Fig. 4); these, moreover, lie in a matrix of undetermined polymer composition and orientation. Distortions of oxygen diffusion and partitioning in this biomembrane therefore have effects on the electrode signal that are in every way as profound as with any implanted device that has been reported (26).

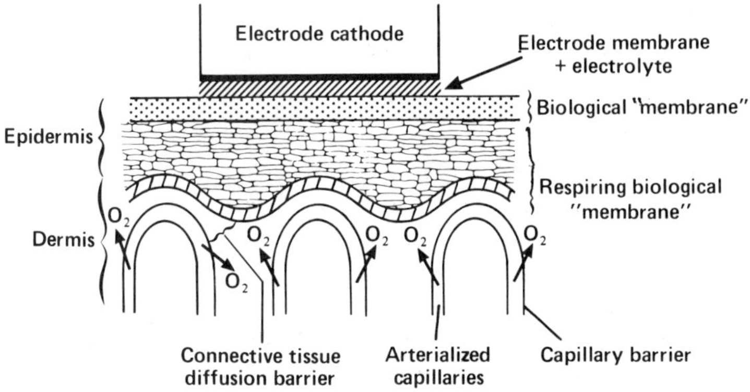

Figure 4 The additional biomembranes that determine the flux of O_2 to a skin surface heated Clark electrode used for transcutaneous monitoring.

Transcutaneous O_2 sensors have seen greatest use in neonatal intensive care monitoring. Le Souef et al. (27) found 44°C to be a satisfactory operating temperature; reliable use has depended on the relatively high O_2 permeability of neonatal skin. A special design feature was the need for a low oxygen consumption by these devices; Huch et al. (28) have used multiple small cathodes, whereas Eberhard et al. (29) employed a low-permeability Mylar membrane. The main value has been in monitoring of neonates with respiratory problems in whom the need for frequent sampling of capillary blood for blood gas analyses has been reduced. Monitoring has revealed large fluctuations of pO_2 in low-birth-weight infants simply as a result of routine nursing care and has highlighted the importance of minimum handling of such babies (30). Use during anesthesia has enabled a more direct control of pO_2 (31).

Transcutaneous pO_2 values are dependent upon skin blood perfusion as well as upon true arterial pO_2. Often this is a disadvantage, but it is possible to exploit the early dissociation between arterial and transcutaneous pO_2 to detect the onset of perfusion failure (shock) (32).

In adults the relation between arterial and transcutaneous pO_2 has proved less clear-cut. The lower vascularity and greater diffusional resistance of adult skin leads to skin surface readings that are 10–20% lower than in arterial blood. Some clinicians have considered this correspondence to be sufficiently good to permit useful trend monitoring (33). Possible applications range from those in the intensive care unit (namely, during mechanical ventilations) to preoperative care in trauma patients or during cardiopulmonary bypass; these aspects have been reviewed by Kram and Shoemaker (34).

The conjunctiva of the eye does not possess a keratinized cell layer; a Clark O_2 electrode attached to the outer surface of a scleral contact lens has therefore permitted oxygen monitoring without surface heating (35,36). Another advantage of this site is that a common blood supply to the brain and conjunctiva enables the latter to specifically reflect cerebral oxygenation; a commercial device is available for clinical monitoring.

Anesthetic Gas Interference

Errors may arise with silver, as well as other cathode electrodes, during anesthesia with nitrous oxide, as the gas is reduced at polarizing voltages often employed for O_2. By limiting polarization voltage to -0.65 V against Ag/AgCl, the interference is largely avoided. Hahn et al. (37) pulsed at -0.6 and -1.4 V, respectively, and were able to obtain oxygen and combined oxygen-nitrous oxide-dependent currents, thus achieving selectivity. Halothane reduction is also readily catalyzed at silver (38), and a current is generated even at -0.4 V; interference can only be prevented with more selective membranes, such as Teflon (14).

CO_2 Measurement

Catheter Sensors

Miniaturization of the Severinghaus CO_2 electrode with its combination of a glass pH sensor, internal fluid electrolyte ($NaHCO_3$), and outer gas-permeable membrane has proved difficult, added to which has been the danger of breakage of the pH-sensitive glass. Parker et al. (39) devised a small CO_2 probe, which also incorporated an O_2 sensor, but response was rather sluggish (~12 min). Coon et al. (40) produced a more robust device by replacing the pH glass with a palladium oxide H^+-sensing wire. The combination of the wire, Ag/AgCl reference electrode, and bicarbonate filling solution was retained behind a solvent-cast silicone-polycarbonate composite membrane that was permeable to CO_2; response times were <30 sec, and the device was used to monitor pCO_2 in patients undergoing neurosurgery (41). A similar device was evaluated in animal experiments (42) but required calibration every 1½ hr as a result of drift, which proved to be a logarithmic function of time. Inclusion of an H^+ ionophore, *p*-octodecyloxy-*m*-chlorophenylhyrazonemesoxalonitrile (OCPH), in the silicone membrane phase of this device has permitted simultaneous pH measurement with a separate external reference electrode (40). A more flexible and safer catheter construction was achieved by Opdycke and Meyerhoff (43), who used a PVC-incorporated ionophore. This H^+-responsive membrane formed part of a PVC tube behind which was the internal reference electrode; silicone rubber tubing formed the outer gas-permeable membrane (Fig. 5). Good in vitro performance was observed, but blood coagulation was a problem during intravascular use, and special buffering was needed to avoid pH electrode drift due to inward diffusion of CO_2 through PVC into the reference electrolyte solution.

A further development has been the substitution of a pH ISFET for the neutral carrier (or glass) electrode (Kuraray Industries, Osaka, Japan). The system is usable with a skin reference electrode.

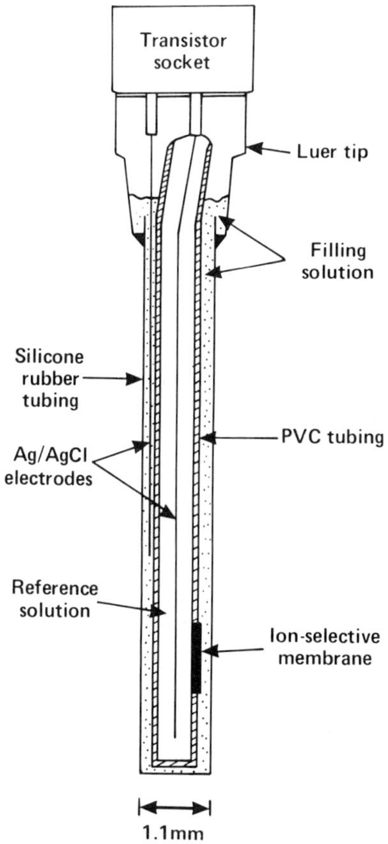

Figure 5 Carbon dioxide catheter sensor based on a neutral carrier membrane for intravascular pCO_2 monitoring. [After Opdyke and Meyerhoff (43).]

Transcutaneous Sensors

A Severinghaus CO_2 electrode applied to the heated skin enables transcutaneous CO_2 monitoring (44). The situation is somewhat different from that for oxygen, since CO_2 diffuses more readily through intact skin and an increase in skin metabolism adds to transcutaneous values; furthermore, the electrode does not consume CO_2. However, temperature affects readings by modifying skin permeability and metabolism with a direct effect on the electrode. The net result is that transcutaneous readings are higher than arterial values, typically by a factor of 1.3 at 44°C (45).

It has been possible to obtain equilibrium transcutaneous readings without heating, but the high diffusional resistance to CO_2 due to the unheated keratinized layer "biomembrane" (Fig. 3), increases response time (46). Circulatory changes have less effect on measurements than they do with transcutaneous O_2; however, in severe shock measured pCO_2 rises (47).

Combined transcutaneous CO_2/O_2 electrodes have also been reported (48). Here, a single gas-permeable membrane and internal electrolytes are used, but a small cathode area is needed to minimize OH^- generation from electrochemical O_2 reduction with resulting drift in internal electrolyte pH.

pH ELECTRODES

Catheter Sensors

Continuous recording of pH has yet to come into routine clinical practice; however, such monitoring could provide a rapid means of gauging tissue hypoxia irrespective of whether the cause is lowered arterial pO_2 or perfusion failure. Several reports point to clinically useful applications. Staehlin et al. (49) used intravascular catheter-mounted glass electrodes to monitor central venous pH in patients during vascular surgery. Catheter construction was simplified by deploying a separate external reference electrode, and a polymer coating over the pH glass improved blood compatibility. The high-impedance output of such devices leads to a high sensitivity to background electrical interference, however, and steps need to be taken to electrically isolate the patients. The fragility of the pH glass also poses a danger. OCPH-doped polymer was used by Cobbe and Poole-Wilson (50) to permit safer pH monitoring; responses were sufficiently fast to follow pH changes during normal respiratory movements, and catheters sited in the venous drainage of the heart demonstrated local blood pH reduction occurring after cardiac pacing (51). Tridodecylamine neutral carrier-based catheter electrodes have been used for animal studies; optical isolation of catheters avoided signal noise, and low-amplitude (<0.02 pH) fluctuations were followed reliably (52).

Metal/metal oxide electrodes generate potentials that are linearly related to pH, and their solid nature and ready miniaturization makes them well suited to invasive monitoring (53). An iridium/iridium oxide-based catheter was thus used as part of a pH-triggered pacemaker. The device increased the pace rate in response to any fall in blood pH and thus was able to respond to the increased needs of exercise. Slow drift occurred over long periods, but the system was designed only to respond to rapid electromagnetic frequency (EMF) changes (54). Physiological considerations indicate that this may prove an important route to feedback pacemakers (55).

Ion-selective field effect transistors (ISFET) provide a solid-state electrode system with in situ conversion of a high-impedance signal to one of low impedance. Furthermore, signal processing and device multiplexing is readily achievable, making ISFET an attractive means of monitoring H^+ and other ions. A pH catheter has been designed for clinical use (56,57). A dual-function pH and pCO_2 electrode was also deployed in the radial artery (58); here local infusion of anticoagulant reduced the dangers of blood clotting. Schepel et al. (59) concluded that, although bioelectrical noise was small with arterial ISFET catheters, a defined electrode stabilization period of 15–60 min was required in vivo. General factors governing in vivo electrode drift have been discussed by Eberhart (60). An ISFET has been used on-line in a flow cell to permit the intermittent determination of pH (as well as Ca^{2+}, Na^{+}, and K^{+}) on continuously sampled venous blood (61); drift correction was achieved by interposing calibrant solutions. There remains a vital need to miniaturize the reference electrode and to improve its hemocompatibility by modifying the liquid junction; such an approach has been attempted using hydrogel at the interface with blood (62).

Tissue Sensors

Early probes used for tissue pH were large, required a long skin incision, and had to be positioned in muscle; Mackenzie et al. (63) reported a convenient antimony/antimony oxide electrode for use in tissue. Routine clinical monitoring has become feasible with the development of a strengthened low electrical resistance glass electrode (64,65). This device has a 1 mm pH-sensitive tip inserted into subcutaneous tissue through a 3 mm stab incision. Most commonly the electrode is used for fetal scalp pH monitoring during labor. Close proximity of the pH glass and reference electrode (Ag/AgCl) minimizes electrical noise, but signals from this superficially applied sensor are subject to movement artifact and firm attachment with two spiral needles is necessary. In a study (66) in which 229 pH-monitoring runs were carried out, tissue pH values were between 7.02 and 7.40 and showed a sufficiently high correlation ($V = 0.71$) with umbilical artery blood pH to be of clinical value. Fetal heart rate patterns have some association with tissue pH (67), and dynamic interrelationships may give added clinical, diagnostic, and prognostic information.

Walters et al. (68) monitored intramyocardial pH using miniature glass electrodes housed in 0.8 mm OD hypodermic needles. Another 8×1 mm glass pH electrode (69) inserted into myocardium during open heart surgery provided early indication of ischemic injury, registered as a decreasing pH (70). Local myocardial damage due to insertion of such probes is inevitable, but this must be balanced against possible clinical benefits. Extracellular deployment of such relatively large electrodes seems valid, however, since changes reflect intracellular pH, particularly when pH is lowered.

FIBEROPTIC SENSORS FOR BLOOD GAS AND pH

The advent of fiberoptic technology has allowed in situ determination of chemical species in inaccessible and optically opaque media; clearly the intravascular compartment is distinguished by both these features. By means of a fluorescent dye immobilized at the tip of optical fibers, Gehrich et al. (71) measured pO_2, pCO_2, and pH. For pO_2, concentration-dependent quenching of fluorescence of an appropriate dye followed; for pH an ionizable pH-sensitive fluorescence dye was employed (hydroxypyrene trisulfonic acid); for pCO_2 the dye monitored bicarbonate solution pH retained behind a silicone CO_2-permeable layer (Fig. 6). An opaque cellulosic covering over this triple probe served to prevent light interaction with blood. A key feature of the system is the relatively smooth surface presented to blood, together with a thromboresistant heparin coating, thus diminishing the risks of thromboembolism and surface fouling. In any case, since the fluorescent optode gives a true equilibrium response the steady-state signal is unaffected by surface fouling. Arterial blood was monitored in humans (72) using a commercial version of the sensor (Cardiovascular Devices, Irvine CA); the probe has a 0.62 mm external diameter (fiber diameter 140 μm), uses a broad-band xenon flash lamp/filtered light source, and gives a response after a 3 min lag. One important practical consideration has been the need to minimize the light transmission distance; this requires the entire optoelectronic monitor assembly to be deployed at the bedside. The basic principles of oxygen probe optimization have been discussed by Opitz and Lübbers (73). An intravascular optode for oxygen based on photoluminescence quenching and tested in animals indicated that stable operation was possible over 8 hr (74).

Siggaard-Andersen et al. (75) undertook detailed evaluation of a corresponding on-line, extracorporeal monitor for pH, pCO_2, and pO_2 and found that, although the overall bias in comparison with in vitro electrochemical measurements was minor, precision was not to the standard required for clinical use.

Despite this move away from electrochemistry, such sensors have not avoided interference from nitrous oxide (affecting all three parameters) or halothane (affecting pO_2).

With regard to future prospects in selected intensive care situations, intravascular optodes may well become an established technique (76). For the clinical environment important features include immunity from electrical noise, added patient safety, lack of need for reference electrodes, and a high stability toward steam sterilization (77), although chemical sterilization is normally used at present.

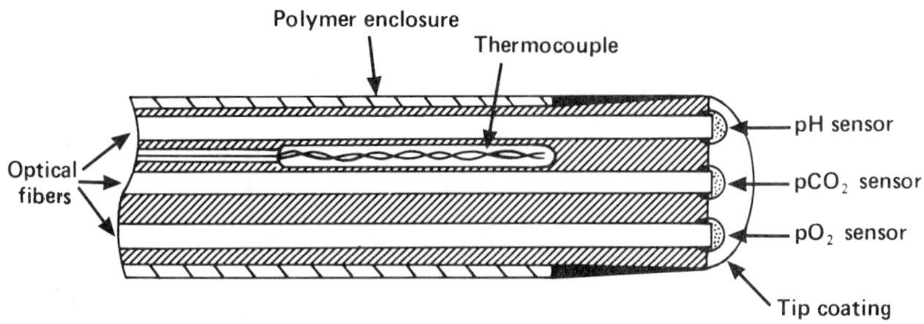

Figure 6 Intravascular blood gas and pH optode based on fluorescent dyes [Gehrich et al. (71)]. A thermistor is required for temperature compensation.

Conjunctival monitoring of CO_2 is possible with fiberoptic conjunctival sensors, again using the Severinghaus electrode principle; a colorimetric pH dye (phenol red) was used to follow bicarbonate solution pH. New clinical data emerging with this device suggest that the close correspondence between tissue and blood pCO_2 may be lost in some situations (78). Colorimetric dye-based sensors can also be used for direct pH monitoring and are readily miniaturized for intravascular use.

ION-SELECTIVE ELECTRODES

An ion-selective electrode membrane develops an interfacial potential that is proportional to the logarithm of activity of the analyte ion in solution. At the heart of the device is a chemically selective membrane component that has an affinity for the analyte via noncovalent interactions. Considerable effort and ingenuity have been exercised to attain high ion selectivity through the synthesis of appropriate ionophores. However, it has proved a formidable task to extend the ion-measuring repertoire of clinical ion-selective electrodes beyond the early systems for Na^+, K^+, Ca^{2+}, Cl^-, and possibly Li^+. Nevertheless, important fundamental insight has been gained from the continuous monitoring of these ions, and the prospects of clinical benefits are high.

Intravascular Electrodes

Catheter tip electrodes have generally been used for intravascular monitoring of ions. Treasure and Band (79) designed a biluminal catheter that contained a Ag/AgCl electrode in each compartment. A PVC membrane doped with valinomycin and supported by a ceramic plug served as a K^+-sensitive surface for one of the compartments, and the complete catheter unit functioned as an in vivo electrochemical cell. A commercially made K^+ electrode was used in clinical studies (Simonsen and Weel, Sidcup, UK). In a later modification (80) the porous plug-mounted compartment formed the main catheter component, and an external sleeve was used to retain an electrolyte film connected to an external reference electrode (Fig. 7). Electrode responses here were followed by means of a telemeter-radioreceiver system to electrically isolate the electrode.

Catheters have been employed for clinical studies (81), as well as for animal experiments in which, for example, K^+ measurement in arterial blood going to the heart has been possible along with that in the venous drainage (82). Band's group (83) also devised a modified sampling system for human studies in which a cannula-mounted electrode monitors K^+ flowing out of an artery into a collection channel.

These basic PVC catheter designs should permit monitoring of other ions, provided the ionophores used have sufficient selectivity (52). Additionally it has been possible to use ISFET with appropriate selective membranes for the inva-

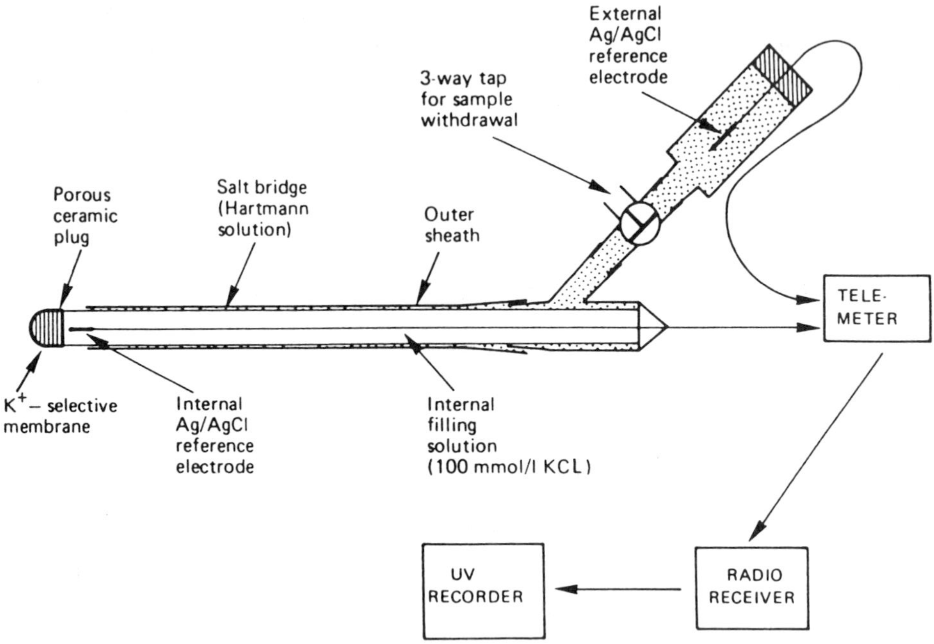

Figure 7 Intravascular K^+ catheter. (From Lim et al. (80).]

sive monitoring of calcium (84) and potassium (85) in blood. As with pH electrodes there is a need for further refinement of the reference electrode configuration and improved hemocompatibility. Tseung and Goffe (86) reviewed the basic electrochemical features needed for a stable in vivo reference electrode and suggested some theoretical advantages of a palladium hydride electrode with low electrical impedance and enhanced resistance to protein fouling.

Tissue Electrodes

The basic catheter K^+ electrode was modified by Hill et al. (87), with the valinomycin membrane strengthened by incorporation of 72% dibutylsebacate. This enabled the device to be used as a "plunge" electrode in myocardial muscle. In animal studies, for example, Coronel et al. (88) showed that during myocardial ischemia a biphasic K^+ change occurs, with a rapid rise followed by a second delayed phase increase associated with irreversible myocardial injury; the anatomic distribution of K^+ changes were also studied. This type of quantitative information about tissue biochemistry will prove invaluable in the design of clinical therapy. Janata's group gave equally relevant insight into blood-tissue interrelationships. Thus implantation of miniature K^+ ISFET showed how local tissue K^+ can rise dramatically in circulatory shock without parallel changes in blood levels (89).

Microelectrodes with a tip diameter up to 200 μm have been developed for fundamental biological investigations. Specialized (90) fabrication techniques and instrumentation are required, but both glass and liquid membrane electrodes can be made (91). The devices are insufficiently robust for clinical application but have been used extensively for dynamic intracellular ion monitoring and also extracellular measurement in some cases.

Extracorporeal Monitoring

Use of electrodes in flow cells has permitted short-term monitoring of ions. Osswald et al. (92) monitored blood K^+ during heart surgery by using a flow assembly in which the potential of a valinomycin electrode was measured against a Ag/AgCl reference; the potential of each was actually determined with respect to a common platinum wire. An extracorporeal circuit was sampled in the study, and the response time of 3 min was sufficient to detect transient surges in blood K^+. A similar flow-through cell arrangement was reported for the measurement of lithium to sodium ion cell concentration ratios (93).

A neutral carrier-based Ca^{2+}-selective electrode was used by Anker et al. (94) in an extracorporeal circuit in a dog: the return of blood here required a change from KCl bridging solution to an isotonic saline to prevent potassium overload. A differential sensor assembly in which heparinized blood and standard flowed past two identical K^+ (or Ca^{2+}) electrodes enabled on-line monitoring with apparently minimal drift (95).

Potassium and sodium electrodes with PVC membranes are susceptible to interference from lipophilic anions in blood. By employing a silicone rubber matrix instead of PVC, selective monitoring of undiluted urine proved possible (96).

Wider acceptability of even the extracorporeal sensors awaits further refinements specifically directed toward achieving operational simplicity. Thus, although the fabrication demands are less stringent than for in vivo sensors, with less need for miniaturization, sterilization requirements obviated, and drift correction possible, there remains the need for high operator skill. It is here that such microfabricated devices as ISFET make the crucial difference between commercial failure and success. With multiple arrays of microfabricated devices in future, there would be in-built redundancy and compensation for signal instability; Ko (97) reported devices with up to 10 sensors. It must be admitted that the individual instability of microchip devices may be greater than with conventional electrodes, even in simple aqueous solutions, and higher tolerances are demanded of the packaging materials used for chemical microsensors generally (98), including the need for electrostatic shielding. One of the key incompatibilities here arise from a need to have total protection from the aqueous environment and indeed, water from vapor for the electronic and input-output connections, and the sensor membrane must retain a degree of ionic permeability and responsiveness to the aqueous phase. One specific interference problem has been that of CO_2 and organic acid diffusion through PVC coatings at ISFET leading to membrane changes and signal drift (99).

DIRECT VOLTAMMETRIC MONITORING

The ready electrochemical oxidation of endogenous, organic, and pharmaceutical agents at polarizing voltages well below those capable of decomposing water can in principle be used for their in vivo monitoring. In practice such an approach is limited by the inability of most standard electrochemical techniques to selectively detect the analyte of interest. A further drawback is the ready tendency of the voltammetric working electrode to be fouled by macromolecules present in biological fluids or to be passivated by the selective adsorption of certain biochemical solutes and their electrochemical reaction products (100). In addition, many so-called inert metal electrodes, such as the platinum electrode, present a variably oxidized surface for electrochemistry depending upon such parameters as pH and the applied external voltage; this can seriously affect the efficiency of surface electron transfer and, therefore, the reliability of voltammetric measurement. In accordance with this Blaedel and Jenkins (101) found platinum to be less reliable for the quantitation of NADH compared to glassy carbon. Adsorption of proteins can both generate "false" currents, showing up as charge-transfer peaks during cyclic voltammetry (102), and also charge electron-transfer efficiency as shown by Ossendorfova et al. (103), who found a damping effect on voltammetric peaks in serum.

In the determination of catecholamines at platinum a sequence of electrooxidation and side group cyclization leads to surface polymer formation, typically with melaninlike reaction products. Lane and Hubbard (104) were able to reduce such film deposition by surface modification using iodide ion. Further advances in electrode modification are anticipated toward the manipulation of surface reactivity and stability (105).

A major evolving area of in vivo voltammetry has been that relating to brain neurotransmitter research. Early studies (106) used graphite paste electrodes solidified with epoxy resin, reducing tip diameters to 50 μm. These allowed in situ detection of a range of catecholamines, but later emphasis has been on the use of carbon fiber electrodes (107) down to 10 μm. These techniques can combine with electrical or pharmacological stimulation methods (108) to give information about basic biology and transmitter diffusion on a real-time basis but are restricted to being probes of extracellular events.

Electrochemical techniques have become more specialized and adapted to monitoring, with reduced perturbation of local neurotransmitter concentrations along with enhanced electrode surface stability (109). Additionally, measurement of brain ascorbate levels has provided useful biological information (110), but in large part the high (millimolar) concentrations of ascorbate have tended to obscure neurotransmitter responses at electrodes (106). Coating of carbon fiber electrodes with the perfluorinated hydrocarbon Nafion, bearing fixed anionic charges, has enabled rejection of the ascorbate anion (111) and permeation of neutral and cationic species (e.g., dopamine). Further selective membrane refinements are needed to achieve reliable speciation, and smaller probes will be needed if the synoptic junction itself is to be studied (112). However, Nafion continues to draw attention, and with optimized deposition techniques membranes of the order of 0.18 μm thickness are being produced that have allowed rapid-scan voltammetry (113). In this way Wightman's group were able to follow dopamine release and uptake over millisecond time scales (114). Multiple monitoring and in situ signal processing may become feasible with the use of integrated circuit (IC) devices (115).

Opportunities for voltammetric intravascular monitoring, at best on an experimental basis, exist but have not yet been extensively examined. Endogenous electrochemical interference is likely, and dynamic trend monitoring of infused drugs should give some information on pharmacodynamics. Acetaminophen, for example, can be readily detected in biological fluids and has been used for in situ calibration of microelectrodes in brain cyclic voltammetry (116). Wang et al. (117) devised a long Teflon catheter with a carbon paste tip to monitor injected acetaminophen in an experimental animal. With differential pulse voltammetry it was possible to avoid blood interference in the anodic range. Nevertheless, standard/blood differences are observed when such direct monitoring of whole blood is carried out (118). The surface oxidation state on carbon, as with platinum, alters electrochemical reactivity, and thus defined anodic pretreatment may help to bring about stabilization for in vivo use (119). With a carbon rod electrode inserted into the aortic arch, Meulemans (120) monitored venous infusions of the antibiotics metronidazole, cefsulodin, chloramphenicol, and nimorazole by differential pulse voltammetry. This study demonstrates the wider applicability of this approach, but with in vitro electrode life spans of several months now reduced in vivo to 3–4 hr, the need to attend to additional interfacial interactions occurring in the biological matrix are paramount.

GLUCOSE ELECTRODES

Two main factors have determined the considerable efforts now being made toward the in vivo monitoring of blood glucose. First, diabetes is a common condition that, despite the most careful and clinically judicious use of insulin, is associated with metabolic instabilities that may cause long-term large vessel and microvascular complications. These instabilities are most readily apparent with regard to glucose, the lability and unpredictability of which not only produce untoward symptoms but, in the extreme, can be life threatening. Clearly, measurement of glucose on a real-time basis would permit better stabilization of its blood levels with some amelioration of the underlying metabolic instability. Second, in the form of an enzyme electrode a glucose sensor is perhaps one of the easiest and most robust of biosensors to be fabricated, due in no small part to the high stability of the glucose oxidase enzyme incorporated into most such systems. Furthermore, the enzyme-catalyzed reaction

$$\text{Glucose} + O_2 + H_2O \longrightarrow \text{gluconic acid} + H_2O_2 \quad (1)$$

can be followed using sensors for O_2, H_2O_2, or H^+, as well as mediator reagents capable of directly interacting with the redox center of the oxidase enzyme. A nonexhaustive survey of the main enzymic and also some nonenzymic electrodes is provided here.

Enzyme Electrodes

O_2-Base Electrodes

Electrodes in which O_2 consumption within the enzyme layer provides the means for following the catalytic degradation of glucose were the focus of early studies. No doubt this was because of the then ready general accessibility of the Clark oxygen electrode. Enshrined within the first description of a functional glucose electrode (121) was at least a partial solution to two fundamental problems with such devices: (1) a dual amperometric input was used by which a second "inactive" enzyme electrode corrected for the background variation in pO_2 in blood; and (2) a small surface area enzyme layer was used by which glucose mass transport approximated to spherical diffusion, thus reducing bulk solution convection (stirring) effects. The dual-electrode principle was then utilized by Layne et al. (122), who devised a galvanic cell for implantation. The electrode output signal was nonlinear, slow (~10 min response), and vulnerable to blood hematocrit and viscosity changes and warranted further optimization. However, the high selectivity conferred by the gas-permeable membranes of these electrodes remains an attractive feature and has more recently stimulated a new geometrical design (123) that enables electrodes to operate under the low oxygen conditions often found in venous blood and tissue. This electrode (Fig. 8) has an axial surface over which both glucose and oxygen can diffuse and a cylindrical surface through which only O_2 can diffuse to the enzyme layer. By thus enhancing the relative diffusion of O_2 to glucose, the glucose operational range for low pO_2 was extended; electrode size affected diffusion distances and therefore response times (124). In an adaptation for long-term in vivo use a three-electrode combination of working, reference, and counterelectrode was proposed for O_2 detection (125). A fully implanted telemetric device for glucose has now been reported (126). Glucose monitoring over long periods requires high enzymic loading to ensure a sufficient reserve of enzyme, but enzyme stability may

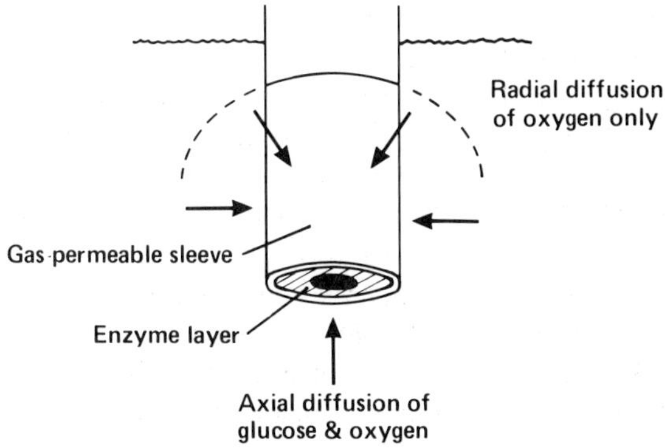

Figure 8 Two-dimensional glucose enzyme electrode with a high oxygen to glucose surface permeability ratio achieved using an O_2-permeable sleeve. [After Gough et al. (123).]

also be enhanced by incorporating catalase to decompose the H_2O_2 product, as this can cause cumulative damage to the enzyme (123).

H_2O_2-Based Electrodes

Detection of the H_2O_2 product of the glucose oxidase reaction at a positively polarized working electrode enables measurement with a single device without O_2 correction. This has been the mode of operation chosen by most workers, despite the possible damaging effect of H_2O_2 on the enzyme.

Considerable practical experience of blood monitoring has been gained over the last two decades through the use of extracorporeal sensors, such as that forming part of a commercial closed-loop system for glucose control, the Biostator (127). The outer membrane typically is a microporous polycarbonate or a cellulosic protein-reflecting semipermeable layer. These membranes have a critical effect on electrode survival in blood since they can ameliorate some of the surface-fouling problems observed with even anticoagulated blood (128). Also of importance is a further selective barrier membrane between the enzyme and the working electrode. This substitutes for the O_2-permeable membrane of a Clark electrode; an equivalent degree of selectivity for the H_2O_2 product is not possible, but thin, dense cellulose acetate membranes of low-molecular-weight cutoff have been adequate to screen out a major proportion of electrochemical interferents in biological fluids (that is, ascorbate, urate, tyrosine, and thiols).

The basic flat-form device for the Biostator resides in a linear flow channel, where it is presented with 1:10 diluted blood at a constant flow rate with facilities for recalibration. A similar sensor (129) forms part of a more recent artificial biofeedback system (Betalike, Esaote Biomedica, Genova) in which exposure is to an ultrafiltrate of blood, reducing direct electrode fouling problems due to blood cells and macromolecules. The precise algorithms to use in such feedback systems remain to be worked out, although the basic biphasic insulin dosage achieved approximates the physiological situation (130). The lag in glucose electrode response, however, conditions the efficiency of the feedback loops and their ability to control glucose without generating high blood insulin levels (131).

A step toward implantation was the use of a perforated gas-permeable outer membrane by Fischer and Abel (132). This reduced O_2 depletion in the enzyme layer, thus extending the electrode linear range and allowing blood measurement without sample dilution. In another approach microporous membranes were treated with organosilane reagent to increase the relative amounts of oxygen reaching the enzyme layer (133). Clark et al. (134) proposed immobilization of the enzyme layer over a silastic drum to furnish an additional path for the diffusion of oxygen.

Practical monitoring with a miniature device was carried out by Shichiri's group (135) in dogs using needle electrodes. Here the enzyme was immobilized in cellulose acetate and a dip-coated low-permeability polyurethane layer reduced glucose diffusion relative to that of O_2. An important consequence of any diffusion-limiting membrane is the relative independence it gives from external solution variables, such as pH, viscosity, and stirring. The basic design of the needle electrode also permitted a rapid response (t_{90} = 16.2 ± 6.2 sec), with low pO_2 dependence of the glucose signal and sufficient in vitro stability (24.8% signal loss in 7 days) to allow short-term in vivo monitoring (136). In practice this electrode system gave a stable glucose output in tissue (pO_2 range 27–72 mmHg), and signal size at 7 days was reduced to 73.5%. A modified form of the needle electrode utilizing three discrete layers is shown in Figure 9; experience with polyarylsulfone indicates that it is a more robust and selective substitute for cellulose acetate (137). Ege (138) refined the membrane deposition procedure so that an aqueous enzyme-polyurethane dispersion is used to coat the needle.

A requirement for covering membranes of greater biocompatibility remains. Shichiri et al. (139) employed a hydrophilic covering layer of polyvinyl alcohol but recently reported that a laminate comprising alginate-polylysine-alginate preserves electrode responsiveness for 14 days (140).

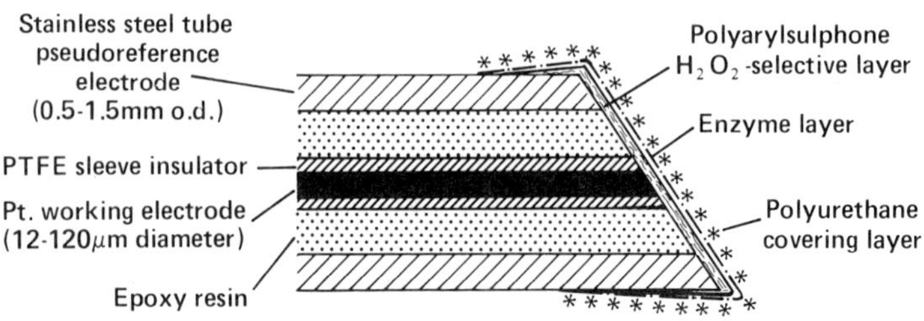

Figure 9 Amperometric needle enzyme electrode for oxidase enzyme substrates. The outer polyurethane layer constitutes a low-permeability diffusion barrier for the substrate.

An important part of electrode construction is the material used for the pseudoreference electrode, normally the needle shaft. Some have relied upon stainless steel (137,138), which, although not appropriate for use in potentiometric devices, appears to have a voltage drift well within the stable plateau region for amperometric H_2O_2 measurement. Silver is also used (135), but Velho et al. (141) found that the potentials of silver (and gold) altered according to the amount of current generated during H_2O_2 detection and only Ag/AgCl provided an invariate potential. Planar integrated-chip glucose enzyme electrodes have also used Ag/AgCl as the reference but employed a platinum counter as well as a working electrode (142,143); the relative merits of two- and three-electrode devices warrants in vivo comparisons.

The greater perceived safety in subcutaneous compared to intravascular implantation of electrodes prompted most research efforts to be targeted toward this latter goal. A fundamental problem arises, however, as to how well tissue glucose levels reflect blood glucose. Indications from implanted chambers used to sample glucose from several extracellular compartments are that tissue levels are lower than those in blood (144). In an alternative sampling technique Abel et al. (145) implanted cotton wicks into subcutaneous tissue; measurement of glucose in the wick specimens indicated close correspondence between blood and tissue glucose (Fig. 10). There was, however, a lag of ~10 min for tissue when blood glucose was rising. A further complicating factor affecting all

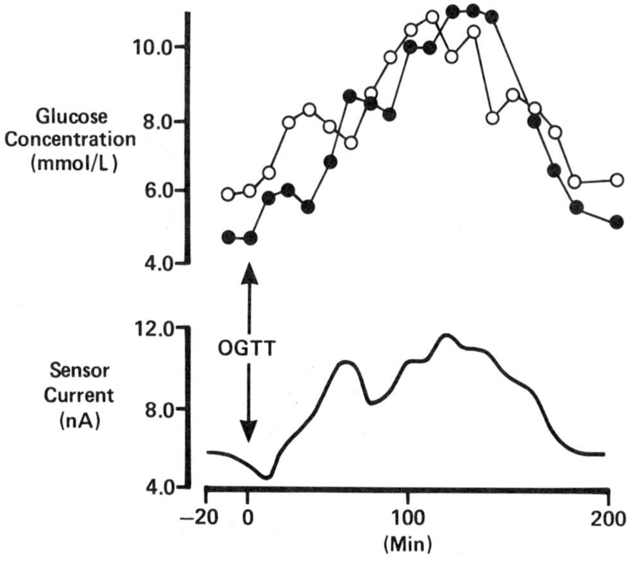

Figure 10 Comparison of venous plasma glucose (○) with wick samples (●) together with a continuous readout from a subcutaneously implanted needle electrode. [After Abel et al. (145).]

sensors is that electrode sensitivity is reduced on implantation; this is due partly to a restricted glucose access to the electrode site and partly to membrane surface fouling (145). The wick study, however, supports the notion that an in vivo calibration of tissue glucose electrodes is valid provided steady-state conditions are achieved. The ideal of a two-point in vivo calibration (146) is attractive but is likely to be difficult to achieve in practice.

Peritoneally implanted electrodes may prove acceptable for in vivo monitoring in the future, but recent glucose clamp experiments have shown long equilibration times with respect to blood (147); thus in vivo calibration here may not be feasible. Certainly for automated closed-loop control the lag period in the subcutaneous tissue compartment is sufficiently short to permit reliable management of the diabetic state (148). Given the small amount of subcutaneous tissue at some implantation sites (e.g., forearm) (149) it may be necessary to modify electrode dimensions or to attempt implantation to a standard depth.

Mediator-Based Electrodes

Exploitation of the electrochemically recycled electron mediator ferrocene for the glucose oxidase reaction furnishes electrodes able to operate without oxygen and with a linear range >20 mmol/L of glucose (150). Construction in a needle form has been feasible using a rigid support to maintain the structural integrity of the exquisite carbon working electrode. With this approach it was possible to monitor subcutaneous glucose in the pig without a lag period observed with respect to blood changes; absolute levels were ~20% of those in blood (151). In a more recent study in humans similar electrodes were calibrated in vivo; during oral glucose tolerance tests substantial (40 min) delays in peak glucose values were seen (152). Polyurethane was used to coat such devices; thus tissue biocompatibility may ultimately be similar to that of H_2O_2-based electrodes. Intravascular catheter electrodes have also been reported; leaching of the mediator from the electrode does not appear to be a problem during in vivo use (153).

Other Glucose-Monitoring Devices

Electrocatalytic Sensors

An implantable glucose sensor, free of the need for enzyme, escapes many of the functional instabilities associated with the use of metastable reagents. Since glucose can be electrochemically oxidized at platinum surfaces, the principle has therefore been exploited for direct glucose assay. In one refinement, the potential jump method (154), the platinum surface is first electrochemically oxidized at a high voltage, the potential then switched to a low value to discharge electrode capacitance, and then finally raised to the working potential (400 mV) for glucose detection. Adjustments to the potential sequence and appropriate choice of covering membranes reduced background electrochemical interference, especially from urea and amino acids. An alternative approach of measuring electrode impedance was proposed (155). Giner's group alternatively refined the signal-processing aspects of cyclic voltammetry for glucose (156). In their "compensated net charge" technique the differences between the charge passed during the cathodic and anodic phases of a potentiometric cycle are computerized. This double layer subtracts out electrode surface oxidation and double-layer charging, leaving a residual difference due to glucose oxidation. Recognition of the glucose component of the cycle confers selectivity, and this approach enabled monitoring of glucose in blood as well as in tissue (157). Whether the electrode and associated data management system can be made sufficiently simple for clinical purposes requires evaluation.

Coated wire electrodes, like ion-selective electrodes, operate on the basis of an ion-dependent interfacial potential. In the absence of an internal electrolyte solution for such electrodes, the mechanism of surface potential registration by the electrode wire remains uncertain. However, Wilkins and Wilkins (158) designed a novel coated wire electrode for glucose that incorporated a barium-glucose adduct in PVC. Changes in ambient glucose apparently generated EMF variations by altering the equilibrium state between the adduct and free Ba^{2+} in the membrane. The application of the coated wire glucose electrode to in vivo monitoring has not been fully assessed at present.

Optical Glucose Sensors

The fiberoptic sensor of Schultz et al. (159) is based on the selective sugar binding affinity of concanavalin A (ConA). This lectin is immobilized on the inner wall of a hollow fiber chamber and binds FITC-labeled dextran (Fig. 11); diffusion of glucose into the chamber displaces some of the dextran, changing the total amount of free labeled dextran present in the chamber. An optical fiber

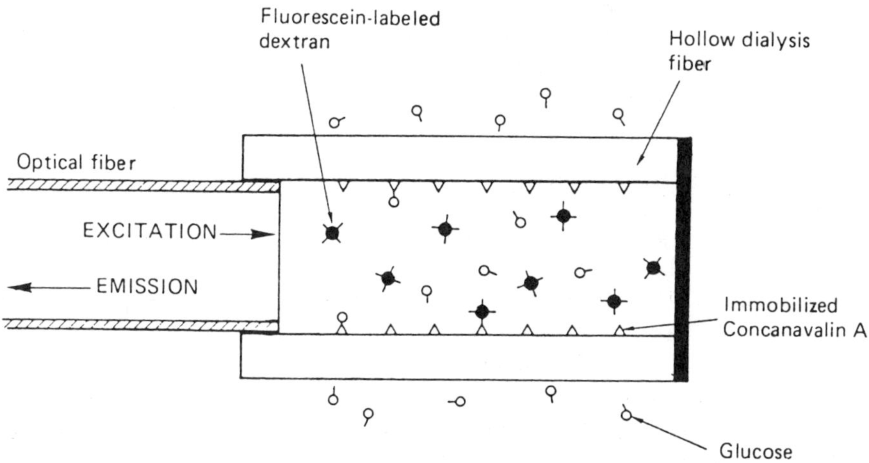

Figure 11 Glucose affinity sensor; a fiberoptic sensor based on selective glucose binding of concanavalin A. [After Schultz et al. (159).]

attached to the end of the hollow fiber is able to detect the change in fluorescence in the axial region of the chamber. The behavior of the device is therefore a function of two competing equilibria:

$$\text{FITC-dextran} + \text{ConA} \longrightarrow \text{FITC-dextran–ConA}$$
$$\text{glucose} + \text{ConA} \longrightarrow \text{glucose-ConA} \qquad (2)$$

An important characteristic of the sensor is its attainment of a true glucose-dependent equilibrium. Therefore, although membrane fouling can retard the speed of response, it should not perturb the eventual steady-state response. The sensor was successfully used for in vivo monitoring in experimental animals. Meadows and Schultz (160) have now described a modification in which ConA is linked to a fluorescence quencher (rhodamine) and is present in free solution in the hollow fiber. In this approach binding of FITC-dextran to ConA leads to a direct loss of fluorescence, avoiding the need to immobilize ConA on the walls of the fiber. With both these fluorescence devices chamber geometry has an important bearing on sensitivity and dynamic response, with slow diffusion through the chamber dominating the speed of response.

Glucose-Responsive Membrane

Polymeric supports are drawing increasing attention as vehicles for in vivo drug delivery, and our knowledge about solute mass transfer through polymer matrices has substantially increased in recent years. The possibility therefore arises of an "intelligent" controlled insulin release system based on the release of insulin from a polymer matrix. Kost et al. (161) studied a range of hydrophilic gels with free amino groups loaded with glucose oxidase. The water content of these gels (60–90%) increased in the presence of glucose since the gluconic acid generated by the glucose oxidase reaction protonated the amino groups, giving an expanded and more permeable polymer network. It is therefore possible that membranes with a glucose-dependent permeability to insulin could thus be engineered and provide a variable in vivo insulin delivery system.

LACTATE ELECTRODES

The end point of anaerobic respiration is lactate, and both hypoxia and cellular derangements, which compromise the cell's ability to utilize oxygen, lead to a lactic acidemia. Thus far, studies of lactate sensors have been mainly restricted to in vitro devices, but work on in vivo electrodes is likely to increase. We reported a needle lactate electrode in which prefabricated enzyme membrane laminates are monitored over the needle tip. These devices have a linear range >10 mmol/L of lactate and operate reliably in undiluted blood in vitro (162). They may prove to be usable in tissue. An intravascular lactate catheter was described by Lim (163). Mascini and coworkers acquired experience of dynamic lactate as well as pyruvate changes in blood using extracorporeal oxidase-based electrodes (164).

BIOCOMPATIBILITY

One final common pathway to in vivo signal deterioration in nearly all biosensors is fouling of the sensor surface. The processes involved are highly complex and have often been neglected in biosensor research. Initial contact between any surface and a protein-containing solution leads to coating of that surface within seconds. In itself this deposited layer has little effect on the signal, but it is important in conditioning the subsequent deposition cascade at that surface. The situation is made more complicated because the protein layer is continually remodeling, surface-bound proteins denature, and there is a continuous, ongoing exchange of protein with the bulk solution. Cuypers et al. (165) showed that from a simple solution fibrinogen deposits as a 50 Å layer within 5 min, irrespective of whether the surface is hydrophilic or hydrophobic. In anticoagulated blood a complex mixture of proteins, including clotting factors, deposits on any surface. The coagulation pathway may then be initiated, coinciding with the arrival of platelets, leukocytes, and red cells in large numbers. A decay in biosensor response is even then only evident when the covering "biolayer" is either thick enough or dense enough to perturb mass transfer. There is undoubtedly a case for much more detailed appraisal and specific engineering of polymeric surfaces used over biosensors. Important material properties that need to be investigated are summarized in Table 2. This list is similar to any that might be used for biomaterials generally, but the polymeric biosensor surface clearly demands especially detailed structural consideration.

With regard to intravascular use, even if the surface makeup is chemically ideally blood compatible fouling is inevitable over a prolonged period. An intravascular device perturbs the normal near laminar flow pattern of blood, and local turbulence remains a potent trigger for enhanced platelet activity (166). Adherence of platelets to the sensor surface eventually leads to thromboembolism with its attendant hazards.

Table 2 Factors Determining Polymer Surface-Biofluid Interactions

Topography	Molecular scale
	Cellular scale
Charge	Density
	Distribution
	Uniformity
Hydrophobic groups	Density
	Chain length
	Mobility
	Uniformity
Polar groups	Density
	Hydration
Crystallinity	Ratio of crystalline and amorphous domains

Tissue poses a rather different interfacing problem to blood. Again, whatever the nominal biocompatibility of an implant, tissue recognizes it as foreign, and a series of reactions is set up designed to degrade and ultimately reject the implant. In large part the processes are akin to wound healing, with an acute inflammatory response giving way to the growth of local vascular tissue, healing tissue, and finally collagen scar formation. The intensity of this response is partly governed by sensor geometry (167); a spherical surface, for example, provokes less response than one that has angular features, and implant size is also an important consideration. Histological techniques are a mainstay of tissue biocompatibility studies, but it should be recognized that a range of hydrolases and other enzymes is involved in the tissue response (168); quantitation of these warrants histochemical evaluation. One specific study was carried out on intraperitoneally implanted glucose enzyme electrodes by Clark et al. (169). These workers found that peritoneal macrophages covered the sensor within a few hours, followed by fibroblasts, leading to a collagen capsule that was glucose permeable. Encroachment of catalase activity into the enzyme layer caused response failure in some cases, but electrodes could also function for over a month. Interestingly, although cellulosic dialysis membranes were associated with an avascular capsule, microporous fluorocarbon membranes stimulated capillary formation and, therefore, an active local circulation. Ertefai and Gough (170) reported a chamber that allows systematic histological evaluation of tissue processes close to an implanted sensor. They were able to quantitate local blood capillary densities and found density to be maximal 20–100 μm from the surface of a silicone rubber implant. Such investigations will increasingly help to interrelate sensor performance and tissue processes, stimulating the development of biocompatible devices.

Potential toxicity problems exist with most implanted materials. Although gold, platinum, and stainless steel are relatively nontoxic metals, for example, intense local toxic problems have been demonstrated with chronic Ag and Ag/AgCl implants (171). In acute experiments Ag/AgCl electrodes also provoked loss of vascular smooth muscle contractility (172). Other electrode components that can have local effects are leached plasticizer, polymer degradation products, reaction products of the biosensor response itself, immunogenic material, and electron mediators. In addition, a systematic strategy has yet to be devised for the sterilization of biosensors containing labile components.

Overall, step-by-step progression toward better biocompatibility is in prospect, rather than any radical solution to interfacing problems emanating from either one single material or any single transducer reaction chemistry.

CONCLUSIONS

The development of miniaturized potentiometric and voltammetric probes aimed at in vivo monitoring continues to be an area of intense research activity. Despite significant progress the in vivo monitoring of important clinical variables is still in its infancy. The technology of biosensor construction has reached an advanced state of development, and the emphasis should now be on exploration of interaction between the biosensor and its immediate external environment or biological matrix.

Development of a sensor for invasive monitoring constitutes a substantially more difficult problem than one for in vitro or extracorporeal use. Not only does the device have to be reliable and function for extended periods, but there must also be no mechanical, toxic, or antigenic effects: hence the emphasis on investigation of more biocompatible materials. Over the past two decades, however, blood gas monitoring, for instance, has progressed from relatively crude bioincompatible and unreliable intravascular electrodes to sophisticated, noninvasive devices.

Although at present there are no commercially available in vivo glucose sensors, recent developments in this area show some promise. For long-term implantation, however, almost drift-free operation is necessary if calibration cannot be guaranteed and a potentially lethal therapeutic agent (such as insulin in a glucose sensor) is to be administered.

Future efforts on in vivo electrochemistry will undoubtedly focus on the development of selective probes for organic drugs and additional analytes of clinical significance, the coupling of enzymatic and immunochemical reactions with in vivo electrodes, the design of new noninvasive and multispecies probes, and the search for means of protecting in vivo electrodes against biological matrix effects.

REFERENCES

1. Solberg, H. E., and Grasbeck, R. (1989). Reference values, *Adv. Clin. Chem., 27*: 1–79.
2. Hunter, K. W. (1989). Biosensors, *Am. J. Clin. Pathol., 91*: S32–S33.
3. Kreuzer, F., and Nessler, C. G. (1958). Method of polarographic in vivo continuous recording of blood oxygen tension, *Science, 128*: 1005–1006.
4. Mindt, W. (1974). Electrochemical methods in diagnosis, *Diagnostik, 7*: 367.
5. Huch, A., Huch, R., Neumayer, E., and Rooth, G. (1972). Continuous intra-arterial pO_2 measurements in infants, *Acta Paediatr. Scand., 61*: 722.
6. Armstrong, R. F., Hutchinson, J. M., Lincoln, C., Ingram, D., and Soutter, L. (1976). Continuous measurement of arterial oxygen tension during one-lung anesthesia, *Br. J. Anaesthesiol., 48*: 1005.
7. Rolfe, P. (1976). Arterial oxygen measurement in the newborn with intravascular transducers. In: *IEE Medical Electronics Monographs*, Nos. 18–22 (D. W. Hill and B. W. Watson, Eds), Perigrinus, London, p. 126.
8. Goddard, P. J., Keith, I., Marcovitch, H., Rolfe, P. J., and Scopes, J. W. (1972). Experience with a catheter-tip transducer for continuous measurement of blood oxygen tension including evaluation in newborn babies, *Arch. Dis. Child, 47*: 675.
9. Kimmich, H. P., and Kreuzer, F. (1969). Catheter pO_2 electrode with low flow dependency and fast response, *Prog. Respir. Res. 3*: 100.

10. Soutter, L. P., Conway, M., and Packer, D. (1975). A system for monitoring arterial oxygen tension in sick newborn babies, *Biomed. Eng., 10*: 257.
11. Conway, M., Durbin, G. M., Ingram, B. A., McIntosch, N., Parker, D., Reynolds, E. O. R., and Soutter, L. P. (1976). Continuous monitoring of arterial oxygen tension using a catheter-tip polarographic electrode in infants, *Pediatrics, 57*: 244.
12. Karube, I., Tarniya, E., Sode, K., Yokoyama, K., Suzuki, H., and Hattori, S. (1988). Integrated microbiosensors for medical use, *Ann. NY Acad. Sci., 542*: 470.
13. Parker, D., Keys, Al, Davies, R., Scopes, J. W., and Markovitch, A. (1971). A disposable catheter-tip transducer for continuous measurement of blood oxygen tension in vivo *Biomed. Eng., 6*: 313.
14. Hahn, C. E. W. (1981). Techniques for measuring the partial pressures of gases in the blood, Part II. In vivo measurements, *J. Phys. E: Sci. Instrumen., 14*: 783.
15. Harris, T. R., and Nugent, M. (1973). Continuous arterial oxygen tension monitoring in the newborn infant, *J. Pediatr., 82*: 929.
16. Strauss, A. W., Escobedo, M., and Goldring, D. (1974). Continuous monitoring of arterial oxygen tension in the newborn infant, *J. Pediatr., 85*: 254.
17. Nilsson, E., Edwall, G., Larsson, R., and Olsson, P. (1981). Polarographic pO_2 sensors with heparinised membranes for in vitro and continuous in vivo registration, *Scand. J. Clin. Lab. Invest., 41*: 557.
18. Oeseburg, B., Zijlstra, W. G., and Ing, T. G. (1968). Uncovered platinum electrodes for the measurement of ΔpO_2 as an aid in diagnostic cardiac catheterization, *Cardiovasc. Res., 2*: 394.
19. Silver, I. A. (1965). Some observations on the cerebral cortex with an ultramicro, membrane covered oxygen electrode, *Med. Electron. Biol. Eng., 3*: 377.
20. Aarnoudes, J. G., Huisjes, H. J., Oeseburg, B., and Zijlstra, W. G. (1981). Subcutaneous oxygen tension in the fetal scalp during labour. Continuous monitoring with a needle electrode, *Br. J. Obstet. Gynaecol., 88*: 517.
21. Schneiderman, G., and Goldstick, T. K. (1978). Oxygen electrode design criteria and performance characteristics: Recessed cathode, *J. Appl. Physiol., 45*: 145.
22. Linsenmeier, R. A., and Yancey, C. M. (1987). Improved fabrication of double-barrelled recessed cathode O_2 microelectrodes, *J. Appl. Physiol., 63*: 2554.
23. van der Kleij, J. A., and de Koning, J. (1981). Tissue oxygen electrode for routine clinical application. In: *Monitoring of Vital Parameters During Extracorporeal Circulation* (H. P. Limmich, Ed), Karger, Basel, p. 95.
24. Silver, I. A. (1987). Microelectrodes in medicine. In: *Biosensors* (M. Akhtar, C. R. Loew, and I. J. Higgins, Eds), Royal Society, London, p. 161.
25. Albanese, R. A. (1971). Use of membrane-covered oxygen cathodes in tissue, *J. Theor. Biol., 33*: 91.
26. Silver, I. A. (1967). Polarography and its biological applications, *Phys. Med. Biol., 12*: 285.
27. Le Souef, P. N., Morgan, A. K., Soutter, L. P., Reynolds, E. O. R., and Parker, D. (1978). Continuous comparison of transuctaneous and arterial oxygen tension in newborn infants with respiratory illness, *Acta. Anaesthesiol. Scand. (Suppl.), 68*: 91.
28. Huch, R., Lübbers, D. W., and Huch, A. (1972). Quantitative continuous measurement of partial oxygen pressure on the skin of adults and newborn babies, *Pflugers Arch., 337*: 185.

29. Eberhard, P., Hammacher, K., and Mindt, W. (1973). Methode zur ratanen messurg des sauerstoffpartialdruckes, *Biomed. Tech.* (Berlin), *12*: 212.
30. Long, J. G., Philip, A. G. S., and Lucey, J. F. (1980). Excessive handling as a cause of hypoxaemia, *Pediatrics, 65*: 203.
31. Shaw, T. J. I., and Bennett, N. R. (1980). Towards safer anaesthesia in sickle cell states, *Anaesthesia, 35*: 394.
32. Buntain, W. L., Conner, E., Emrico, J., and Cassaday, G. (1979). Transcutaneous (tc pO_2) measurements as an aid to fluid therapy in necrotizing enterocolitis, *J. Pediatr. Surg., 14*: 728.
33. Tremper, K. K., Waxman, K., and Shoemaker, W. C. (1981). Use of transcutaneous oxygen sensors to titrate PEEP, *Ann. Surg., 193*: 206.
34. Kram, H. B., and Shoemaker, W. C. (1989). Transcutaneous, conjunctival and organ pO_2 and pCO_2 monitoring in the adult. In: *Textbook of Critical Care Medicine* (W. C. Shoemaker, S. Ayres, A. Grenvik, P. R. Holbrook, and W. L. Thompson, Eds), W. B. Saunders, Philadelphia, p. 283.
35. Kram, H. B. (1985). Non-invasive tissue oxygen monitoring in surgical and critical care medicine, *Surg. Clin. North Am., 65*: 1005.
36. Shoemaker, W. C., and Lawner, P. M. (1983). Method for continuous conjunctival oxygen monitoring during carotid artery surgery, *Crit. Care Med., 11*: 946.
37. Hahn, C. E. W., Brooks, W. N., Albery, W. J., and Rolfe, P. (1979). O_2 and NO_2 analysis with a single intravascular catheter electrode. An in vitro study, *Anaesthesia, 34*: 263.
38. Bates, M. L., Feigold, A., and Gold, M. I. (1975). The effects of anaesthetics on an in vivo oxygen electrode, *Am. J. Clin. Pathol., 64*: 448.
39. Parker, D., Delpy, D., and Lewis, M. (1978). Catheter-tip electrode for continuous measurement of pO_2 and pCO_2, *Med. Biol. Eng. Comput., 16*: 599.
40. Coon, R. L., Lai, N. C. J., and Kampine, J. P. (1976). Evolution of a dual-function pH and pCO_2 in vivo sensor, *J. Appl. Physiol., 40*: 625.
41. Neumark, J., Bardeen, A., Sulzer, E., and Kampine, J. P. (1975). Miniature intravascular pCO_2 sensors in neurosurgery, *J. Neurosurg., 43*: 172.
42. Olievier, C. N., Berkenbosch, A., and Quanjer, P. L. H. (1978). In vivo measurement of carbon dioxide tension with a miniature electrode, *Pfluegers Arch., 373*: 269.
43. Opdycke, W. N., and Meyerhoff, M. E. (1986). Development and performance of tubular polymer membrane electrode based carbon dioxide catheters, *Anal. Chem., 58*: 950.
44. Huch, A., Huch, R., Seiler, D., Galster, H., Meinzer, K., and Lübbers, D. W. (1977). Transcutaneous pCO_2 measurement with a miniaturized electrode, *Lancet, 1*: 982.
45. Hansen, T. N., and Tooley, W. H. (1979). Skin surface carbon dioxide tension in sick infants, *Pediatrics, 64*: 942.
46. Cabal, L., Hodgman, J., and Siassi, B. (1981). Factors affecting heated transcutaneous pO_2 and unheated transcutaneous pCO_2 in pre term infants, *Crit. Care Med., 9*: 298.
47. Tremper, K. K., Mentelos, R. A., and Shoemaker, W. C. (1981). Clinical and experimental transcutaneous pCO_2 monitoring, *J. Clin. Eng., 6*: 143.
48. Bhat, R., Diaz-Blanco, J., Chaudry, U., and Vidyasagar, D. (1985). Trans-

cutaneous oxygen and carbon dioxide monitoring in sick neonates using a combined sensor, *Chest, 88*, 890.
49. Staehlin, H. B., Carlsen, E. N., Hinshaw, D. B., and Smith, L. L. (1968). Continuous blood pH monitoring using an indwelling catheter, *Am. J. Surg., 116*: 280.
50. Cobbe, S. M., and Poole-Wilson, P. A. (1980). Catheter tip pH electrodes for continuous intravascular recording, *J. Med. Eng. Technol., 4*: 122.
51. Cobbe, S. M., and Poole-Wilson, P. A. (1982). Continuous coronary sinus and arterial monitoring during pacing-induced ischaemia in coronary artery disease, *Br. Heart J., 47*: 369.
52. Simon, W., Ammann, D., Anker, P., Oesch, U., and Bond, D. M. (1984). Ion-selective electrodes and their clinical application in the continuous ion monitoring, *Ann. NY Acad. Sci., 428*: 279.
53. Gaub, W. T., and King, L. H. (1980). Palladium-palladium oxide pH electrodes, *Anal. Chem., 52*: 270.
54. Cambilli, L., Alcidi, L., Papeschi, G., Wiechmann, V., Padeletti, L., and Grassi, G. (1978). Preliminary experience with the pH triggered pacemaker, *Pace, 1*: 448.
55. Editorial (1988). Pacemakers to make the heart beat faster with exercise, *Lancet, 2*: 23.
56. Bergveld, P. (1985). Research and development of biosensors for chemical determination in the human body, *Tijdschr. Ned. Ver Klin. Chem., 10*: 153.
57. Bergveld, P. (1986). The development and application of FET-based biosensors, *Biosensor, 2*: 15.
58. Kohama, A., Nakamara, Y., Nakamara, M. (1984). Continuous monitoring of arterial and tissue pCO_2 with sensors based on the pH-ISFET, *Crit. Care Med., 12*: 940.
59. Schepel, S. J., de Rooij, N. F., Koning, G., Oeseburg, B., and Zijlstra, W. G. (1984). In vivo experiments with a pH ISFET electrode, *Med. Biol. Eng. Comput., 22*: 6.
60. Eberhart, R. C. (1985). Indwelling blood compatible chemical sensors, *Surg. Clin. North Am. 65*: 1025.
61. Sibbald, A., Covington, A. K., and Carter, D. F. (1985). On line patient monitoring system for the simultaneous analysis of blood K^+, Ca^{2+}, Na^+ and pH using a quadruple function Chem FET integrated circuit sensor, *Med. Biol. Eng. Comput., 23*: 329.
62. Margules, G. S., Hunter, C. M., and MacGregor, D. C. (1983). Hydrogel based in vivo reference electrode catheter, *Med. Biol. Eng. Comput., 21*: 1.
63. Mackenzie, J. W., Salkind, A. J., and Topaz, S. R. (1974). Intramuscular pH needle electrode, a preliminary report, *J. Surg. Res., 16*: 632.
64. Stamm, O., Lat Scha, U., Janecek, P., and Campana, A. (1976). Development of a special electrode for continuous subcutaneous pH measurement in the infant scalp, *Annu. J. Obstet. Gyecol., 124*: 193.
65. Mindt, W., Maurer, H., and Mueller, W. (1978). Principle and characteristics of the ROCHE tissue pH electrode, *Arch. Gynaecol., 226*: 9.
66. Weber, T., and Nickelsen, C. (1988). Tissue pH/pCO_2 and continuous base excess monitoring in the human fetus, *J. Perinat. Med., 16* (Suppl.): 151.

67. Young, B. K. (1981). Continuous fetal tissue monitoring in labor, *J. Perinat. Med.*, *9*: 189.
68. Walters, F. J. M., Wilson, G. J., Steward, D. J., Domenech, R. J., and MacGregor, D. C. (1979). Intramyocardial pH as an index of myocardial metabolism during cardiovascular surgery, *J. Thorac. Cardiovasc. Surg.*, *78*: 319.
69. Lange, R., Kloner, R. A., Zierler, B. A., Carlson, N., Seiler, M. D., and Khuri, S. F. (1983). Time course of ischaemic alterations during normothermic and hypothermic arrest and its reflection by on-line monitoring of tissue pH, *J. Thorac. Cardiovasc. Surg.*, *86*: 418.
70. Khuri, S. F., Josa, M., Marston, W., Braunwald, N. S., Smith, B., Tow, D. T., Van Cisim, M., and Barasamian, E. M. (1983). First report of intramyocardial pH in man, *J. Thorac. Cardiovasc. Surg.*, *86*: 667.
71. Gehrich, J. L., Lübbers, D. W., Opitz, N., Hansmann, D. R., Miller, W. W., Tusa, J. K., and Yafuso, M. (1986). Optical fluorescence and its application to an intravascular blood gas monitoring system, *IEEE-BME, 33*: 117.
72. Shapiro, B., Cane, R. D., Chamka, C. M., Bandala, L. E., and Peruzzi, W. T. (1989). Preliminary evaluation of an intra-arterial blood gas system in dogs and humans, *Crit. Care Med.*, *17*: 455.
73. Opitz, N., and Lübbers, D. W. (1987). Theory and development of fluorescent-based optochemical oxygen sensors: Oxygen optodes, *Int. Anesthesiol. Clin.*, *25:3* 177.
74. Baker, S. J., Tremper, K. K., and Heitzman, H. A. (1987). Continuous fiber-optic oxygen tension in dogs, *Crit. Care Med.*, *15*: 403.
75. Siggaard-Andersen, O., Gothgen, I. H., Wimberley, P. D., Rasmussen, J. P., and Fogh-Andersen, N. (1988). Evaluation of the Gas-STAT^R fluorescence sensors for continuous measurement of pH, pCO_2 and pO_2 during cardiopulmonary bypass and hypothermia, *Scand. J. Clin. Lab. Invest.*, *48*: 77.
76. Shapiro B. (1989). pH and blood gas monitoring measurements: Discerning innovation from sophistication, *Crit. Care Med.*, *17*: 967.
77. Abraham, E., Markle, D. R., Pinhoster, G., and Fink, S. E. (1986). Non-invasive measurement of conjunctival pCO_2 with a fiberoptic sensor, *Crit. Care Med.*, *14*: 138.
78. Kram, H. B., Fink, S., Tsang, M., Markle, D., Appel, P. L., and Shoemaker, W. C. (1988). Non-invasive measurement of tissue carbon dioxide tension using a fiberoptic conjunctival sensor: effects of respiratory and metabolic alkalosis and acidosis, *Crit. Care Med.*, *16*: 280.
79. Treasure, T., and Band, D. M. (1977). A catheter tip potassium selective electrode, *J. Med. Eng. Technol.*, *1*: 271.
80. Lim, M., Linton, R. A. F., and Band, D. M. (1982). Continuous intravascular monitoring of epinephrine-induced changes in plasma potassium, *Anesthesiology*, *57*: 272.
81. Love, J. W., Jahnke, E. J., Wilcox, A. A., and Kater, J. A. (1982). Continuous monitoring of intravascular potassium: A new technique, *Cardiovasc. Med.*, *1*: 273.
82. Ellingsen, O., Vengen, U. A., and Lebekk, A. I. (1987). Myocardial and potassium uptake during α- and β-adrenoceptor stimulation, *Am. J. Physiol.*, *253*: H 799.

83. Linton, R. A. F., Lim, M., Wolff, C. B., Wilmshurst, P., and Band, D. M. (1984). Arterial plasma potassium measured continuously during exercise in man, *Clin. Sci., 67*: 427.
84. McKinley, B. A., Wong, K. C., Janata, J., Jordan, W. S., and Westenkow, D. R. (1981). In vivo continuous monitoring of ionised calcium in dogs using ion sensitive field effect transistors, *Crit. Care Med., 9*: 333.
85. McKinley, B. A., Saffle, J., Jordan, W. S., Janata, J., Moss, S. D., and Westenskow, D. R. (1980). In vivo continuous monitoring of K^+ in animals using ISFET probes, *Med. Instrum., 14*: 93.
86. Tseung, A. C. C., and Goffe, R. A. (1978). In vivo reference electrodes, *Med. Biol. Eng. Comput., 16*: 677.
87. Hill, J. L., Gettes, L. S., Lynch, M. R., and Hebert, N. C. (1978). Flexible valinomycin electrodes for on-line determination of intravascular and myocardial $[K^+]$, *Am. J. Physiol., 235*: H 455.
88. Coronel, R., Fiolet, J. W. T., Wilms-Schopman, T. J. G., Schaapherder, A. F. M., Johnson, T. A., Gettes, L. S., and Janse, M. J. (1988). Distribution of extracellular potassium and its relation to electrophysiological changes during acute myocardial ischaemia in the isolated perfused porcine heart, *Circulation, 77*: 1125.
89. McKinley, B. A., Houtchens, B. A., and Janata, J. (1981). Continuous monitoring of interstitial fluid potassium during haemorrhagic shock in dogs, *Crit. Care Med., 9*: 845.
90. Hinke, J. A. M. (1959). Glass micro-electrodes for measuring intracellular activities of sodium and potassium, *Nature, 184*: 1257.
91. Walker, J. L. (1971). Ion specific liquid ion exchanger microelectrodes, *Anal. Chem., 43*: 89A.
92. Osswald, H. F., Asper, R., Dimai, W., and Simon, W. (1974). On-line continuous potentiometric measurement of potassium concentration in whole blood during open heart surgery, *Clin. Chem., 38*: 79.
93. Metzger, E., Dohner, R., Simon, W., Vanderschmitt, D. J., and Gautschi, K. (1987). Lithium/sodium ion concentration ratio measurements in blood serum with lithium and sodium ion selective liquid membrane electrodes, *Anal. Chem., 59*: 1600.
94. Anker, P., Amman, D., Meier, P.C., and Simon, W. (1984). Neutral carrier electrode for continuous measurement of blood Ca^{2+} in the extracorporeal circulation, *Clin. Chem., 30*: 454.
95. Fogt, E. J., Eddy, A. R., Clemens, A. H., Fox, J., and Heath, H. (1980). Use of electrochemical sensors for on-line monitoring of ionised calcium, potassium and glucose in whole blood of living dogs, *Clin. Chem., 26*: 1425.
96. Dutsch, S., Jennye, H. B., Schlatter, K. J., Perisset, P. M. J., Wolff, G., Clerc, J. T., Pretsch, E., and Simon, W. (1985). Microprocessor controlled ex vivo monitoring of sodium and potassium concentrations in undiluted urine with ion-selective electrodes, *Anal. Chem., 57*: 578.
97. Ko, W. H. (1986). Solid state transducer for biomedical applications, *Trans. Am. Soc. Artif. Intern. Organs, 32*: 629.
98. Smith, R. L., and Collins, S. D. (1988). Micromachined packaging for chemical microsensors, *IEEE Trans. Electron., 35*: 787.

99. Li, X., Verpoorte, E. M. J., and Harrison, D. J. (1988). Elimination of neutral species interference at the ion-selective membrane/semiconductor device interface, *Anal. Chem., 60*: 493.
100. Adams, R. N. (1969). *Electrochemistry at Solid Electrodes*, Marcel Dekker, New York.
101. Blaedel, W. J., and Jenkins, R. A. (1975). Study of the electrochemical oxidation of reduced nicotinamide adenine dinucleotide, *Anal. Chem., 47*: 1337.
102. Ramasamy, N., Srinivasen, S., and Sawyer, P. V. (1974). Electrochemical behaviour of blood coagulation proteins—factors V and VIII, *Electrochim. Acta, 19*: 137.
103. Ossendorfova, J., Pradac, J., Pradacova, J., and Koryta, J. (1975). Cyclic voltammetric response of blood serum, *J. Electroanal. Chem., 58*: 255.
104. Lane, R. F., and Hubbard, A. T. (1976). Differential double pulse voltammetry at chemically modified platinum electrodes for in vivo determination of catecholamines, *Anal. Chem., 48*: 1287.
105. Murray, R. W., Ewing, A. G., and Durst, R. A. (1987). Chemically modified electrodes. Molecular design for electroanalysis, *Anal. Chem., 59*: 379A.
106. Adams, R. N. (1976). Probing brain chemistry with electroanalytic techniques, *Anal. Chem., 48*: 1126A.
107. Armstrong-James, M., and Millar, J. (1979). Carbon fibre microelectrodes, *J. Neurosci. Methods, 1*: 279.
108. Marsden, C. A., Conti, J., Strape, E., Curzon, G., and Adams, R. N. (1979). Monitoring 5-hydroxytryptamine release in the brain of the freely moving unanaesthetised rat using in vivo voltammetry, *Brain Res., 73*: 15.
109. Wightman, R. M. (1981). Microvoltammetric electrodes, *Anal. Chem., 53*: 1125A.
110. Stamford, J. A., Kruk, Z. L., and Millar, J. (1984). A double cycle high speed voltammetric technique allowing direct measurement of irreversibly oxidised species: Characterisation and application to the temporal measurement of ascorbate in the rat central nervous system, *J. Neurosci. Meth., 10*: 107.
111. Rice, M. E., Oke, A. F., Bredberry, C. W., and Adams, R. N. (1985). Perfluorinated cation exchange membrane (Nafion), *Brain Res., 340*: 151.
112. Wightman, R. M., May, L. J., and Michael, A. C. (1988). Detection of dopamine dynamics in the brain, *Anal. Chem., 60*: 769A.
113. Kristensen, E. W., Kuhr, W. G., and Wightman, R. M. (1987). Temporal characterisation of perfluorinated ion exchange coated microvoltammetric electrodes for in vivo use, *Anal. Chem., 59*: 1752.
114. Wightman, R. M., Amature, C., Engstrom, R. C., and May, L. J. (1988). Real-time characterisation of dopamine overflow and uptake in the rat striatum, *Neuroscience, 25*: 513.
115. Pickard, R. J. (1979). A review of printed circuit microelectrodes and their production, *J. Neurosci. Methods, 1*: 301.
116. Morgan, M. E., and Freed, C. R. (1981). Acetaminophen as an internal standard for calibrating in vivo electrochemical electrodes, *J. Pharmacol. Exp. Ther., 219*: 49.
117. Wang, J., Hutchins, L. D., Selim, S., and Cummins, L. B. (1984). Voltammetric microelectrode for in vivo monitoring of acetaminophen in primate species, *Bioelectrochem. Bioeng., 12*: 193.

118. Wang, J., Hutchins, L. D., Selim, S., and Cumming, L. B. (1983). Catheter microelectrode assembly for in vivo and in vitro voltammetric analysis of body fluids, *Talanta, 30*: 121.
119. Feng, J. X., Brazell, M., Renner, K., Kasser, R., and Adams, R. N. (1987). Electrochemical pretreatment of carbon fibers for in vivo electrochemistry: Effects on sensitivity and response time, *Anal. Chem., 59*: 1863.
120. Meulemans, A. (1987). Measurement of electroactive antibiotic drugs in the bloodstream of rats with a catheter electrode, *Anal. Chem., 59*: 1872.
121. Updike, S. J., and Hicks, G. P. (1967). The enzyme electrode, *Nature, 214*: 986.
122. Layne, E. C., Schultz, R. D., Thomas, L. J., Slama, G., Sayler, D. F., and Bessman, S. P. (1976). Continuous extracorporeal monitoring of animal blood using the glucose electrode, *Diabetes, 25*: 81.
123. Gough, D. A., Lucisano, J. Y., and Tse, P. H. S. (1985). Two dimensional enzyme electrode sensor for glucose, *Anal. Chem., 57*: 2351.
124. Lucisano, J. Y., and Gough, D. A. (1988). Transient response of the two dimensional glucose sensor, *Anal. Chem., 60*: 1272.
125. Lucisano, J. Y., Armour, J. C., and Gough, D. A. (1987). In vitro stability of an oxygen sensor, *Anal. Chem., 59*: 736.
126. McKean, B. D., and Gough, D. A. (1988). A telemetry-instrumentation system for chronically implanted glucose and oxygen sensors, *IEE Trans. Biomed. Eng., 35*: 526.
127. Clarke, W. L., and Santiago, J. V. (1977). The characteristics of a new glucose sensor for use in an artificial pancreatic beta cell, *Artif. Organs, 1*: 78.
128. Hanning, I., Vadgama, P., Covington, A. K., and Alberti, K. G. M. M. (1986). Improved blood compatibility at a glucose enzyme electrode used for extracorporeal monitoring, *Anal. Lett., 19*: 461.
129. Fisher, M. E., and Teo, K. L. (1989). Optimal insulin infusion resulting from a mathematical model of blood glucose dynamics, *IEEE Trans. BME, 36*: 479.
130. Pfeiffer, E. F. (1987). On the way to the automated (blood) glucose regulation in diabetes: The dark past, the grey present and the rosy future, *Diabetologia, 30*: 51.
131. Albisser, A. M., Leibel, B. S., Ewart, G., Davidovac, Z., Botz, C. K., and Zingg, W. (1974). An artificial endocrine pancreas, *Diabetes, 23*: 389.
132. Fischer, U., and Abel, P. (1982). A membrane combination for implantable glucose sensors. Measurements in undiluted biological fluids, *Trans. Am. Soc. Artif. Intern. Organs, 28*: 245.
133. Mullen, W. H., Keedy, W. F., Churchouse, S. J., and Vadgama, P. (1986). Glucose enzyme electrode with extended linearity; Application to undiluted blood measurement, *Anal. Chim. Acta, 183*: 59.
134. Clark, L. C., Noyes, L. K., Spokane, R. B., Sudan, R., and Miller, M. L. (1987). Design and long-term performance of surgically implanted electroenzymatic glucose sensors, *Ann. NY Acad. Sci., 501*: 534.
135. Shichiri, M., Kawamori, R., Yamasaki, Y., Hakui, N., and Abe, H. (1982). Wearable-type artificial endocrine pancreas with needle type glucose sensor, *Lancet, 2*: 1129.
136. Shichiri, M., Kawamori, R., and Yamasaki, Y. (1988). Needle-type glucose sensor, *Methods Enzymol., 137*: 326.
137. Churchouse, S. J., Mullen, W. H., Battersby, C., and Vadgama, P. (1986). Needle enzyme electrodes for biological studies, *Biosensors, 2*: 325.

138. Ege, H. (1989). Construction and testing of a needle glucose electrode with a membrane formed from a polyurethane dispersion. In: *2nd Workshop on Biomedical Engineering of the EEC on Chemical Sensors for In Vivo Monitoring*, Firenze.
139. Shichiri, M., Yamasaki, Y., Nao, K., Sekiya, M., and Ueda, N. (1988). In vivo characteristics of needle-type glucose sensor—measurements of subcutaneous glucose concentrations in human volunteers, *Horm. Metab. Res. Suppl., 20*: 17.
140. Shichiri, M., Fukushima, H, Yamaguchi, K, Kawamori, R., Yamasaki, Y., Ueda, N., and Kamada, T. (1989). Membrane design for extending the long-life of an implantable glucose sensor, *Diabetes Nutr. Metab., 2*: 309.
141. Velho, G., Froguel, P., Sternberg, R., Thevenot, D. R., and Reach, G. (1989). In vitro and in vivo stability of electrode potentials in needle type glucose sensors, *Diabetes, 38*: 164.
142. Gernet, S., Koudelka, M., and de Rooij, N. F. (1989). A planar glucose enzyme electrode. *Sens. Act., 17*: 537.
143. Sansen, W., and Lambrechts, M. (1985). Glucose sensor with telemetry system. In: *Implantable Sensors for Closed-Loop Prostetic Systems* (W. Ko. Ed), Futura Publishing, New York, p. 167.
144. Wolfson, S. K., Tokarsky, J. F., and Krupper, M. A. (1982). Glucose concentration at possible sensor sites, *Diabetes Care, 5*: 162.
145. Abel, P., Fischer, U., Brunstein, E., and Ertle, R. (1988). The GOD-H_2O_2-electrode as an approach to implantable glucose sensors. *Horm. Metab. Res. Supp., 20*: 26.
146. Velho, G., Froguel, P., Thevenot, D. R., and Reach, G. (1988). In vivo calibration of a subcutaneous glucose sensor for determination of subcutaneous glucose kinetics, *Diabetes Nutr. Metab., 1*: 227.
147. Velho, G., Frognel, P., and Reach, G. (1989). Determination of peritoneal glucose kinetics in rats: Implications for the peritoneal implantation of closed-loop insulin delivery system, *Diabetologia, 32*: 331.
148. Rebrin, K., Fischer, U., Woedtke, T. V., Abel, P., and Brunstein, E. (1989). Automated feedback control of subcutaneous glucose concentration in diabetic dogs, *Diabetologia, 32*: 573.
149. Bolinder, J., and Anders, F. (1989). Ultrasonic measurement of forearm subcutaneous adipose tissue. Thickness suitable for monitoring of subcutaneous glucose concentration? *Diabetes Care, 12*: 305.
150. Pickup, J. C., and Claremont, D. J. (1985). A potentially implantable glucose sensor with direct electron transfer, *Diabetes Res. Clin. Pract., Suppl. 1*: 447.
151. Claremont, D. J., Sambrook, I. E., Penton, C., and Pickup, J. C. (1986). Subcutaneous implantation of ferrocene mediated glucose sensor in pigs, *Diabetologia, 29*: 817.
152. Pickup, J. C., Shaw, G. W., and Claremont, D. J. (1989). In vivo molecular sensing in diabetes mellitus: An implantable glucose sensor with direct electron transfer, *Diabetologia, 32*: 213.
153. Matthews, D. R., Bown, E., Beck, T. W., Plotkin, E., Lock, L., Gosden, E., and Wickham, M. (1988). An amperometric needle-type glucose sensor tested in rats and man, *Diabetic Med., 5*: 248.
154. Gebherdt, U., Luft, G., Richter, G. J., and van Strum, F. (1978). Development of an implantable electrocatalytic glucose sensor, *Bioelectrochem. Bioeng., 5*: 607.

155. Lucadon, G. I., Luft, G., Preidel, W., and Richter, G. J. (1988). The electrocatalytic glucose sensor, *Horm. Metab. Res. Suppl., 20*: 41.
156. Lerner, H., Giner, J., Soeldner, J. S., and Colton, C. K. (1984). Implantable electrochemical glucose sensor, *Ann. NY Acad. Sci., 428*: 263.
157. Sarangapani, S., Giner, J., Soldner, J. S., Colton, C. K., Picha, G., Mayhan, K. G., and Drake, R. F. (1988). Electrocatlytic glucose sensor, *Horm. Metab. Res. Suppl., 20*: 43.
158. Wilkins, E. S., and Wilkins, M. G. (1988). The coated wire electrode glucose sensor, *Horm. Metab. Res. Suppl., 20*: 50.
159. Schultz, J. S., Mansouri, S., and Goldstein, I. J. (1982). Affinity sensor: A new technique for developing implantable sensors for glucose and other metabolites, *Diabetes Care, 5*: 245.
160. Meadows, D., and Schultz, J. S. (1988). Fiberoptic biosensors based on fluorescence energy transfer, *Talanta, 35*: 145.
161. Kost, J., Horbett, T. A., Ratner, B. D., and Singh, M. (1985). Glucose sensitive membranes containing glucose oxidase: Activity, swelling and permeability studies, *J. Biomed. Mater. Res., 19*: 1117.
162. Battersby, C. M., and Vadgama, P. (1988). Needle lactate electrode for whole blood measurement, *Diabetes Nutr. Metab., 1*: 87.
163. Lim, S. P. (1988). A catheter tip polarographic lactic acid and lactate sensor for extended use in vivo, *Clin. Res., 36*: A926.
164. Mascini, M., Fortunati, S., Moscone, D., Palleschi, G., Massi-Benedetti, M., and Fabietti, P. (1985). An L-lactate sensor with immobilised enzyme for use in in vivo studies with an endocrine artificial pancreas, *Clin. Chem., 31*: 451.
165. Cuypers, P. A., Hermens, W. T., and Henker, H. C. (1977). Ellipsometric study of protein film on chromium, *Ann. NY Acad. Sci., 283*: 77.
166. Goldsmith, H. L. (1974). Blood flow and thrombosis, *Thromb. Diath. Haemorrh., 34*: 35.
167. Woodward, S. C. (1982). How fibroblasts and giant cells encapsulate implants: Consideration of design in glucose sensors, *Diabetes Care, 5*: 278.
168. Salthouse, T. W. (1976). Cellular enzymic activity at the polymer tissue interface: A review, *J. Biomed. Mater. Res., 10*: 197.
169. Clark, L. C., Noyes, L. K., Spokane, R. B., Sudan, R., and Miller, M. L. (1988). Long term implantation of voltammetric oxidase/peroxide glucose sensors in the rat peritoneum, *Methods Enzymol., 137*: 68.
170. Ertefai, S., and Gough, D. A. (1989). Physiological preparation for studying the response of subcutaneously implanted glucose and oxygen sensors, *J. Biomed. Eng., 11*: 362.
171. Dymond, A. M., Kaechele, L. E., Jurist, J. M., and Crandall, P. H. (1970). Brain tissue reaction to some chronically implanted metals, *J. Neurosurg., 33*: 574.
172. Jackson, W. F., and Duling, B. R. (1983). Toxic effects of silver-silver chloride electrodes on vascular smooth muscle, *Circ. Res., 53*: 105.

14

Trends and Prospects

Loïc J. Blum and Pierre R. Coulet

CNRS-Université Claude Bernard Lyon 1
Villeurbanne, France

INTRODUCTION

The dual nature of biosensors, that is, a biological element associated with a physical transducer, undoubtedly confers to such devices great potential provided the inherent fragility of the bioelement can be overcome.

As a matter of fact, the number of commercially available biosensors is still low. Amperometric enzyme electrodes mainly for glucose (and to a lesser extent for lactate) based on O_2, H_2O_2, or mediated systems are practically the only types of biosensors that have been launched until now with some success. This contrasts singularly with the abundant literature published for more than two decades on the subject.

The persisting strong interest in biosensors can be explained, however, by the urgent need to monitor reliable parameters in real time in domains as different as

Industrial bioprocesses
Critical care units in hospitals
Environment control
Defense

Obviously, considering the very high selectivity that can be obtained, the sensing layer is the essential part of the biosensor but also its Achilles' heel because of the natural fragility and often low stability of biological material.

Several directions from which improvements can be expected, for both the sensing layer and the transducer (Fig. 1), are briefly developed here.

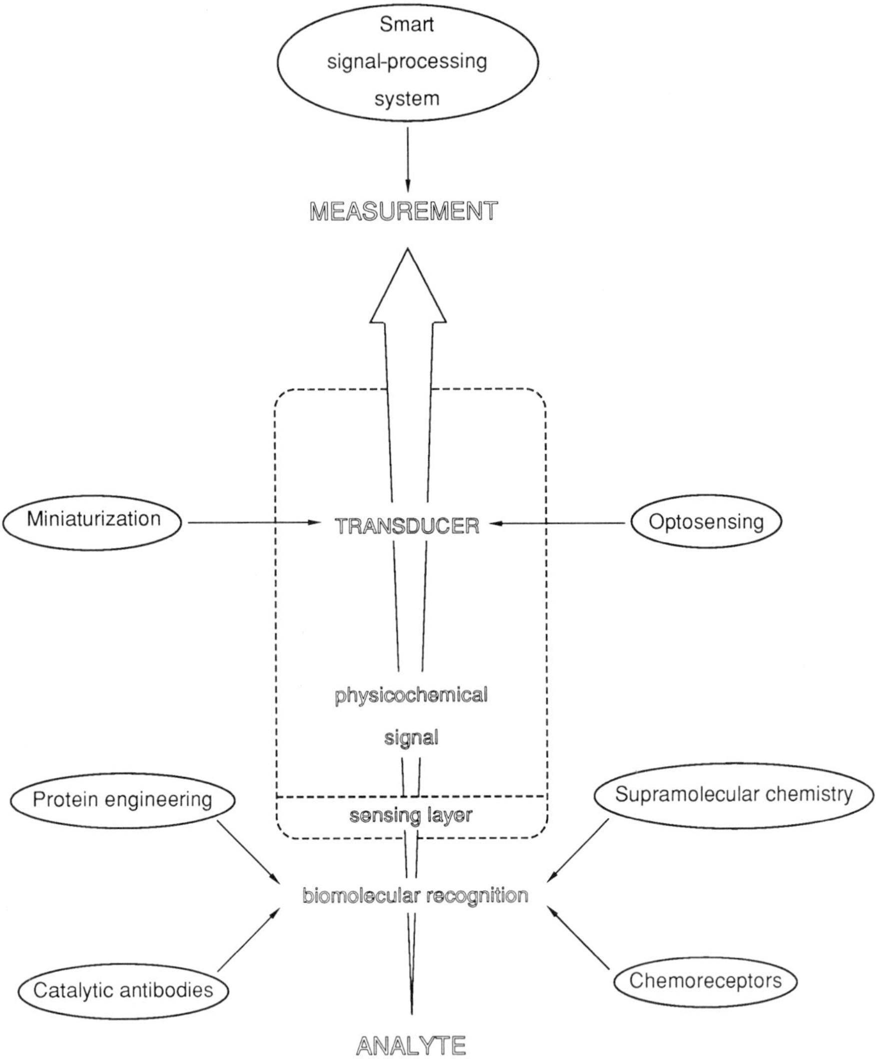

Figure 1 Trends for improving biosensor potentialities.

SENSING LAYERS

Higher selectivity, improved stability, and better sensitivity, as well as new specificities, are the present requirements for better acceptance of biosensors by potential users. Some trends have been identified:

Protein engineering
Catalytic antibodies

Chemoreceptors
Supramolecular chemistry for sensing

Protein engineering with site-directed mutagenesis that allows amino acid substitutions will lead to a better understanding of protein stability and thus to the production of more stable biocatalysts (1). This approach, although still in its infancy, will be the most promising in the short term for designing highly stable and reliable enzyme-based devices (2).

Catalytic monoclonal antibodies are new biologically engineered tools capable of both molecular recognition and catalytic properties. For their preparation it is required that a molecule that resembles the transition state in catalysis is stable enough to be used as an hapten for inducing antibody formation. The complementary determining region of the antibody may then behave as an enzymelike active site. These tailor-made biocatalysts appear very promising when target molecules cannot be sensed by available enzymes (3,4).

Chemoreceptors located in biologic membranes, especially neuroreceptors, appear particularly attractive for the selective recognition of toxins or drugs (5,6). Tedesco et al. (7) pointed out that one of the central problems with the use of biological membrane receptors was the transduction mechanism. In living organisms the nicotinic receptor (ion permeability mechanism) and the β-adrenergic receptor system (with cyclic AMP acting as a messenger within the cell) are the two major classes involved in complex sequences. Incorporating reconstituted molecular assemblies, with the natural receptor as a key component in a suitable bioactive layer associated with a transducer, still appears extremely difficult, taking into account not only the extraction and purification of the receptor itself but also its stability and functionality after reconstitution.

Extreme sensitivity has been attained by using the intact chemosensing structures of antenules of the blue crab (8) connected to a suction electrode utilized in electrophysiology. A signal was obtained for kainic acid and quisqualic acid concentrations as low as 10^{-15} M.

Supramolecular chemistry for sensing. It is possible to design artificial receptor molecules containing intramolecular cavities into which a substrate may fit, leading to cryptates that are inclusion complexes. Molecular behavior at the supramolecular level held promise not only for molecular recognition but also for mimicking biocatalysis and transport. This appears the first age of "chemionics," with the expected development of molecular electronics, photonics, or ionics (9).

Cyclodextrins are crownlike structures that have distinctive properties as a result of their hydrophobic cavity, which is able to trap and release "guest" molecules. Cyclodextrins can be functionalized by grafting chemical groups to mimic biocatalysis, as has been done for a chymotrypsinlike system (10). Functional groups (hydroxyl, imidazolyl, and the carboxylate ion) attached to the cyclodextrin ring lead to artificial biocatalysis with approximately the same trapped within cyclodextrins, and rate enhancement for their hydrolysis could be obtained (11,12).

The exploration of these areas does not exclude the search for affinity biomolecules from new natural sources, such as microorganisms living under extreme conditions, hot abyssal springs for example, which present characteristics that may confer improved performance to biosensors.

TRANSDUCERS

An alternative to overcome the relatively short lifetime of any biological sensing layer coated onto a conventional transducer (*e.g.*, a potentiometric or amperometric electrode) is to use disposable and cheap tips, easily replaceable after each assay if necessary. The pen-sized glucose sensor that was developed at Cranfield is a good model (13), and this principle will certainly gain in popularity in coming years.

Miniaturization

With miniaturization, mass production used for microfabrication of integrated circuits is also certainly a good approach, which furthermore allows miniaturization to be achieved together with the possibility of preparing arrays of sensors with different specificities, leading to multifunctional devices. Such miniaturization requires a corresponding evolution of the methods of preparation of the sensing layers. In this case, the bioactive membrane cannot be prepared separately and further associated with the transducer but a direct coating of the sensitive part must be achieved. Microlithographic techniques for depositing ultrathin layers have been developed for FET-based and mass-sensitive biosensors. The potential of Langmuir-Blodgett films mimicking biological membranes and incorporating enzyme antibodies or new engineered proteins or supramolecular structures begins to be explored as well as the potential of molecular wires for direct electron transfer between a protein and a transducer.

Optosensing

A growing interest is devoted to optical sensors, especially those involving fiberoptics. Either the waveguide is used only to transport light to and from an immobilized reagent phase or a change in the characteristics of the waveguide itself is involved.

A new generation is now emerging based on the evanescent waves generated at the interface of two different optical media. Special coating capable of molecular recognition located at this interface will change optical properties upon interaction with an analyte. Internal reflectance spectroscopic (IRS) techniques have been applied to the development of fiberoptic chemical sensors and novel optical biosensors based on ATR (attenuated total reflection), TIRF (total internal reflection fluorescence), and SPR (surface plasmon resonance) appear promising (14,15).

NEEDS AND NEW AREAS

Improvements in the treatment of the signal for a better interpretation of measurement data will favor the use of biosensors by unskilled personnel. Periodic autotests and automatic calibrations are among the requirements to be satisfied, especially when biosensors must be used outside the laboratory. Under these conditions the sampling procedure must be as simple as possible and biosensors must be capable of operation with crude samples without pretreatment or dilution. As a matter of fact, biosensors may prove to be useful when samples cannot be sent to central laboratories. Mobile laboratories will surely gain in importance with the need of having facilities on the site where immediate analysis are needed, for example, in accidental pollution and emergency situations. Multivariate and miniaturized reliable biosensors could be of a great interest because of the limited space available for an optimized measuring capacity of this type of laboratory.

At present a great deal of effort is being expended throughout the world to designing automatic control stations for pollution monitoring in water and air. Such systems must be robust, low in energy consumption, and autonomous with little maintenance. If reliable, multifunctional, and self-contained biosensors can be constructed, a decisive step will be achieved in environmental control. Biosensors able to work in gaseous or organic phases would certainly contribute to such a goal.

If in vivo implantable sensors are still a long-term challenge, the requirement for feedback regulation in extracorporeal systems could be fulfilled by reliable biosensors, not only for organic compounds but also for ions. Ambulatory monitoring for periods shorter than 3 days with miniaturized systems can also be expected.

In some cases sensitivity will be the main factor to be considered. For instance, the detection of cancer markers in biological fluids requires a reliable measurement below the nanomolar level. A biosensor able to give an unambiguous response in the early stages of the disease would undoubtedly boost the biosensor concept, not only in the biomedical engineering area but also in other expanding domains.

In conclusion, as stated many improvements can be imagined to make these analytical tools widely available, and the huge amount of interdisciplinary knowledge accumulated since the first description of an enzyme electrode makes us truly optimistic for the future of biosensors and bioelectronics.

REFERENCES

1. Nosoh, Y., and Sekiguchi, T. (1990). Protein engineering for thermostability, *Trends Biotechnol.*, *8*: 16.
2. Ulmer, K. M. (1987). Engineering proteins for biosensor applications. In: *Biosensors, International Workshop 1987*, GBF monographs Vol. 10 (R. D. Schmid, G. G. Guilbault, I. Karube, H.-L. Schmidt, and L. B. Wingard, Eds), VCH, Weinheim, p. 245.
3. Schultz, P. R. (1989). Catalytic antibodies, *Angew. Chem. Int. Ed. Engl.*, *28*: 1283.
4. Green, B. S., and Tawfik, D. S. (1989). Catalytic monoclonal antibodies: Tailor-made enzyme-like catalysts for chemical reactions, *Trends Biotechnol.*, *7*: 304.
5. Belli, S. L., Buch, R. M., and Rechnitz, G. A. (1987). Response model for chemoreceptor biosensors, *Anal. Lett.*, *20*: 327.
6. Wingard, L. B. Jr. (1987). Possibilities for biosensors based on neuroreceptors. In: *Biosensors, International Workshop 1987*, GBF monographs Vol. 10 (R. D. Schmid, G. G. Guilbault, I. Karube, H.-L. Schmidt, and L. B. Wingard, Eds), VCH, Weinheim, p. 133.
7. Tedesco, J. L., Krull, U. J., and Thompson, M. (1989). Molecular receptors and their potential for artificial transduction, *Biosensors*, *4*: 135.
8. Buch, R. M., and Rechnitz, G. A. (1989). Intact chemoreceptor-based biosensors: Extreme sensitivity to some excitatory aminoacids, *Anal. Lett.*, *22*: 2685.
9. Lehn, J.-M. (1988). Supramolecular chemistry-scope and perspectives, molecules, supermolecules and molecular devices (Nobel lecture), *Angew. Chem. Int. Ed. Engl.*, *27*: 89.
10. D'Souza, V. T., Hanabusa, K., O'Leary, T., Gadwood, R. C., and Bender, M. L. (1985). Synthesis and evaluation of a miniature organic model of chymotrypsin, *Biochem. Biophys. Res. Commun.*, *129*: 727.
11. Rebeck, J., Jr. (1988). Molecular recognition: Model studies with convergent functional groups, *J. Mol. Recog.*, *1*: 1.
12. Thomas, J. M. (1989). Advanced catalysts: Interfaces in the physical and biological sciences, *Angew. Chem. Adv. Mater.*, *101*: 1105.
13. Matthews, D. R., Holman, R. R., Bown, E., Steemson, A., Watson, A., Hughes, S., and Scott, D. (1987). Pen-sized digital 30-second blood glucose meter, *Lancet*, *1*: 778.
14. Hall, E. A. H. (1988). Developing biosensors. In: *Analytical Uses of Immobilized Biological Compounds for Detection, Medical and Industrial Uses* (G. G. Guilbault and M. Mascini, Eds), D. Reidel Publishing, Dordrecht, The Netherlands, p. 227.
15. Dessy, R. E. (1989). Waveguides as chemical sensors, *Anal. Chem.*, *61*: 1079A.

Index

Absorbance, 167, 182, 195, 207
Acetaminophen, 319
  interference, 65
Acetic acid, 277
  sensor, 282
Acetylcholinesterase, 53
Acrylic beads, 89
Adenosine
  deaminase, 53, 67
  diphosphatase, 91
  diphosate (ADP), 29, 91, 96
  triphosphatase, 91
  triphosphate (ATP), 91, 92, 96, 213, 214, 269, 280
    bioluminescent measurement of, 214, 217, 233, 234, 239
    converting reactions, 218
    sensor, 20, 233, 238
ADP (*see* Adenosine diphosphate)
Aerosol, 114
Aflatoxin B, 293
Agar entrapment of microbial cells, 273
Albumin
  bovine, 117, 132, 146, 183
  human, 120, 122, 260
Alcohol, 4, 90, 163
  biosensors, 178
  dehydrogenase, 16, 59, 236
  oxidase, 4, 90, 91, 178
  primary, 236
  secondary, 236
  sensor, 284
Alginate, 98
  -polylsine-alginate, 322
Ambulatory monitoring, 343
Amino acid, 4, 70, 77
  interference, 324
  oxidase, 4
L-Amino acid, 17
Ammonia, 114
  electrode, 291
  gas sensor, 283
  optode (*see* Optode)
Ammonium dihydrogen phospate, 109
Ampicillin, 101
Amplification
  chemical, 96
  cofactor, 59
  enzymatic, 96
α-Amylase, 28
Analyzer
  coated piezoelectric crystal, 113
  immobilized enzyme flow-enthalpimetric, 84
  microprocessor-based, 31
  two-electrode-based, 31
Anesthetic gas interferences, 310
Anode (*see* Electrodes)
Antibiotics, 218, 319

Antibody, 47, 99, 107, 115, 172, 249
  catalase-labeled, 256
  catalytic, 340, 341
  -coated crystal, 132
  enzyme-labeled, 48
  monoclonal, 67, 208, 341
  polyclonal, 67
  solid-phase, 48
  Texas red-labeled, 207
Antigen, 47, 107, 115, 172
  chromophoric, 195
  -coated crystal, 117
  enzyme-labeled, 67, 99
  fluorescein-labeled, 207
  fluorescent, 195
Apyrase, 91
Arginine, 70
Ascorbate, 10, 91, 188
  brain level, 319
  interference of, 321
  oxidase, 91, 180, 188
Ascorbic acid (*see* ascorbate)
Asparaginase, 67
Aspartame, 72
Aspregillus ochraceus, 256
Assimilable sugar sensor, 282
ATP (*see* Adenosine triphosphate)
Attenuated total reflection (ATR), 343

Bacillus cereus, 93
Bacillus subtilis, 127, 293
Bacterial infection, 218
Bacteriophages, 218
Bacterium cadaveris, 70
Barium titanate, 109
Benzopyrenetetraol, 208
Benzoquinone, 93
Beverages, 91, 93
Bifunctional reagents (*see also* glutaraldehyde), 9
Biocatalysis, 63
Biocompatibility, 304, 322, 327–328
Bioconjugates, 2
Bioelectronics, 343
Biological oxygen demand, 6
  sensor, 285
Bioluminescence, 213
  bacterial, 215, 216, 218, 279
  firefly, 214
  reactions, 214–215
Biomass, 218
Biooptode (*see* Fiberoptic biosensor)
Biopolymer, 70
Bioprocess, 339
Bioreactors, 70
Biosensors
  amperometric, 7–45, 52
    commercial devices, 32–35
    for enzyme immunoassay, 47–61
  -based analyzers, 7, 32–35
  calorimetric, 87 (*see also* Enzyme thermistor)
  definition of, 1–7
  FET-based, 139–162, 342
  fiberoptic
    bioluminescence- and chemiluminescence-based, 213–247
    chemically mediated, 163–194
    fluorophore- and chromophore-based, 195–211
  in vivo, 303–338
  mass-sensitive, 342
  microbial, 267–301
  monofunctional, 233
  multifunctional, 238–241, 343
  NADH-based, 197–204, 233, 237
  piezoelectric crystal, 107–138
  potentiometric, 52, 63–82
    commercial devices, 77
  single-chip, 149
  whole-cell, 203
Biotin-avidin, 184
Blood, 28, 91
  arterial, monitoring of, 314
  carbon dioxide measurement in, 310
  constituents, interference of, 55
  lipids, 94
  monitoring, 321
  oxygen measurement in, 305–308
  pH measurement in, 312–313
  typing, 254
BOD (*see* Biological oxygen demand)

Boronate, 18
Brevibacterium lactofermentum, 282
Bromothymol blue, 182
Buffer capacity, 64, 153, 182, 183, 184
1-Butanol, 236
Butyric acid, 178
Butyrylcholine, 95
  esterase, 76

Calcium
  ion-selective field effect transistor, 316
  monitoring, 317
Calorimetry, 83, 85, 270
Cancer, 27
Candida boidinii, 90
Captan, 293
Carbodiimide, 18, 174
Carbon black, 166
Carbon
  dioxide, 172
    arterial partial pressure, 304
    conjunctival monitoring of, 315
  optode (*see* Optode)
Carbonic anhydrase, 96
Carcinogen, 293
Carcinogenicity, 256, 293
Cardiac infection, 27
L-Carnitine, 16
Casein, 280
Catalase, 67, 90, 92, 93, 94, 96, 97, 176, 178, 255, 321
Catecholamines, 318
Catheter
  electrodes, 315
  sensors, 305, 310, 312
Cathode (*see* Electrodes)
Cefsulodin, 319
Cellobiose, 91
Cephalosporin, 92
  C, 290
  sensor, 289
Cephaloridine, 290
Cephalosporinase, 289
Cephalosporium acremonium, 290
Cephalothin, 290
Cetyltrimethylammonium bromide (CTAB), 231, 232
Charge transfer, 13
Chemical kinetics, 13
Chemiluminescence, 213
  electrogenerated, 216, 230
  reactions, 215–216
Chemionics, 341
Chemoreception, 63
Chemoreceptors, 341
Chloramphenicol, 319
Chlorophyll, 203
Cholesterol, 92, 94, 163
  biosensor, 180
  esterase, 92
  esters, 94
  oxidase, 28, 92, 180
Choline
  esterase, 32, 67, 95
  oxidase, 32, 94
Chromophore, 195
  detection, 196
α-Chymotrypsin, 97
Chymotrypsin-like system, 341
Cirrhosis, 27
Citrobacter freundii, 289
Cladding, 167
Clark oxygen electrode, 1, 12, 13, 14, 21, 26, 256, 274, 305, 320
  miniature, 307
Closed-loop system, 159
  for glucose control, 321
Cocaine, 129
Codeine, 259
Coenzymes, 28
  recycling, 96
  regeneration, 97
Cofactor (*see also* Coenzymes)
  amplification for immunoassay, 59
  flavin-based, 204
  regeneration, 18, 280
Collagen, 272
Concanavalin A, 91, 95, 185, 206, 249, 325
  -conjugated Sepharose, 91, 95

Controlled pore glass (CPG), 89, 92, 93, 94, 95
Convection, 13
Core, 167
CPG (*see* Controlled pore glass)
CPPO (*see* bis-2, 4, 5-Trichloro-6 pentoxycarbonyl oxalate)
Creatine kinase, 20, 29
Creatinine, 76, 77, 92
  deiminase, 92
Cresol green, 182
Critical care units, 339
Cross-reactivity, 134
Cryptates, 341
Current
  anodic, 11
  background, 13
  cathodic, 11, 12
Cyanide, 95, 203
Cyanuric chloride, 18
Cyclodextrins, 341, 342

Darlington transistors, 86
Decacyclene, 180
Decanal, 214
Decarboxylase, 175
Deoxyribonucleic acid (DNA)
  hybridization, 121
  immobilization, 115
Dextran, 325
  dopamine-modified, 18
Diabetes, 319
Dichlorofluorescein, 172
Diffusion, 13
Digoxin, 55
2,4-Dinitrophenol, 67
2,4-Dinitrophenyl-aminocaproic acid, 56
1,2-Dioxetanedione, 216
Diphenylanthracene, 216
Direct voltametric monitoring, 318
Disaccharide, 94, 96
DNA (*see* Deoxiribonucliec acid)
Dopamine
  detection, 20

Drain, 140
  voltage, 141
Drugs, 99, 341

Effector, determination of, 95, 98
Electroactivity, 11, 12
Electrochemical detection, 7
Electrochemical interferents in biological fluids, 321
Electrodes (*see also* Sensors)
  aluminum, 111, 274
  amperometric, 7–45, 47–61
    enzyme activity determinations with, 27
    substrate determinations with, 27
  antimony oxide, 65
  carbon, 16, 19, 55, 65
    fiber, 319
  carbon dioxide sensing, 289
  enzyme-mediated, 20
  fluoride, 77
  galvanic, 270, 271
  gas, 66, 271
  gold, 12, 14, 65, 111
  immunotype, 249
  impedance, 324
  ion-selective, 64, 271, 315
    cation-selective, 65
  mediated-based, 324
  metal-metal oxide, 65
  modified, 17
  nickel, 111
  normal hydrogen (NHE), 11
  organic salt, 59
  palladium oxide, 65
  pH-glass, 64, 65
  platinum, 12, 14, 19, 65, 274
  polarographic, 270
  potentiometric, 2, 63–82
  redox, 65
  reference, 11, 66
    Ag/AgCl, 11, 13, 148, 315, 323
    saturated calomel (SCE), 15, 148
  rotating ring-disk, 12
  silver, 111

solid-state pH, 65
survival in blood, 321
titanium, 65
Endotoxin, 127
Enterobacter cloacae, 92
Enthalpy
change, 83, 84, 85, 97, 270
protonation, 96
Environment control, 339, 343
Enzyme
activity, 7
-based amperometric biosensors, 7–45,
catalysis in organic solvents, 97
in chemically mediated fiberoptic biosensors, 173
electrode, 2, 7
amperometric, 7–45, 339
design, 11
L-glutamine, 10
potentiometric, 63–82
hydrogen peroxide generating, 213, 218
immobilization, 2, 8, 69, 174
immunoassay, 2, 17, 47–51
intrinsic fluorescence of, 204
labels, 255
-linked immunosorbent assays, 49
thermometric, 99
stability, 178, 236
synthetic, 77
thermistors, 83–105
commercially available devices, 102
substrate determination with, 90
Eosin, 172
Escherichia coli, 70, 101, 289, 293, 297
genetically engineered, 99
Estriol, 259
Ethanol, 16, 101, 178, 180, 201, 202, 235, 285
Ethylenediamine tartrate, 109
Evanescent fied, 207
Extracorporeal monitoring, 304, 317
Extracorporeal sensors, 321
FAD (*see* Flavine adenine dinucleotide)
Fermentation, 91
broth, 267
process control, 280
Ferricyanide, 15, 59, 216
Ferrocene, 11, 16, 324
continuous-flow amperometric detection, 259
derivatives, 341
-lidocaine complex, 55
-linked substrate, 55
FET (*see* Field effect transistor)
α-Fetoprotein, anti-, 256
FIA (*see* Flow injection analysis)
Fiber
bifurcated, 167, 196
bundle, 167, 196, 233
optical (*see* Fiberoptic)
quartz, 199
Fiberoptic, 342
biosensors, 163
antibody-based, 195, 206–208
bioluminescence- and chemiluminescence-based, 213–247
chemically mediated, 163–194
fluorophore- and chromophore-based, 195-211
intrinsic, 163, 164
immunosensors, 257, 260, 261
probe, 5
sensor, 325
for blood gas and pH, 313
Fibrinogen
deposition on electrodes, 327
gelation time of, 127
Fibroblasts, 328
Field effect transistor (FET), 139, 254
-based biosensors, 139–162
Films (*see also* Membranes)
Mylar, 66
polyethylene, 12
polypropylene, 12
polyvinylchloride, 12
PTFE, 12
rubber, 12

Films (*Continued*)
silicon rubber, 66
Teflon, 66
Firefly, 213
FITC (*see* Fluoresceineisothiocyanate)
Flavine
adenine dinucleotide (FAD), 50, 56, 204
mononucleotide (FMN), 204, 269
reduced ($FMNH_2$), 214, 217, 280
Flow analysis
with bioluminescence-based fiberoptic sensors, 237
with enzyme thermistors, 83–105
with fluorophore-based fiberoptic biosensors, 205
with microbial biosensors, 277
Flow injection analysis (FIA), 176, 188, 279
fiberoptic biosensors in, 187
Fluorescamine, 183
Fluoresceinamine, 183
Fluoresceine
derivative, 185
-labeled antigen, 207
Fluoresceineisothiocyanate (FITC), 184
labeled dextran, 185, 186, 187, 325
Fluorescence, 165, 182, 195, 207, 326
energy-transfer, 172, 187, 207
intrinsic, 166
quenching of, 180
Fluorophore, 172, 195, 209, 251
detection, 197
FMN (*see* Flavine mononucleotide)
Formaldehyde, 129
Fragrance meter, 129
Frequency
electrical oscillation, 108
resonant, 108, 110, 126
measurement of, 112–113
shift, 117, 123, 126, 133
vibrational, 108
D-Fructose, 186, 280, 283

Galactose, 17, 92
oxidase, 28, 92
β-Galactosidase, 28, 49, 50, 93, 96
β-Galactosylumbelliferone, 50
Gas-phase analysis, 128
Gate electrode, 141, 254
Gel entrapping, 272–273
Gelatin, 69
beads, 99
Glass beads, 184
Glucoamylase, 28
D-Gluconate, 20
dehydrogenase, 20
Gluconobacter oxydans, 98
β-Glucosidase, 91
Glucoprocesseur, 31
Glucose, 1, 7, 18, 77, 84, 91, 92, 93, 101, 126, 157, 163, 185, 187, 200, 204, 231, 232, 276, 280, 282, 321, 325, 339
analyzer, 28
dehydrogenase, 200
electrodes, 319
in vivo monitoring of, 319
monitoring in human blood, 157
oxidase, (GOD), 1, 4, 9, 15, 16, 17, 18, 19, 28, 29, 31, 49, 50, 55, 67, 92, 93, 96, 97, 182, 204, 231, 232, 255, 259, 324
apoenzyme, 56
holoenzyme, 56
stability, 320
sensor, 16, 17, 155, 176, 182, 280
implantable, 324
pen-sized, 342
Glucose-6-phosphate, 29
dehydrogenase, 49, 58
α-Glucosidase, 28
Glutamate, 31, 202
decarboxylase, 288
oxidase, 31
Glutamic acid sensor, 288
Glutamine, 72
Glutaraldehyde, 9, 10, 18, 19, 70, 89, 115, 121, 144, 146, 174, 180, 183
Glutathione, 182
Glycerokinase, 29
Glycerol, 98, 236

Glycerol-3-phosphate oxidase, 30
Glycolate, 17
GOD (*see* Glucose oxidase)
Graphite disks, 11

H_2O_2 (*see* Hydrogen peroxide)
Heat capacity, 85
Heavy metals, detection of, 95
Hemin, 216
Hemocyanin, 120
Hepatitis, 27
Herbicides, 203
Hexacyanoferrate, 15, 16
Hexokinase, 92, 96, 126
High-performance liquid chromatography (HPLC), 92, 93
Histidine, 70
Hollow fiber, 183, 325
Hormones, 99
Horseradish peroxidase (*see* Peroxidase)
HPTS (*see* 1-Hydroxypyrene-3,6,8-trisulfonate)
Human gonadotropin hormone, 254
Hydrocarbons
  aromatic, 114
Hydrogen
  chloride, 114
  peroxide, 8, 11, 96, 175, 178, 213, 215, 216, 217, 230, 250, 339
    chemiluminescent determination of, 215–216, 232, 233
    detection, 13, 20, 93, 321, 323
    dismutation, 13
    reduction, 13
  sulfide, 114
Hydrolase, 8, 28
1-Hydroxypyrene-3,6,8-trisulfonate (HPTS), 171, 182, 183, 314
  fatty acid ester of, 205
Hygrometer, 113
Hypoxanthine, 94

IgA, IgE, IgG, IgM (*see* Immunoglobulin A, E, G, M)
Immobilization
  of biocatalysts on piezoelectric crystals, 115
  of bioluminescence enzymes, 219–220
  of enzymes, 8, 69
  of microorganisms, 271
  of proteins, 69
Immunoassays, 2, 47 (*see also* Enzyme immunoassays)
  electrochemical, 54, 259
  enyme-amplified, 59
  enzyme-linked, 47, 50, 99
  fluorescent labels in, 50
  heterogeneous, 255–260
  hmogeneous, 260–263
  kinetic, 125
  labels, 49
  nonisotopic, 47
  potentiometric, 67
  sandwich, 120
Immunobiosensors (*see* Immunosensors)
ImmunoFET, 140
Immunoglobulin
  A, 120
  E, 120
  G, 55, 120, 121, 124, 125, 126, 128, 262
  M, 120
Immunoinhibition, 29
Immunoreaction, 63, 108
Immunosensors, 249–266
  amperometric, 47–61
  potentiometric, 67
Inflammatory response, 328
Inhibitor, determination of, 95
Insulating layer, 141
  capacitance, 141
Insulin
  bovine, 99, 258
  dosage, 322
  in diabete, 319
  porcine, 258
  release of, 326
Interfacial mass changes, 110
Internal reflectance spectroscopy (IRS), 342

Intravascular
electrodes, 315
optode, 314
space, 304
Invertase, 28, 94
In vivo monitoring, 304
Ionophore, 67
Ion-selective field effect transistor (ISFET), 63, 139, 310, 312, 315
-based biosensor, 140
silicon on sapphire, 142
ISFET (*see* Ion-selective field effect transistor)

Kainic acid, 341
Klebsiella pneumoniae, 70
Klebsiella sp., 127

Labels
electrochemically active, 259
electrochemiluminescence, 261
enzyme, 255
fluorescence, 260
liposome, 260
β-Lactamase, 92, 93, 97
β-Lactams, 93
L-lactate, 16, 29, 93, 96, 101
Lactate, 163, 188, 199, 201, 204, 326, 339
biosensor, 177
dehydrogenase, 15, 20, 28, 29, 30, 31, 93, 96, 99, 197
electrode, 326
monooxygenase, 29, 93, 177, 204
oxidase, 29, 93, 96, 99, 178, 188
Lactic acidemia, 326
Lactobacillus plantarum, 99
Lactose, 77, 93, 96
Langmuir-Blodgett
films, 342
technique, 19, 190
Laser, 167
Leucocytes, 327
Lidocaine, 17, 55, 260
Light
emission, 214, 216
emitting diode, 170
exciting, 165
guide (*see* Fiberoptic)
scattering, 207
source, 165
pulsed, 166
Lipoprotein lipase, 94, 97
Liposome, 251
Liver damage, 27
Lophine, 215
Luciferase, 214, 256
bacterial, 214, 217, 280
firefly, 214, 216, 217, 238
Luciferin, 214, 256
Luminescence, 5
reagents, 213
Luminescent reactions, 5, 213–215
Luminol, 215, 220, 230, 256, 259, 261
chemiluminescence, 215, 233
micellar mediated, 232
Lysine, 70, 77
decarboxylase, 70

Malate, 93
dehydrogenase, 30, 236
Malic enzyme, 93
Maltose, 77
D-Mannose, 187
Mass transport, 13, 52, 64, 326
Mediated detection, 15
Mediated systems, 339
Mediators, 15, 16, 59
Membranes (*see also* Films)
anion-exchange, 205
bacteria-collagen, 275
cellophane, 10
cellulose acetate, 5, 10, 11, 272, 283, 321, 322
collagen, 10, 30, 290, 291
crystal, 65
cuprophane, 171

dialysis, 12, 31, 69, 174
fluorocarbon, 328
gas-permeable, 66, 172, 285, 291, 297, 322
hydrophobic, 178
ion-selective, 63
ion-sensitive, 141, 142
nitrocellulose, 272
nylon, 199
perm-selective, 9, 202
polyamide, 231, 233
polycarbonate, 5, 10, 321
polypropylene, 66
polyvinyl chloride, 249, 317
preactivated, 5, 10, 233
silicon-polycarbonate composite, 310
silicon rubber, 166
Teflon, 66, 169, 185, 202, 274, 277, 283, 285, 293
ultrafiltration, 202
Mercury vapors, 114
Metal ions, detection of, 96
Methane sensor, 286
Methanol, 180, 285
Methionine, 77
Methotrexate, 207
N-Methyl-*N*-nitro-*N*-notrosoguanidine, 293
Methylomonas flagellata, 286
N-Methylphenazinium, 59
Metronidazole, 319
Microacalormeter, 83
Microelectrodes, 17
carbon fiber, 20
Mithomycin, 293
Molasses, 276, 282
Morphine, 259
Multibiosensor, 147, 155
Multienzyme systems, 21, 235, 238–241
Mutagen, 293
Mutagenesis, site-directed, 341
Mutagenicity, 293
Mutagen sensor, 293
Mycobacterium smegmatis, 93
Myocardial damage, 27, 313
NAD (*see* Nicotinamide adenine dinucleotide)
NADH (*see* Nicotinamide adenine dinucleotide, reduced)
NADP (*see* Nicotinamide adenine dinucleotide phosphate)
NADPH (*see* Nicotinamide adenine dinucleotide phosphate, reduced)
Needle electrode
for glucose, 322
for lactate, 326
Needle oxygen electrode, 307
Nephropathy, 256
Nerve gas, 98
Neuroreceptors, 341
Neurotransmitter, 318
NHE (*see* Electrode, normal hydrogen)
Nicotinamide adenine dinucleotide (NAD), 59
dehydrogenase, 17
macromolecular derivative, 201
polyethylene glycol conjugate, 202
reduced (NADH), 17, 198, 213, 217, 280, 318
amperometric measurement, 58
bioluminescent measurement of, 215, 233, 235, 239, 269
fluorimetric detection of, 198
intracellular, 203
oxidase, 59
sensor, 233, 236, 238
voltammetric detection, 20
Nicotinamide adenine dinucleotide phosphate (NADP), 59, 203
reduced (NADPH), 213, 215, 217
Nicotinic acid, 99
Nimorazole, 319
Nitrobacter sp., 283, 296, 297
Nitrosomonas sp., 283
Nitrifying bacteria, 283, 292
Nocardia erythropolis, 92
Nonactin-polyvinylchloride, 65
Nylon
coils, 5
membranes, 199
nets, 10

Ochratoxin A, 256
On-line monitoring, 267
Optode (*see also* Sensors, optical)
  ammonia, 168, 172, 185
  carbon dioxide, 168, 172
  fluorescent, 314
  oxygen, 168–170, 175
  pH, 168, 171–172, 182
Optosensing, 342
Orosomucoid, 55
Oscillator circuits, 112
O_2 (*see* Oxygen)
Oxalate, 93, 101
  decarboxylase, 93
  derivatives, 215
  oxidase, 93
Oxalic acid esters, 216
Oxaloacetate, 235
  decarboxylase, 31
Oxidase, 8, 28, 90, 180, 218
Oxidoreductase, 217
  NAD(P)H-FMN, 215
Oxygen, 1, 8, 11, 214, 339
  consumption, 92, 175
    by bacteria, 276, 297
  detection, 12, 20, 168–170
  electrode (*see also* Clark oxygen electrode), 2, 12, 249, 252, 256, 270, 275, 277, 283, 286, 291, 293, 297
  diffusion coefficient, 21
  optode (*see* Optode)
  partial pressure, 11, 12, 175, 182
    arterial, 304
    in blood, 320
Oxygenase, 180
Ozone, 114

Pancreas, artificial, 159
Parathion, 129
Parker's law, 165
Passive agglutination method, 117
Pd-MOSFET, 140
Pediococcus pseudomonas, 93
Peltier elements, 84, 86
Penicillinase, 93, 183, 184
Penicillin, 163
  biosensor, 183, 184
  G, 93, 94
  V, 93, 94
Penicillium viridicatum, 256
Peritoneal cavity, 304
Peroxidase, 16, 17, 19, 49, 96, 97, 99, 216, 219, 232, 233, 255, 256
Peroxyoxalate chemiluminescence, 216
  micellar enhanced, 231
Perylene, 178, 216, 231
Pesticides, 76, 77, 95, 203
  organophosphorus, 76, 114
Phenol red, 171, 172
7-Phenylacetylamidodesacetoxysporanic acid, 290
Phenylalanine, 70, 77
Phenytoin, 58
Phosgene, 114
Phosphatase
  acid, 59
  alkaline, 29, 49, 55, 59, 99, 196
Phosphatidylcholine, 19
Phospholipase
  A, 120
  D, 94
Phospholipids, 94
Photinus pyralis, 214, 233, 238
Photobacterium leiognathi, 214, 219
Photobacterium phosphoreum, 214, 279
Photoluminescence quenching, 314
Photomicrobial detector, 279
Photodiodes, 217, 269
Photomultiplier tube, 196, 217, 230, 259, 269, 279
Pichia pastoris, 91
Piezoelectric
  crystal, 107, 111–112
    avidin-coated, 122
    commercially available instrumentation, 113–115
    biosensors, 107–138
  effect, 108–109

material, 255
oscillators, 4
Platelets, 327
Polarography, 13
Pollutant, 95
gases, 114, 296
Pollution
of air, 295
monitoring, 343
Polyacrylamide
gel, 2, 5, 9, 69, 171, 174, 183, 220, 272
matrix, 126
Polyacrylic acid, 219
Polyalcohols, 70
Polyarylsulfone, 322
Polynucleotide hybridization, 121
Polyurethane, 324
Polyvinyl alcohol, 9, 322
Polypeptides, 70
Potassium
ion-selective field effect transistor, 316
measurement in arterial blood, 315
monitoring, 317
Potential jump method, 324
Probes (*see* also Sensors)
thermal enzyme, 84
Proinsulin, human, 99
1-Propanol, 236
2-Propanol, 236
Propylene glycol dinitrate, 114
Protein A, 125, 126
Protein engineering, 340, 341
Pseudomonas fluorescens, 275
2-Pyridine, aldoxime, 77
Pyrrole, 19
Pyruvate, 20, 28, 29, 96, 199, 201
kinase, 29, 96
oxidase, 31

Quantum yield, 165, 216, 217
fluorescence, 166
light detector, 217
luminescent reaction, 216
Quartz
crystal, 108, 109
optical waveguide, 207
oscilating, 110
wafer, 111
Quisqualic acid, 341

Radioimmunoassay, 47, 132
Radioisotope, 47
Rayleigh wave propagation, 111
Reactors, electroenzymatic, 18
Receptor
β-adrenergic, 341
nicotinic, 341
Reflectance, 167, 182
Refractive index, 167
Reusability, 6
Rhodamine, 326
B, 187
Rhodanese, 95
Ribonucleic acid (RNA)
hybridization, 121
RNA (*see* Ribonucleic acid)
Rochelle salt, 109

Saccharides, 70
Saccharomyces cerevisiae, 127, 280
Salmonella typhimurium, 121, 293
Salting-out, 14
Sauerbrey equation, 111, 125
SCE (*see* Electrode, saturated calomel)
Sensors (*see also* Biosensors)
ammonia, 66
ammonium, 65
amperometric, 7–45, 164
carbon dioxide, 66
chloride, 65
cyanide, 65
electrocatalytic, 324
electrogravimetric, 109
flow through, 28
fluoride, 65
in vivo, 70, 343
iodide, 65

Sensors (*see also* Biosensors)
optical, 164, 165, 342
potentiometric, 164
Serum, 28, 92, 93
Severinghaus carbon dioxide electrode, 310, 311
Sewage, 95
Silicon
dioxide, 109, 111
photodiode, 217
p-type, 140
rubber, 9, 180
Sodium nitrite sensor, 295
Sorbitol, 235
Source, 140
Stability
of biological materials, 339
operational, 70, 236, 237
Starch gel, 9, 69
Steinhart-Hart equation, 86
Stern Volmer equation, 168, 176
Subcutaneous tissue, 304
Sucrose, 94, 101, 186, 283
Sulfur dioxide, 114
Supramolecular chemistry, 341
Surface acoustic wave (SAW), 4, 111, 124
Surface plasmon resonance (SPR), 254, 343
Surface voltage, 142
Synzymes (*see* Enzymes, synthetic)
Syphilis, 254

TCPO (*see* bis-2,4,6-Trichlorophenyl oxalate)
TELISA (*see* Thermometric enzyme-linked immunosorbent assay)
7,7,8,8-Tetracyanoquinodimethane, 59
Theophylline, 17
thermistors, 84, 86
with immobilized enzymes, 83–105
with immobilized cells, 97–101, 280
determination of enzyme activity with, 100
measurements of metabolic activity with, 100–101
Thermometric enzyme-linked immunosorbent assay (TELISA), 99
Thermopiles, 86
Thermus aquaticus, 59
Thiols
interference of, 321
Threshold voltage, 142
Thromboembolism, 327
Tissue
electrodes, 307, 316
sensors, 313
Total internal reflection fluorescene (TIRF), 343
Tourmaline, 109
Toxins, 341
Transaminases
glutamate oxaloacetate, 30, 31
glutamate-pyruvate, 30, 31, 201
Transcutaneous sensors, 308–310, 311
Transducer, 1, 342
amperometric, 8
-based fiberoptic biosensors, 175–185
optical, 165
temperature, 86
Transmembrane potential, 249, 252
bis-2,4,6-Trichlorophenyl oxalate (TCPO), 216, 231
bis-2,4,5-Trichloro-6-pentoxycarbonyl oxalate (CPPO), 216
Trichosporon brassicae, 277, 285
Trichorosporon cutaneum, 285
Triglycerides, 94, 97
Tryptophane, 72
Tyrosinase, 180
Tyrosine, 70, 77
interference of, 321

Universal Sensors, 77
Urea, 65, 75, 77, 84, 94, 101, 139, 163, 187, 291

biosensor, 184, 185
interference of, 65, 324
Urease, 4, 10, 49, 53, 65, 67, 75, 94, 95, 140, 145, 146, 149, 187, 291
Uric acid, 94
interference of, 65, 321
Uricase, 10, 94
Urine, 39, 93, 284

Valinomycin, 315
Venom honeybee, 120
Vibrio fischeri, 214, 217, 218, 233, 235, 236, 238
Vibrio harveyi, 214, 217, 233, 237, 238, 297
Vitamins, 98
Voltametry
cyclic, 19, 20, 55, 318, 325
differential pulse, 319
square-wave, 19, 20

Waveguide, 207
Wheastone bridge, 87
Whole blood, 28 (*see also* Blood)

Xanthine, 94
oxidase, 94

Yellow Springs Instrument, 31